T0140116

Studies in Big Data

Volume 10

Series editor

Janusz Kacprzyk, Polish Academy of Sciences, Warsaw, Poland
e-mail: kacprzyk@ibspan.waw.pl

About this Series

Studies in Big Data

The series "Studies in Big Data" (SBD) publishes new developments and advances in the various areas of Big Data- quickly and with a high quality. The intent is to cover the theory, research, development, and applications of Big Data, as embedded in the fields of engineering, computer science, physics, economics and life sciences. The books of the series refer to the analysis and understanding of large, complex, and/or distributed data sets generated from recent digital sources coming from sensors or other physical instruments as well as simulations, crowd sourcing, social networks or other internet transactions, such as emails or video click streams and other. The series contains monographs, lecture notes and edited volumes in Big Data spanning the areas of computational intelligence incl. neural networks, evolutionary computation, soft computing, fuzzy systems, as well as artificial intelligence, data mining, modern statistics and Operations research, as well as self-organizing systems. Of particular value to both the contributors and the readership are the short publication timeframe and the world-wide distribution, which enable both wide and rapid dissemination of research output.

More information about this series at http://www.springer.com/series/11970

Witold Pedrycz · Shyi-Ming Chen

Editors

Granular Computing and Decision-Making

Interactive and Iterative Approaches

 Springer

Editors
Witold Pedrycz
Department of Electrical and Computer
 Engineering
University of Alberta
Edmonton, Alberta
Canada

Shyi-Ming Chen
Department of Computer Science and
 Information Engineering
National Taiwan University of
 Science and Technology
Taipei
Taiwan

ISSN 2197-6503   ISSN 2197-6511   (electronic)
Studies in Big Data
ISBN 978-3-319-36490-2   ISBN 978-3-319-16829-6   (eBook)
DOI 10.1007/978-3-319-16829-6

Springer Cham Heidelberg New York Dordrecht London

Printed on acid-free paper

Springer International Publishing AG Switzerland is part of Springer Science+Business Media
(www.springer.com)

Preface

Fuzzy decision-making - including its underlying methodology, plethora of algorithmic developments and a rich and diversified slew of application studies form a cornerstone of fuzzy sets. In spite of the ongoing research and a long list of accomplishments, there are still a number of open and intriguing issues worth pursuing. In particular, there is a genuine need to come to grip as to the dominant aspects of the methodology and resulting technology along with their systematic and coherent usage.

The overarching theme of this edited volume dwells upon the two key features of decision making processes that become quite commonly present throughout numerous studies and applications, and deal with *interactive* and *iterative* nature of these processes. Decision-making is inherently interactive. Fuzzy sets help realize human-machine communication in an efficient way by facilitating a two-way interaction in a friendly and transparent manner. This interaction is carried out at a suitable level of information granularity not confining the user to the use of precise numeric data. Subsequently any decision support system releases non-numeric findings, which become helpful to comprehend the results and assess potential impact of decisions being made. Human-centric interaction is of paramount relevance as a leading guiding design principle of decision support systems. The facet of interaction comes hand in hand with iterative processes of decision-making. Resulting recommendations issued by the system can be assessed and may trigger an iterative scheme of gathering additional evidence, refining goals and constraints, assessing the findings and, if necessary, proceeding with the next iterative loop. For instance, an iterative nature of decision-making becomes apparent in consensus building where a series of iterations (sessions) among the participants is commonly observed.

The facets of interactivity and the iterative way in which decisions are formed, give rise to the title of this volume - I^2 Fuzzy Decision Making, for brief. We are of opinion that even though these aspects of the decision processes are quite common and vividly present in numerous studies, they have not been fully and systematically exploited and supported by the technology of fuzzy sets. The objective of this volume is to revisit the underlying paradigm and practice of fuzzy decision-making emphasizing pivotal facets of their realization in an interactivity and iterative way.

The ultimate objectives of the proposed edited volume is to provide the reader with an updated, in-depth material on the conceptually appealing and practically sound methodology and practice of I^2 Fuzzy Decision Making.

The book engages a wealth of methods of fuzzy sets, brings new concepts, architectures and practice of fuzzy decision making. The chapters cover a wealth of ideas, algorithms, and applications. The growing role of information granules, Granular Computing and Computing with Words (CWW) as well as type-2 fuzzy sets is fully reflected in the book where a number of contributions are directly devoted to this subject either by offering some methodological insights or presenting interesting applications. There is a group of chapters dealing sequential decision -making, consensus building and group of decision making. Representative application studies including air quality, economics, library management, and image processing are a testimony of the leading role of the technology of fuzzy sets and information granules.

Given the leading theme of this undertaking, the book is aimed at a broad audience of researchers and practitioners. Owing to the nature of the material being covered and a way the main threads have been organized, the volume will appeal to the well-established communities including those active in various disciplines in which decision-making processes play a pivotal role and serve as a vehicle to produce solutions to existing problems. Those involved in operations research, management, various branches of engineering, social sciences, logistics, and economics will benefit from the exposure to the subject matter.

In virtue of the way in which the edited volume has been arranged, this book may serve as a useful and timely reference material for graduate students and senior undergraduate students in courses such as those on decision-making, Computational Intelligence, operations research, pattern recognition, risk management, and knowledge-based systems.

We would like to take this opportunity to express our deep thanks to the contributors to the volume for sharing results of their advanced and original research and delivering their views at the rapidly expanding areas of fundamental and applied research. The reviewers deserve our thanks for their constructive and timely input. We greatly appreciate a continuous support and encouragement coming from the Editor-in-Chief, Professor Janusz Kacprzyk whose leadership and vision make this book series a unique vehicle to disseminate the most recent, highly relevant and far-fetching publications in Computational Intelligence and decision-making and their various applications. The editorial staff at Springer has done a meticulous job and working with them was a pleasant experience.

We hope that the readers will find this volume of genuine interest and the innovative ideas put forward in this volume will become instrumental in fostering progress in research, education, and numerous practical endeavors in this exciting domain.

Witold Pedrycz
Shyi-Ming Chen

Contents

Granularity Helps Explain Seemingly Irrational Features of Human Decision Making

Joe Lorkowski and Vladik Kreinovich

Abstract. Starting from well-known studies by Kahmenan and Tversky, researchers have found many examples when our decision making seems to be irrational. In this chapter, we show that this seemingly irrational decision making can be explained if we take into account that human abilities to process information are limited; as a result, instead of the exact *values* of different quantities, we operate with *granules* that contain these values. On several examples, we show that optimization under such granularity restriction indeed leads to observed human decision making. Thus, granularity helps explain seemingly irrational human decision making.

Keywords: Decision making, Granularity, Seemingly irrational behavior.

1 Seemingly Irrational Human Decision Making: Formulation of the Problem

In the Ideal World, People Should make Perfect Decisions. In many real-life situations, we know what is best for us, and we know the exact consequences of each of our actions. In this case, a rational person should select an action that leads to the best possible outcome.

This assumption underlies basic (idealized) economic models: in these models, our decision making may hurt others but every person is interested in selecting a decision which is the best for him/herself.

In the Real World, People's Decisions are not Perfect. In the perfect world, people should make perfect decisions. It is well known, however, that our world is not perfect, and that many people make decisions which are not in their own best

J. Lorkowski · V. Kreinovich
Department of Computer Science, University of Texas at El Paso, 500 W. University, El Paso, Texas 79968, USA
e-mail: lorkowski@computer.org, vladik@utep.edu

W. Pedrycz and S.-M. Chen (eds.), *Granular Computing and Decision-Making*
Studies in Big Data 10, DOI: 10.1007/978-3-319-16829-6_1

interests. People eat unhealthy food, fail to exercise, get drunk, smoke, take drugs, gamble, and do many other things which – as they perfectly know – are bad for their health and bad for their wallets.

Such Imperfect Decisions can Still be Described in Optimization Terms. People engage in all kinds of unhealthy and asocial decision making because they get a lot of positive emotions from this engagement. A drug addict may lose his money, his family, his job, his health – but he gets so much pleasure from his drugs that he cannot stop. A gambler may lose all his money, but the pleasure of gambling is so high that he continues gambling (and losing) money until no money is left.

These examples of bad decisions are bad from the viewpoint of a person's health or wealth or social status. In all these examples, people clearly know what they want – e.g., more pleasure from drugs or from gambling – and they select a decision which is the "best" from this viewpoint.

On Top of this Well-known Abnormal Decision Making, there are also Many Examples When a Seemingly Rational Decision is Actually Irrational. It is well known that people make seemingly irrational decisions as described above, i.e., that they are optimizing objective functions which lead to their physical and ethical ruin.

Somewhat less known for the general public – but well known in psychology – is the fact that many quite rational people, people who are not addicted to drugs or gambling, people who normally lead a reasonably rational life, often make decisions which, at first glance, may seem reasonable but which, on deeper analysis, are irrational. This was first discovered in the studies of the Nobel Prize Winner Daniel Kahneman and his coauthor Amos Tversky; see, e.g., [8].

What they discovered is that sometimes people behave in such a way that no optimization can explain. Let us give a simple example. When a person is given two alternatives A and B, the person is usually able to conclude that, e.g., to him or her, A is better than B. We may disagree – as is the case of drug addiction – that A is an unhealthy choice, but A is what a person prefers. In this case, if now we offer the same person yet another alternative C, he or she may stick to A, or he or she may switch to C – but we do not expect this person to select B (since we already know that to this person, B is worse than another available alternative – namely, the alternative A). We do not expect to observe such a weird choice – but in some situations, this is exactly what has been observed; an example will be given later in this section.

This is a relatively clear example of a seemingly irrational decision making; we will show later that there are more subtle ones, where irrationality is not as easy to explain – but is clearly present; see examples in the next section.

So Why Irrational Decision Making? The fact that normal, reasonable people often make seemingly irrational decisions is puzzling. We humans come from billions years of improving evolution, we have flown to the Moon, we have discovered secrets of the Universe, we have learned to predict weather and to cure previously fatal diseases – this victorious image of Human Beings with a capital H does not seem to fit with simple decision mistakes, when the same person prefers A to B and then B to A, without even realizing that he/she is inconsistent.

Yes, we are not perfect, there are wars, crimes, and exploitations, but the common wisdom seems to indicate that most of our problems are caused by our selfishness – a criminal robs people because he wants to maximize his gain and he does not care if other people get hurt; a manufacturer of unhealthy foods maximizes his profit and does not care that people's health deteriorates as a result, etc. In all these cases, we blame the "evil" person for his selfish and vicious preferences, implicitly assuming that this person looks for what is best for him/her.

We can understand a person following a bad-for-society optimization criterion, but it is difficult to perceive a person whose decision making does not follow any optimization at all – and often, such a person is us.

How are Such Examples of Seemingly Irrational Decision Making Explained Now: The Idea of Bounded Rationality. An established explanation for the seemingly irrational decision making is that we humans have a limited ability to process information – especially when the decision needs to be made urgently. On the qualitative level, this idea of *bounded rationality* is in good accordance with the observations; for example, usually, the more time we are given to make decisions, the more rational our decisions become; see, e.g., [8].

The Existing Explanations Explain the Very Fact of Seemingly Irrational Decision Making, but not its Observed Specifics. The main limitation of the existing explanation is that it explains the *fact* that our decisions are sometimes not rational. In principle, under bounded resources, we can make different decisions, so we should observe various types of seemingly irrational decision making.

In many situations, however – for example, in the two situations described above – different decision makers exhibit the same deviations from the rational decision making. How can we explain these consistent deviations?

Main Idea behind Our Explanations: Optimization under Granularity. When we do not have enough time to take all the information into account, a natural idea is to use partial information. For example, when a man sees an animal in the jungle, it could be a predator, so an immediate decision needs to be made on whether to run away or not. Ideally, we should take into account all the details of an animal image, but there is no time for that, a reasonable reaction is to run way if an animal is sufficiently large.

So, instead of considering each data set separately, we, in effect, combine these data sets into "granules" corresponding to the partial information that is actually used in decision making; see, e.g., [22]. In the above example, instead of using the animal's size, we only take into account whether this size is greater than a certain threshold s_0 or not. In effect, this means that we divide the set of all possible values of size into two granules:

- a granule consisting of small animals, whose size is smaller than s_0, and
- a granule consisting of large (and thus, potentially dangerous) animals, whose size is larger than or equal to s_0.

What we plan to show is that in many cases, if we take into account only algorithms that process such granular information, then the observed human decision making can be shown to be *optimal* among such granular algorithms – although, of course, if we could take into account all available information, we would be able to make a better decision.

Comment. Some results from this paper were first presented at major fuzzy conferences [12, 13, 14]; several other results appear in this chapter for the first time.

Structure of the Paper. Section 2 provides detailed description of two examples of seemingly irrational decision making. In Section 3, we show that granularity helps explain the first of these examples; the second example is explained in Section 4. Sections 5 and 6 show that similar granularity-based arguments can also explain why utility grows as square root of money (Section 5) and how fuzzy techniques can be reconciled with the traditional decision making (Section 6). Section 7 contains conclusions and future work.

2   Examples of (Seemingly) Irrational Decision Making

Examples of Irrational Decision Making. Kahneman's book *Thinking, Fast and Slow* [8] has many examples of seemingly irrational decision making. In this chapter, we will concentrate on two examples. We selected these examples because they are, in our opinion, the easiest to explain without getting into the details of decision making theory and mathematical optimization. Let us describe these two examples in detail.

First Example of Seemingly Irrational Decision Making: Compromise Effect. The first example comes from simple decisions that most of us do very frequently: decisions on what to buy.

A customer shopping for an item usually has several choices. Some of these choices have better quality, leading to more possibilities, etc. – but are, on the other hand, more expensive. For example, a customer shopping for a photo camera has plenty of choices ranging from the cheapest ones (that work only in good lighting) to professional cameras that enable the users to make highest-quality photos even under complex circumstances. A traveller planning to spend a night in a new city has a choice from the cheapest motels (which provide a place to sleep) to luxurious hotels providing all kinds of comfort, etc. A customer selects one of the alternatives by taking into account the additional advantages of more expensive choices versus the need to pay more money for these choices.

In many real-life situations, customers face numerous choices. As usual in science, a good way to understand complex phenomena is to start by analyzing the simplest cases. In line with this reasoning, researchers provided customers with two alternatives and recorded which of these two alternatives a customer selected. In many particular cases, these experiments helped better understand the customer's selections – and sometimes even predict customer selections.

At first glance, it seems like such pair-wise comparisons are all we need to know: if a customer faces several choices $a_1, a_2, \ldots, a_n$, then a customer will select an alternative a_i if and only if this alternative is better in pair-wise comparisons that all other possible choices. To confirm this common-sense idea, in the 1990s, several researchers asked the customers to select one of the three randomly selected alternatives.

The experimenters expected that since the three alternatives were selected at random, a customers would:

- sometimes select the cheapest of the three alternative (of lowest quality of all three),
- sometimes select the intermediate alternative (or intermediate quality), and
- sometimes select the most expensive of the three alternatives (of highest quality of all three).

Contrary to the expectations, the experimenters observed that in the overwhelming majority of cases, customers selected the intermediate alternative; see, e.g., [25, 26, 29]. In all these cases, the customer selected an alternative which provided a compromise between the quality and cost; because of this, this phenomenon was named *compromise effect*.

Why is this Irrational? At first glance, selecting the middle alternative is reasonable. Let us show, however, that such a selection is not always rational.

For example, let us assume that we have four alternative $a_1 < a_2 < a_3 < a_4$ ordered in the increasing order of price and at the same time, increasing order of quality. Then:

- if we present the user with three choices $a_1 < a_2 < a_3$, in most cases, the user will select the middle choice a_2; this means, in particular, that, to the user, a_2 is better than the alternative a_3;
- on the other hand, if we present the user with three other choices $a_2 < a_3 < a_4$, in most cases, the same user will select the middle choice a_3; but this means that, to the user, the alternative a_3 is better than the alternative a_2.

If in a pair-wise comparison, a_2 is better, then the second choice is wrong. If in a pair-wise comparison, the alternative a_3 is better, then the first choice is wrong. In both cases, one of the two choices is irrational.

This is not Just an Experimental Curiosity, Customers' Decisions have been Manipulated this Way. At first glance, the above phenomena may seem like one of optical illusions or logical paradoxes: interesting but not that critically important. Actually, it is serious and important, since, according to anecdotal evidence, many companies have tried to use this phenomenon to manipulate the customer's choices: to make the customer buy a more expensive product.

For example, if there are two possible types of a certain product, a company can make sure that most customers select the most expensive type – simply by offering, as the third option, an even more expensive type of the same product.

Manipulation Possibility has Been Exaggerated, but Mystery Remains. Recent research showed that manipulation is not very easy: the compromise effect only happens when a customer has no additional information – and no time (or no desire) to collect such information. In situations when customers were given access to additional information, they selected – as expected from rational folks – one of the three alternatives with almost equal frequency, and their pairwise selections, in most cases, did not depend on the presence of any other alternatives; see, e.g., [28].

The new experiment shows that the compromise effect is not as critical and not as wide-spread as it was previously believed. However, in situation when decisions need to be made under major uncertainty, this effect is clearly present – and its seemingly counterintuitive, inconsistent nature is puzzling.

Second Example of Seemingly Irrational Decision Making: Biased Probability Estimates. In the first example, we considered situations with a simple choice, in which we have several alternatives, and we know the exact consequences of each alternative. In many practical cases, the situation is more complicated: for each decision, depending on how things go, we may face different consequences. For example, if a person invests all his retirement money in the stock market, the market may go up – in which case, he will gain – or it may go down, in which case he will lose a big portion of his savings. A person who takes on a potentially dangerous sport (like car racing) will probably gain a lot of pleasure, but there is also a chance of a serious injury.

To make a decision in such situations, it is important to estimate the probability of different outcomes. In some cases we know these probabilities – e.g., in the state-run lotteries, probabilities of winning are usually disclosed. In other cases, a person has to estimate these probabilities.

Of course, based on the limited information, we can get only approximate estimates of the corresponding probabilities. However, we expect that these estimates are unbiased, i.e., that, on average, they should provide a reasonably accurate estimate. Indeed, if we systematically overestimate small probabilities, then we would overestimate our gain in a lottery and, on average, lose. Similarly, if we systematically underestimate small probabilities, then, in particular, we will underestimate the probability of a disaster and invest in too risky stocks – and also lose on average.

This is what we expect: unbiased estimates, but this is not what we observe. What we observe is that small probabilities are routinely *over*estimated, while probabilities close to 1 are routinely *under*estimated. This is not just an occasional phenomenon: for each actual probability, the estimated probability is consistently different. For different actual probabilities p_i, the corresponding estimated probabilities $\tilde{p}_i$ are given in [8] (see also references therein):

p_i	0	1	2	5	10	20	50	80	90	95	98	99	100
$\tilde{p}_i$	0	5.5	8.1	13.2	18.6	26.1	42.1	60.1	71.2	79.3	87.1	91.2	100

Why Biased Estimates? As we have mentioned, biased estimates are contrary to rational decision making: overestimating a small probability of success may get a decision maker involved in risky situations where, on average (and thus, in the long run) the decision maker will lose. On the other hand, overestimating a small probability of a disaster will make a decision maker too cautious and prevent him/her from making a rational risky decision.

3 Explaining the First Example of Seemingly Irrational Human Decision making: Granularity Explains the Compromise Effect

Compromise Effect: Reminder. We have three alternative a, a' and a'':

- the alternative a is the cheapest – and is, correspondingly, of the lowest quality among the given three alternatives;
- the alternative a' is intermediate in terms of price – and is, correspondingly, intermediate in terms of quality;
- finally, the alternative a'' is the most expensive – and is, correspondingly, of the highest quality among the given three alternatives.

We do not know the exact prices, we just know the order between them; similarly, we do not know the exact values of quality, we just know the order between them. In this situation, most people select an alternative a'.

Let us Describe the Corresponding Granularity. The "utility" of each alternative comes from two factors:

- the first factor comes from the quality: the higher the quality, the better – i.e., larger the corresponding component u_1 of the utility;
- the second factor comes from price: the lower the price, the better for the user – i.e., the larger the corresponding component u_2 of the utility.

The fact that we do not know the exact value of the price means, in effect, that we consider three possible levels of price and thus, three possible levels of the utility u_1:

- low price, corresponding to high price-related utility;
- medium price, corresponding to medium price-related utility; and
- high price, corresponding to low price-related utility.

In the following text, we will denote "low" by L, "medium" by M, and "high" by H. In these terms, the above description of each alternative by the corresponding pair of utility values takes the following form:

- the alternative a is characterized by the pair (L, H);
- the alternative a' is characterized by the pair (M, M); and
- the alternative a'' is characterized by the pair (H, L).

Natural Symmetries. We do not know a priori which of the two utility components is more important. As a result, it is reasonable to treat both components equally. In order words, the selection should be the same if we simply swap the two utility components – i.e., we should select the same of three alternatives before and after swap:

- if we are selecting an alternative based on the pairs (L,H), (M,M), and (H,L),
- then we should select the exact same alternative if the pairs were swapped, i.e., if:

  - the alternative a was characterized by the pair (H,L);
  - the alternative a' was characterized by the pair (M,M); and
  - the alternative a'' was characterized by the pair (L,H).

Similarly, there is no reason to a priori prefer one alternative or the other. So, the selection should not depend on which of the alternatives we mark as a, which we mark as a', and which we mark as a''. In other words, any permutation of the three alternatives is a reasonable symmetry transformation. For example, if, in our case, we select an alternative a which is characterized by the pair (L,H), then, after we swap a and a'' and get the choice of the following three alternatives:

- the alternative a which is characterized by the pair (H,L);
- the alternative a' is characterized by the pair (M,M); and
- the alternative a'' is characterized by the pair (L,H),

then we should select the same alternative – which is now denoted by a''.

What can be Conclude Based on These Symmetries. Now, we can observe the following: that if we *both* swap u_1 and u_2 *and* swap a and a'', then you get the exact same characterization of all alternatives:

- the alternative a is still characterized by the pair (L,H);
- the alternative a' is still characterized by the pair (M,M); and
- the alternative a'' is still characterized by the pair (H,L).

The only difference is that:

- now, a indicates an alternative which was previously denoted by a'', and
- a'' now denotes the alternative which was previously denoted by a.

As we have mentioned, it is reasonable to conclude that:

- if in the original triple selection, we select the alternative a,
- then in the new selection – which is based on the exact same pairs of utility values – we should also select an alternative denoted by a.

But this "new" alternative a is nothing else but the old a''. So, we conclude that:

- if we selected a,
- then we should have selected a different alternative a'' in the original problem.

This is clearly a contradiction:

- we started by assuming that, to the user a was better than a'' (because otherwise a would not have been selected in the first place), and
- we ended up concluding that to the same user, the original alternative a'' is better than a.

This contradiction shows that, under the symmetry approach, we cannot prefer a.

Similarly:

- if in the original problem, we preferred an alternative a'',
- then this would mean that in the new problem, we should still select an alternative which marked by a''.

But this "new" a'' is nothing else but the old a. So, this means that:

- if we originally selected a'',
- then we should have selected a different alternative a in the original problem.

This is also a contradiction:

- we started by assuming that, to the user a'' was better than a (because otherwise a'' would not have been selected in the first place), and
- we ended up concluding that to the same user, the original alternative a is better than a''. This contradiction shows that, under the symmetry approach, we cannot prefer a''.

We thus conclude that out of the three alternatives a, a', and a'':

- we cannot select a, and
- we cannot select a''.

This leaves us only once choice: to select the intermediate alternative a'.

This is exactly the compromise effect that we planned to explain.

Conclusion. Experiments show when people are presented with three choices $a < a' < a''$ of increasing price and increasing quality, and they do not have detailed information about these choices, then in the overwhelming majority of cases, they select the intermediate alternative a'.

This "compromise effect" is, at first glance, irrational: selecting a' means that, to the user, a' is better than a'', but in a similar situation when the user is presented with $a' < a'' < a'''$, the same principle would indicate that the user will select a'' – meaning that a'' is better than a'.

Somewhat surprisingly, a natural granularity approach explains this seemingly irrational decision making.

4 Explaining the Second Example of Seemingly Irrational Human Decision making: Granularity Explains Why Our Probability Estimates Are Biased

Main Idea. Probability of an event is estimated, from observations, as the frequency with which this event occurs. For example, if out of 100 days of observation, rain occurred in 40 of these days, then we estimate the probability of rain as 40%. In general, if out of n observations, the event was observed in k of them, we estimate the probability as the ratio $\dfrac{k}{n}$.

This ratio is, in general, different from the actual (unknown) probability. For example, if we take a fair coin, for which the probability of head is exactly 50%, and flip it 100 times, we may get 50 heads, but we may also get 47 heads, 52 heads, etc.

Similarly, if we have the coin fall heads 50 times out of 100, the actual probability could be 50%, could be 47% and could be 52%. In other words, instead of the exact value of the probability, we get a *granule* of possible values. (In statistics, this granule is known as a *confidence interval*; see, e.g., [27].)

In other words:

- first, we estimate a probability based on the observations; as a result, instead of the exact value, we get a granule which contains the actual (unknown) probability; this granule is all we know about the actual probability;
- then, when a need comes to estimate the probability, we produce an estimate based on the granule.

Let us analyze these two procedures one by one.

Probability Granules: Analysis of the First Procedure and the Resulting Formulas. It is known (see, e.g., [27]), that the expected value of the frequency is equal to p, and that the standard deviation of this frequency is equal to

$$\sigma = \sqrt{\frac{p \cdot (1-p)}{n}}.$$

It is also known that, due to the Central Limit Theorem, for large n, the distribution of frequency is very close to the normal distribution (with the corresponding mean p and standard deviation σ).

For normal distribution, we know that with a high certainty all the values are located within 2-3 standard deviations from the mean, i.e., in our case, within the interval $(p - k_0 \cdot \sigma, p + k_0 \cdot \sigma)$, where $k_0 = 2$ or $k_0 = 3$: for example, for $k_0 = 3$, this is true with confidence 99.9%. We can thus say that the two values of probability p and p' are (definitely) distinguishable if the corresponding intervals of possible values of frequency do not intersect – and thus, we can distinguish between these two probabilities just by observing the corresponding frequencies.

In precise terms, the probabilities $p < p'$ are distinguishable if

$$(p - k_0 \cdot \sigma, p + k_0 \cdot \sigma) \cap (p' - k_0 \cdot \sigma', p' + k_0 \cdot \sigma') = \emptyset,$$

where

$$\sigma' \stackrel{\text{def}}{=} \sqrt{\frac{p' \cdot (1-p')}{n}},$$

i.e., if $p' - k_0 \cdot \sigma' \geq p + k_0 \cdot \sigma$. The smaller p', the smaller the difference $p' - k_0 \cdot \sigma'$. Thus, for a given probability p, the next distinguishable value p' is the one for which

$$p' - k_0 \cdot \sigma' = p + k_0 \cdot \sigma.$$

When n is large, these value p and p' are close to each other; therefore, $\sigma' \approx \sigma$. Substituting an approximate value σ instead of σ' into the above equality, we conclude that

$$p' \approx p + 2k_0 \cdot \sigma = p + 2k_0 \cdot \sqrt{\frac{p \cdot (1-p)}{n}}.$$

If the value p corresponds to the i-th level, then the next value p' corresponds to the $(i+1)$-st level. Let us denote the value corresponding to the i-th level by $p(i)$. In these terms, the above formula takes the form

$$p(i+1) - p(i) = 2k_0 \cdot \sqrt{\frac{p \cdot (1-p)}{n}}.$$

The above notation defines the value $p(i)$ for non-negative integers i. We can extrapolate this dependence so that it will be defined for all non-negative real values i. When n is large, the values $p(i+1)$ and $p(i)$ are close, the difference

$$p(i+1) - p(i)$$

is small, and therefore, we can expand the expression $p(i+1)$ in Taylor series and keep only linear terms in this expansion:

$$p(i+1) - p(i) \approx \frac{dp}{di}.$$

Substituting the above expression for $p(i+1) - p(i)$ into this formula, we conclude that

$$\frac{dp}{di} = \text{const} \cdot \sqrt{p \cdot (1-p)}.$$

Moving all the terms containing p into the left-hand side and all the terms containing i into the right-hand side, we get

$$\frac{dp}{\sqrt{p \cdot (1-p)}} = \text{const} \cdot di.$$

Integrating this expression and taking into account that $p = 0$ corresponds to the lowest 0-th level – i.e., that $i(0) = 0$ – we conclude that

$$i(p) = \text{const} \cdot \int_0^p \frac{dq}{\sqrt{q \cdot (1-q)}}.$$

This integral can be easily computed if introduce a new variable t for which $q = \sin^2(t)$. In this case,

$$dq = 2 \cdot \sin(t) \cdot \cos(t) \cdot dt,$$

$1 - p = 1 - \sin^2(t) = \cos^2(t)$ and therefore,

$$\sqrt{p \cdot (1-p)} = \sqrt{\sin^2(t) \cdot \cos^2(t)} = \sin(t) \cdot \cos(t).$$

The lower bound $q = 0$ corresponds to $t = 0$ and the upper bound $q = p$ corresponds to the value t_0 for which $\sin^2(t_0) = p$ – i.e., $\sin(t_0) = \sqrt{p}$ and $t_0 = \arcsin\left(\sqrt{p}\right)$. Therefore,

$$i(p) = \text{const} \cdot \int_0^p \frac{dq}{\sqrt{q \cdot (1-q)}} = \text{const} \cdot \int_0^{t_0} \frac{2 \cdot \sin(t) \cdot \cos(t) \cdot dt}{\sin(t) \cdot \cos(t)} =$$

$$\int_0^{t_0} 2 \cdot dt = 2 \cdot \text{const} \cdot t_0.$$

We know how t_0 depends on p, so we get

$$i(p) = 2 \cdot \text{const} \cdot \arcsin\left(\sqrt{p}\right).$$

We can determine the constant from the condition that the largest possible probability value $p = 1$ should correspond to the largest level $i = m$. From the condition that $i(1) = m$, taking into account that

$$\arcsin\left(\sqrt{1}\right) = \arcsin(1) = \frac{\pi}{2},$$

we conclude that

$$i(p) = \frac{2m}{\pi} \cdot \arcsin\left(\sqrt{p}\right).$$

Thus,

$$\arcsin\left(\sqrt{p}\right) = \frac{\pi \cdot i}{2m},$$

hence

$$\sqrt{p} = \sin\left(\frac{\pi \cdot i}{2m}\right)$$

and thus,

$$p(i) = \sin^2\left(\frac{\pi \cdot i}{2m}\right).$$

Thus, probability granules are formed by intervals $[p(i), p(i+1)]$. Each empirical probability is represented by the granule i to which is belongs.

From Granules to Probability Estimates: Analysis of the Second Procedure. As we have mentioned, instead of the actual probabilities, we have probability *labels* corresponding to m different granules:

- the first label corresponds to the smallest certainty,
- the second label corresponds to the second smallest certainty,
- etc.,
- until we reach the last label which corresponds to the largest certainty.

We need to produce some estimates of the probability based on the granule. In other words, for each i from 1 to m, we need to assign, to each i-th label, a value p_i in such a way that labels corresponding to higher certainty should get larger numbers: $p_1 < p_2 < \ldots < p_m$.

Before we analyze how to do it, let us recall that one of the main objectives of assigning numerical values is that we want computers to help us solve the corresponding decision problems, and computers are not very good in dealing with granules; their natural language is the language of numbers. From this viewpoint, it makes sense to consider not all theoretically possible exact real numbers, but only computer-representable real numbers.

In a computer, real numbers from the interval $[0,1]$ are usually represented by the first d digits of their binary expansion. Thus, computer-representable numbers are $0, h \stackrel{\text{def}}{=} 2^{-d}, 2h, 3h, \ldots$, until we reach the value $2^d \cdot h = 1$.

In our analysis, we will assume that the "machine unit" $h > 0$ is fixed, and we will this assume that only multiples of this machine units are possible values of all n probabilities p_i.

For example, when $h = 0.1$, each probability p_i takes 11 possible values: $0, 0.1, 0.2, 0.3, 0.4, 0.5, 0.6, 0.7, 0.8, 0.9$, and 1.0.

In the modern computers, the value h is extremely small; thus, whenever necessary, we can assume that $h \approx 0$ – i.e., use the limit case of $h \to 0$ instead of the actual small "machine unit" h.

For each h, we consider all possible combinations of probabilities $p_1 < \ldots < p_m$ in which all the numbers p_i are proportional to the selected step h, i.e., all possible combinations of values $(k_1 \cdot h, \ldots, k_m \cdot h)$ with $k_1 < \ldots < k_m$.

For example, when $m = 2$ and $h = 0.1$, we consider all possible combinations of values $(k_1 \cdot h, k_2 \cdot h)$ with $k_1 < k_2$:

- For $k_1 = 0$ and $p_1 = 0$, we have 10 possible combinations $(0, 0.1), (0, 0.2), \ldots, (0, 1)$.
- For $k_1 = 1$ and $p_1 = 0.1$, we have 9 possible combinations $(0.1, 0.2), (0.1, 0.3), \ldots, (0.1, 1)$.
- $\ldots$
- Finally, for $k_1 = 9$ and $p_1 = 0.9$, we have only one possible combination $(0.9, 1)$.

For each i, for different possible combinations $(p_1, \ldots, p_m)$, we get, in general, different value of the probability p_i. According to the complete probability formula, we can obtain the actual (desired) probability P_i if we combine all these value p_i with the weights proportional to the probabilities of corresponding combinations:

$$P_i = \sum_{p_1 < \ldots < p_m} p_i \cdot \mathrm{Prob}(p_1, \ldots, p_m).$$

Since we have no reason to believe that some combinations $(p_1, \ldots, p_m)$ are more probable and some are less probable, it is thus reasonable to assume that all these combinations are equally probable. Thus, P_i is equal to the arithmetic average of the values p_i corresponding to all possible combinations $(p_1, \ldots, p_m)$.

For example, for $m = 2$ and $h = 0.1$, we thus estimate P_1 by taking an arithmetic average of the values p_1 corresponding to all possible pairs. Specifically, we average:

- ten values $p_1 = 0$ corresponding to ten pairs $(0, 0.1), \ldots, (0, 1)$;
- nine values $p_1 = 0.1$ corresponding to nine pairs $(0.1, 0.2), \ldots, (0.1, 1)$;
- $\ldots$
- and a single value $p_1 = 0.9$ corresponding to the single pair $(0.9, 1)$.

As a result, we get the value

$$P_1 = \frac{10 \cdot 0.0 + 0 \cdot 0.1 + \ldots + 1 \cdot 0.9}{10 + 9 + \ldots + 1} = \frac{16.5}{55} = 0.3.$$

Similarly, to get the value p_2, we average:

- a single value $p_2 = 0.1$ corresponding to the single pair $(0, 0.1)$;
- two values $p_2 = 0.2$ corresponding to two pairs $(0, 0.2)$ and $(0.1, 0.2)$;
- ...
- ten values $p_2 = 1.0$ corresponding to ten pairs $(0, 1), \ldots, (0.9, 1)$.

As a result, we get the value

$$P_2 = \frac{1 \cdot 0.1 + 2 \cdot 0.2 + \ldots + 10 \cdot 1.0}{1 + 2 + \ldots + 10} = \frac{37.5}{55} = 0.7.$$

The probability p_i of each label can take any of the equidistant values $0, h, 2h, 3h$, ..., with equal probability. In the limit $h \to 0$, the resulting probability distribution tends to the uniform distribution on the interval $[0, 1]$.

In this limit $h \to 0$, we get the following problem:

- we start with m independent random variable $v_1, \ldots, v_m$ which are uniformly distributed on the interval $[0, 1]$;
- we then need to find, for each i, the conditional expected value

$$E[v_i \,|\, v_1 < \ldots < v_m]$$

of each variable v_i under the condition that the values v_i are sorted in increasing order.

Conditional expected values are usually more difficult to compute than unconditional ones. So, to solve our problem, let us reduce our problem to the problem of computing the unconditional expectation.

Let us consider m independent random variables each of which is uniformly distributed on the interval $[0, 1]$. One can easily check that for any two such variables v_i and v_j, the probability that they are equal to each other is 0. Thus, without losing generality, we can safely assume that all m random values are different. Therefore, the whole range $[0, 1]^m$ is divided into $m!$ sub-ranges corresponding to different orders between v_i. Each sub-range can be reduced to the sub-range corresponding to $v_1 < \ldots < v_m$ by an appropriate permutation in which v_1 is swapped with the smallest $v_{(1)}$ of m values, v_2 is swapped with the second smallest $v_{(2)}$, etc.

Thus, the conditional expected value of v_i is equal to the (unconditional) expected value of the i-th value $v_{(i)}$ in the increasing order. This value $v_{(i)}$ is known as an *order statistic*, and for uniform distributions, the expected values of all order statistics are known (see, e.g., [1, 2, 5]): $P_i = \dfrac{i}{m+1}$.

So, if all we know is that our degree of certainty is expressed by i-th label on an m-label scale of granules, then it is reasonable to assign, to this case, the probability $P_i = \dfrac{i}{m+1}$.

Let us Now Combine the Two Procedures. In the first procedure, based on the empirical frequency p, we find a label i for which

$$p \approx \sin^2 \left(\frac{\pi \cdot i}{2m} \right).$$

Based on this label, we then estimate the probability as $P_i = \dfrac{i}{m+1}$. For large m, we have $P \approx \dfrac{i}{m}$. Substituting P instead of $\dfrac{i}{m}$ into the formula for p, we conclude that

$$p \approx \sin^2 \left(\frac{\pi}{2} \cdot P \right).$$

Based on this formula, we can express the estimate P in terms of the actual probability p, as

$$P \approx \frac{1}{\pi} \cdot \arcsin(\sqrt{p}).$$

Comparing Our Estimates P with Empirical Probability Estimates $\widetilde{p}_i$: First Try. Let us compare the probabilities p_i, Kahneman's empirical estimates $\widetilde{p}_i$, and the estimates $P_i = \dfrac{1}{\pi} \cdot \arcsin(\sqrt{p_i})$ computed by using the above formula:

p_i	0	1	2	5	10	20	50	80	90	95	98	99	100
$\widetilde{p}_i$	0	5.5	8.1	13.2	18.6	26.1	42.1	60.1	71.2	79.3	87.1	91.2	100
P_i	0	6.4	9.0	14.4	20.5	29.5	50.0	70.5	79.5	85.6	91.0	93.6	100

The estimates P_i are closer to the empirical probability estimates $\widetilde{p}_i$ than the original probabilities, but the relation does not seem very impressive.

We Will Show that the Fit is Much Better than it Seems at First Glance. At first glance, the above direct comparison between the observed estimates $\widetilde{p}_i$ and the values P_i seems to make perfect sense. However, let us look deeper.

The observed estimates come from the fact that users select an alternative a that maximizes the expected gain $u(a) = \sum w_i(a) \cdot u_i$, where $w_i(a) \overset{\text{def}}{=} p_i(a)$. It is easy to observe that if we multiply all the weights by the same positive constant $\lambda > 0$, i.e., consider the weights $w'_i(a) = \lambda \cdot w_i(a)$, then for each action, the resulting value of the weighted gain will also increase by the same factor:

$$w'(a) = \sum w'_i(a) \cdot u_i = \sum \lambda \cdot w_i(a) \cdot u_i = \lambda \cdot \sum w_i(a) \cdot u_i = \lambda \cdot w_i(a).$$

The relation between the weighted gains of two actions a and a' does not change if we simply multiply both gains by a positive constant:

- if $w_i(a) < w_i(a')$, then, multiplying both sides of this inequality by λ, we get $w_i'(a) < w_i'(a')$;
- if $w_i(a) = w_i(a')$, then, multiplying both sides of this equality by λ, we get $w_i'(a) = w_i'(a')$;
- if $w_i(a) > w_i(a')$, then, multiplying both sides of this inequality by λ, we get $w_i'(a) > w_i'(a')$.

All we observe is which of the two actions a person selects. Since multiplying all the weights by a constant does not change the selection, this means that based on the selection, we cannot uniquely determine the weights: an empirical selection which is consistent with the weights w_i is equally consistent with the weights $w_i' = \lambda \cdot w_i$.

This fact can be used to *normalize* the empirical weights, i.e., to multiply them by a constant so as to satisfy some additional condition.

In [8], to normalize the weights, the authors use the requirement that the weight corresponding to probability 1 should be equal to 1. Since for $p = 1$, the corresponding value P is also equal to 1, we get a perfect match for $p = 1$, but a rather lousy match for probabilities intermediate between 0 and 1.

Instead of this normalization, we can select λ so as to get the best match "on average".

How to Improve the Fit: Details. A natural idea is to select λ from the Least Squares method, i.e., select λ for which the relative mean squares difference

$$\sum_i \left(\frac{\lambda \cdot P_i - \widetilde{p}_i}{P_i} \right)^2$$

is the smallest possible. Differentiating this expression with respect to λ and equating the derivative to 0, we conclude that

$$\sum_i \left(\lambda - \frac{\widetilde{p}_i}{P_i} \right) = 0,$$

i.e., that

$$\lambda = \frac{1}{m} \cdot \sum_i \frac{\widetilde{p}_i}{P_i}.$$

Resulting Match. For the above values, this formula leads to $\lambda = 0.910$.

Result. The resulting values $P_i' = \lambda \cdot P_i$ are much closer to the empirical probabilities $\widetilde{p}_i$:

p_i	0	1	2	5	10	20	50	80	90	95	98	99	100
$\widetilde{p}_i$	0	5.5	8.1	13.2	18.6	26.1	42.1	60.1	71.2	79.3	87.1	91.2	100
$P_i' = \lambda \cdot P_i$	0	5.8	8.2	13.1	18.7	26.8	45.5	64.2	72.3	77.9	82.8	87.4	91.0

For most probabilities p_i, the difference between the values P'_i and the empirical probability estimates $\widetilde{p}_i$ is so small that it is below the accuracy with which the empirical weights can be obtained from the experiment.

Thus, granularity ideas indeed explain Kahneman and Tversky's observation of biased empirical probability estimates.

Conclusion. Kahneman and Tversky showed that when people make decisions, then instead of – as should be rational – weighting outcomes with weights proportional to probabilities of different outcomes – they use *biased* weights, overestimating the importance of low-probability events and underestimating the importance of high-probability events. In this section, we show that this observable bias can be explained if we take into account granularity – imposed by our limited rationality (i.e., our limited ability to process information).

5 Granularity Explains Why Utility Grows as Square Root of Money

What We do in this Section. In this section, we provide another example when granularity explains observed decision making. To explain this example, we need to recall the traditional decision theory and the corresponding notion of utility.

Main Assumption behind the Traditional Decision Theory. Traditional approach to decision making is based on an assumption that for each two alternatives A' and A'', a user can tell:

- whether the first alternative is better for him/her; we will denote this by $A'' < A'$;
- or the second alternative is better; we will denote this by $A' < A''$;
- or the two given alternatives are of equal value to the user; we will denote this by $A' = A''$.

Towards a Numerical Description of Preferences: The Notion of Utility. Under the above assumption, we can form a natural numerical scale for describing preferences. Namely, let us select a very bad alternative A_0 and a very good alternative A_1. Then, most other alternatives are better than A_0 but worse than A_1.

For every probability $p \in [0, 1]$, we can form a lottery $L(p)$ in which we get A_1 with probability p and A_0 with probability $1 - p$.

- When $p = 0$, this lottery coincides with the alternative A_0: $L(0) = A_0$.
- When $p = 1$, this lottery coincides with the alternative A_1: $L(1) = A_1$.

For values p between 0 and 1, the lottery is better than A_0 and worse than A_1. The larger the probability p of the positive outcome increases, the better the result:

$$p' < p'' \text{ implies } L(p') < L(p'').$$

Thus, we have a continuous scale of alternatives $L(p)$ that monotonically goes from $L(0) = A_0$ to $L(1) = A_1$. We will use this scale to gauge the attractiveness of each alternative A.

Due to the above monotonicity, when p increases, we first have $L(p) < A$, then we have $L(p) > A$, and there is a threshold separating values p for which $L(p) < A$ from the values p for which $L(p) > A$. This threshold value is called the *utility* of the alternative A:

$$u(A) \stackrel{\text{def}}{=} \sup\{p : L(p) < A\} = \inf\{p : L(p) > A\}.$$

Then, for every $\varepsilon > 0$, we have

$$L(u(A) - \varepsilon) < A < L(u(A) + \varepsilon).$$

We will describe such (almost) equivalence by $\equiv$, i.e., we will write that $A \equiv L(u(A))$.

How to Elicit the Utility from a User: A Fast Iterative Process. Initially, we know the values $\underline{u} = 0$ and $\overline{u} = 1$ such that $A \equiv L(u(A))$ for some $u(A) \in [\underline{u}, \overline{u}]$.

On each stage of this iterative process, once we know values $\underline{u}$ and $\overline{u}$ for which $u(A) \in [\underline{u}, \overline{u}]$, we compute the midpoint u_{mid} of the interval $[\underline{u}, \overline{u}]$ and ask the user to compare A with the lottery $L(u_{\text{mid}})$ corresponding to this midpoint. There are two possible outcomes of this comparison: $A \leq L(u_{\text{mid}})$ and $L(u_{\text{mid}}) \leq A$.

- In the first case, the comparison $A \leq L(u_{\text{mid}})$ means that $u(A) \leq u_{\text{mid}}$, so we can conclude that $u \in [\underline{u}, u_{\text{mid}}]$.
- In the second case, the comparison $L(u_{\text{mid}}) \leq A$ means that $u_{\text{mid}} \leq u(A)$, so we can conclude that $u \in [u_{\text{mid}}, \overline{u}]$.

In both cases, after an iteration, we decrease the width of the interval $[\underline{u}, \overline{u}]$ by half. So, after k iterations, we get an interval of width 2^{-k} which contains $u(A)$ – i.e., we get $u(A)$ with accuracy 2^{-k}.

How to Make a Decision Based on Utility Values. Suppose that we have found the utilities $u(A')$, $u(A'')$, ..., of the alternatives A', A'', ... Which of these alternatives should we choose?

By definition of utility, we have:

- $A \equiv L(u(A))$ for every alternative A, and
- $L(p') < L(p'')$ if and only if $p' < p''$.

We can thus conclude that A' is preferable to A'' if and only if $u(A') > u(A'')$. In other words, we should always select an alternative with the largest possible value of utility. So, to find the best solution, we must solve the corresponding optimization problem.

Before We Go Further: Caution. We are *not* claiming that people estimate probabilities when they make decisions: we know they often don't. Our claim is that when people make *definite* and *consistent* choices, these choices *can* be described by probabilities. (Similarly, a falling rock does not solve equations but follows Newton's equations $ma = m\dfrac{d^2x}{dt^2} = -mg$.) In practice, decisions are often *not* definite (i.e., uncertain) and *not* consistent.

How to Estimate Utility of an Action. For each action, we usually know possible outcomes $S_1, \ldots, S_n$. We can often estimate the probabilities $p_1, \ldots, p_n$ of these outcomes.

By definition of utility, each situation S_i is equivalent to a lottery $L(u(S_i))$ in which we get:

- A_1 with probability $u(S_i)$ and
- A_0 with the remaining probability $1 - u(S_i)$.

Thus, the original action is equivalent to a complex lottery in which:

- first, we select one of the situations S_i with probability p_i: $P(S_i) = p_i$;
- then, depending on S_i, we get A_1 with probability $P(A_1 \mid S_i) = u(S_i)$ and A_0 with probability $1 - u(S_i)$.

The probability of getting A_1 in this complex lottery is:

$$P(A_1) = \sum_{i=1}^{n} P(A_1 \mid S_i) \cdot P(S_i) = \sum_{i=1}^{n} u(S_i) \cdot p_i.$$

In this complex lottery, we get:

- A_1 with probability $u = \sum_{i=1}^{n} p_i \cdot u(S_i)$, and
- A_0 with probability $1 - u$.

So, the utility of the complex action is equal to the sum u.

From the mathematical viewpoint, the sum defining u coincides with the expected value of the utility of an outcome. Thus, selecting the action with the largest utility means that we should select the action with the largest value of expected utility $u = \sum p_i \cdot u(S_i)$.

How Uniquely Determined is Utility. The above definition of utility u depends on the selection of two fixed alternatives A_0 and A_1. What if we use different alternatives A_0' and A_1'? How will the new utility u' be related to the original utility u?

By definition of utility, every alternative A is equivalent to a lottery $L(u(A))$ in which we get A_1 with probability $u(A)$ and A_0 with probability $1 - u(A)$. For simplicity, let us assume that $A_0' < A_0 < A_1 < A_1'$. Then, for the utility u', we get $A_0 \equiv L'(u'(A_0))$ and $A_1 \equiv L'(u'(A_1))$. So, the alternative A is equivalent to a complex lottery in which:

- we select A_1 with probability $u(A)$ and A_0 with probability $1 - u(A)$;
- depending on which of the two alternatives A_i we get, we get A_1' with probability $u'(A_i)$ and A_0' with probability $1 - u'(A_i)$.

In this complex lottery, we get A_1' with probability

$$u'(A) = u(A) \cdot (u'(A_1) - u'(A_0)) + u'(A_0).$$

Thus, the utility $u'(A)$ is related with the utility $u(A)$ by a linear transformation $u' = a \cdot u + b$, with $a > 0$. In other words, utility is defined modulo a linear transformation.

Traditional Approach Summarized. We assume that

- we know possible actions, and
- we know the exact consequences of each action.

Then, we should select an action with the largest value of expected utility.

Empirical Fact. It has been experimentally determined that for situations with monetary gain, utility u grows with the money amount x as $u \approx x^\alpha$, with $\alpha \approx 0.5$, i.e., approximately as $u \approx \sqrt{x}$; see, e.g., [8] and references therein.

What We do in this Section. In this section, we explain this empirical dependence.

Main Idea Behind Our Explanation. Money is useful because one can buy goods and services with it. The more goods and services one buys, the better. In the first approximation, we can say that the utility increases with the increase in the number of goods and service.

In these terms, to estimate the utility corresponding to a given amount of money, we need to do two things:

- first, we need to estimate how many goods and services a person can buy for a given amount of money;
- second, we need to estimate what value of utility corresponds to this number of goods and services.

Step 1: Estimating How Many Goods and Services a Person Can Buy. Different goods and services have different costs c_i; some are cheaper, some are more expensive. We know that all the costs c_i are bounded by some reasonable number C, so they are all located within an interval $[0, C]$. Let us sort the costs of different items in increasing order: $c_1 < c_2 < \ldots < c_n$.

In these terms, the smallest amount of money that we need to buy a single item is c_1. The smallest amount of money that we need to buy two items is $c_1 + c_2$, etc. In general, the smallest amount of money that we need to buy k items is $c_1 + c_2 + \ldots + c_k$.

How does this amount depends on k? We do not know the exact costs c_i, all we know is that these costs are sorted in increasing order. Similarly to the previous section, we can therefore consider all possible combinations $c_1 < \ldots < c_n$, and take, as an estimate C_i for c_i, the average value of c_i over all such combinations. Similarly to the previous section, we can conclude that $C_i = C \cdot \dfrac{i}{n+1}$.

In these terms, the expected amount of money needed to buy k items is equal to

$$C_1 + C_2 + \ldots + C_k = \frac{C}{n} \cdot (1 + 2 + \ldots + k) = \frac{C}{2n} \cdot k \cdot (k+1) \approx \text{const} \cdot k^2.$$

Step 2: Estimating the Utility Corresponding to k Items. Let u_k denote the utility corresponding to k items. We know that all the values u_k are bounded by some reasonable number U, so they are all located within an interval $[0, U]$. Clearly the more items, the better, i.e., the larger utility. Thus, we conclude that $u_1 < u_2 < \ldots < u_n$.

We do not know the exact values of u_k, all we know is that these utility values are sorted in increasing order. We can thus consider all possible combinations $u_1 < \ldots < u_n$, and take, as an estimate U_k for u_k, the average value of u_k over all such combinations. Similarly to the previous section, we can conclude that

$$U_k = U \cdot \frac{k}{n+1} = \text{const} \cdot k.$$

Let us Combine These Two Estimates. What is the utility corresponding to the amount of money x? To answer this question, first, we estimate the number of items k that we can buy with this amount. According to our estimates, $x = \text{const} \cdot k^2$, so we conclude that $k = \text{const} \cdot \sqrt{x}$. Then, we use this value k to estimate the utility $U \approx U_k$. Substituting $k = \text{const} \cdot \sqrt{x}$ into the formula $U \approx U_k = \text{const} \cdot k$, we conclude that $U \approx \text{const} \cdot \sqrt{x}$.

Since, as we have mentioned, utility is defined modulo a linear transformation, we can thus conclude that $U \approx \sqrt{x}$, which is exactly what we wanted to explain.

Conclusion. Thus, granularity indeed explains an interesting difficult-to-explain empirical fact – that utility grows as square root of money amount.

6 Granularity Helps Reconcile Traditional Decision Making with Fuzzy Techniques

Fuzzy Uncertainty and Fuzzy-Based Decision Making. In addition to applying traditional decision theory, another very successful way of making decisions under uncertainty is to use techniques based on fuzzy logic and fuzzy uncertainty.

Fuzzy logic (see, e.g., [10, 21, 30]) has been designed to describe imprecise ("fuzzy") natural language properties like "big", "small", etc. In contrast to "crisp" properties like $x \le 10$ which are either true or false, experts are not 100% sure whether a given value x is big or small. To describe such properties P, fuzzy logic proposes to assign, to each possible value x, a degree $\mu_P(x)$ to which the value x satisfies this property:

- the degree $\mu_P(x) = 1$ means that we are absolutely sure that the value x satisfies the property P;
- the degree $\mu_P(x) = 0$ means that we are absolutely sure that the value x does not satisfy the property P; and
- intermediate degrees $0 < \mu_P(x) < 1$ mean that we have *some* confidence that x satisfies the property P but we also have a certain degree of confidence that the value x does not satisfy this property.

How do we elicit the degree $\mu_P(x)$ from the expert? One of the usual ways is to use granules, i.e., more specifically, a *Likert scale*, i.e., to ask the expert to mark his or her degree of confidence that the value x satisfies the property P by one of the labels $0, 1, \ldots, n$ on a scale from 0 to n. If an expert marks m on a scale from 0 to n, then we take the ratio m/n as the desired degree $\mu_P(x)$. For example, if an expert marks her confidence by a value 7 on a scale from 0 to 10, then we take $\mu_P(x) = 7/10$.

For a fixed scale from 0 to n, we only get $n + 1$ values this way: $0, 1/n, 2/n, \ldots,$ $(n-1)/n = 1 - 1/n$, and 1. If we want a more detailed description of the expert's uncertainty, we can use a more detailed scale, with a larger value n.

Problem: How to Reconcile Traditional Decision Making Theory with Fuzzy Techniques? The traditional decision theory describes rational human decision making, it has many practical applications. On the other hand, fuzzy techniques are also very successful in many application problems, in particular, in control and in decision making (see, e.g., [10, 21]).

It is therefore desirable to combine these two techniques, so that we would able to capitalize on the successes of both types of techniques. To enhance this combination, it is desirable to be able to describe both techniques in the same terms. In particular, it is desirable to describe fuzzy uncertainty in terms of traditional decision making.

How do We Select a Mark on a Likert Scale? In Section 3, we simply used the labels marked by people on a Likert scale. But how do people select which labels to mark? To understand this, let us recall how this marking is done. Suppose that we have a Likert scale with $n + 1$ labels $0, 1, 2, \ldots, n$, ranging from the smallest to the largest.

Then, if the actual value of the quantity x is very small, we mark label 0. At some point, we change to label 1; let us mark this threshold point by x_1. When we continue increasing x, we first have values marked by label 1, but eventually reach a new threshold after which values will be marked by label 2; let us denote this threshold by x_2, etc. As a result, we divide the range $[\underline{X}, \overline{X}]$ of the original variable into $n + 1$ intervals $[x_0, x_1], [x_1, x_2], \ldots, [x_{n-1}, x_n], [x_n, x_{n+1}]$, where $x_0 = \underline{X}$ and $x_{n+1} = \overline{X}$:

- values from the first interval $[x_0, x_1]$ are marked with label 0;
- values from the second interval $[x_1, x_2]$ are marked with label 1;
- ...
- values from the n-th interval $[x_{n-1}, x_n]$ are marked with label $n - 1$;
- values from the $(n + 1)$-st interval $[x_n, x_{n+1}]$ are marked with label n.

Then, when we need to make a decision, we base this decision only on the label, i.e., only on the interval to which x belongs. In other words, we make n different decisions depending on whether x belongs to the interval $[x_0, x_1]$, to the interval $[x_1, x_2], \ldots,$ or to the interval $[x_n, x_{n+1}]$.

Decisions Based on the Likert Discretization are Imperfect. Ideally, we should take into account the exact value of the variable x. When we use Likert scale, we only take into account an interval containing x and thus, we do not take into account part of the original information. Since we only use part of the original information about x, the resulting decision may not be as good as the decision based on the ideal complete knowledge.

For example, an ideal office air conditioner should be able to maintain the exact temperature at which a person feels comfortable. People are different, their temperature preferences are different, so an ideal air conditioner should be able to maintain any temperature value x within a certain range $[\underline{X}, \overline{X}]$. In practice, some air conditioners only have a finite number of settings. For example, if we have setting corresponding to 65, 70, 75, and 80 degrees, then a person who prefers 72 degrees will probably select the 70 setting or the 75 setting. In both cases, this person will be somewhat less comfortable than if there was a possibility of an ideal 72 degrees setting.

How do We Select a Likert Scale: Main Idea. According to the general ideas of traditional (utility-based) approach to decision making, we should select a Likert scale for which the expected utility is the largest.

To estimate the utility of decisions based on each scale, we will take into account the just-mentioned fact that decisions based on the Likert discretization are imperfect. In utility terms, this means that the utility of the Likert-based decisions is, in general, smaller than the utility of the ideal decision.

Which Decision should We Choose within Each Label? In the ideal situation, if we could use the exact value of the quantity x, then for each value x, we would select an optimal decision $d(x)$, a decision which maximizes the person's utility.

If we only know the label k, i.e., if we only know that the actual value x belongs to the $(k+1)$-st interval $[x_k, x_{k+1}]$, then we have to make a decision based only on this information. In other words, we have to select one of the possible values $\widetilde{x}_k \in [x_k, x_{k+1}]$, and then, for all x from this interval, use the decision $d(\widetilde{x}_k)$ based on this value.

Which Value $\widetilde{x}_k$ should We Choose: Idea. According to the traditional approach to decision making, we should select a value for which the expected utility is the largest.

Which Value $\widetilde{x}_k$ Should We Choose: Towards a Precise Formulation of the Problem. To find this expected utility, we need to know two things:

- we need to know the probability of different values of x; these probabilities can be described, e.g., by the probability density function $\rho(x)$;
- we also need to know, for each pair of values x' and x, what is the utility $u(x', x)$ of using a decision $d(x')$ in the situation in which the actual value is x.

In these terms, the expected utility of selecting a value $\widetilde{x}_k$ can be described as

$$\int_{x_k}^{x_{k+1}} \rho(x) \cdot u(\widetilde{x}_k, x)\, dx. \tag{5.1}$$

Thus, for each interval $[x_k, x_{k+1}]$, we need to select a decision $d(\widetilde{x}_k)$ corresponding to the value $\widetilde{x}_k$ for which the expression (5.1) attains its largest possible value. The resulting expected utility is equal to

$$\max_{\widetilde{x}_k} \int_{x_k}^{x_{k+1}} \rho(x) \cdot u(\widetilde{x}_k, x)\, dx. \tag{5.2}$$

How to Select the Best Likert Scale: General Formulation of the Problem. The actual value x can belong to any of the $n+1$ intervals $[x_k, x_{k+1}]$. Thus, to find the overall expected utility, we need to add the values (5.2) corresponding to all these intervals. In other words, we need to select the values $x_1, \ldots, x_n$ for which the following expression attains its largest possible value:

$$\sum_{k=0}^{n} \max_{\tilde{x}_k} \int_{x_k}^{x_{k+1}} \rho(x) \cdot u(\tilde{x}_k, x) \, dx. \tag{5.3}$$

Equivalent Reformulation in Terms of Disutility. In the ideal case, for each value x, we should use a decision $d(x)$ corresponding to this value x, and gain utility $u(x, x)$. In practice, we have to use decisions $d(x')$ corresponding to a slightly different value, and thus, get slightly worse utility values $u(x', x)$. The corresponding decrease in utility $U(x', x) \stackrel{\text{def}}{=} u(x, x) - u(x', x)$ is usually called *disutility*. In terms of disutility, the function $u(x', x)$ has the form

$$u(x', x) = u(x, x) - U(x', x),$$

and thus, the optimized expression (5.1) takes the form

$$\int_{x_k}^{x_{k+1}} \rho(x) \cdot u(x, x) \, dx - \int_{x_k}^{x_{k+1}} \rho(x) \cdot U(\tilde{x}_k, x) \, dx.$$

The first integral does not depend on $\tilde{x}_k$; thus, the expression (5.1) attains its maximum if and only if the second integral attains its minimum. The resulting maximum (5.2) thus takes the form

$$\int_{x_k}^{x_{k+1}} \rho(x) \cdot u(x, x) \, dx - \min_{\tilde{x}_k} \int_{x_k}^{x_{k+1}} \rho(x) \cdot U(\tilde{x}_k, x) \, dx. \tag{5.4}$$

Thus, the expression (5.3) takes the form

$$\sum_{k=0}^{n} \int_{x_k}^{x_{k+1}} \rho(x) \cdot u(x, x) \, dx - \sum_{k=0}^{n} \min_{\tilde{x}_k} \int_{x_k}^{x_{k+1}} \rho(x) \cdot U(\tilde{x}_k, x) \, dx.$$

The first sum does not depend on selecting the thresholds. Thus, to maximize utility, we should select the values $x_1, \ldots, x_n$ for which the second sum attains its smallest possible value:

$$\sum_{k=0}^{n} \min_{\tilde{x}_k} \int_{x_k}^{x_{k+1}} \rho(x) \cdot U(\tilde{x}_k, x) \, dx \to \min. \tag{5.5}$$

Let is Recall that are Interested in the Membership Function. For a general Likert scale, we have a complex optimization problem (5.5). However, we are not interested in general Likert scales per se, what we are interested in is the use of Likert scales to elicit the values of the membership function $\mu(x)$.

As we have mentioned earlier, in an n-valued scale:

- the smallest label 0 corresponds to the value $\mu(x) = 0/n$,
- the next label 1 corresponds to the value $\mu(x) = 1/n$,
- ...
- the last label n corresponds to the value $\mu(x) = n/n = 1$.

Thus, for each n:

- values from the interval $[x_0, x_1]$ correspond to the value $\mu(x) = 0/n$;
- values from the interval $[x_1, x_2]$ correspond to the value $\mu(x) = 1/n$;
- ...
- values from the interval $[x_n, x_{n+1}]$ correspond to the value $\mu(x) = n/n = 1$.

The actual value of the membership function $\mu(x)$ corresponds to the limit $n \to \infty$, i.e., in effect, to very large values of n. Thus, in our analysis, we will assume that the number n of labels is huge – and thus, that the width of each of $n+1$ intervals $[x_k, x_{k+1}]$ is very small.

Let us Take into Account that Each Interval is Narrow. Let us use the fact that each interval is narrow to simplify the expression $U(x', x)$ and thus, the optimized expression (5.5).

In the expression $U(x', x)$, both values x' and x belong to the same narrow interval and thus, the difference $\Delta x \overset{\text{def}}{=} x' - x$ is small. Thus, we can expand the expression $U(x', x) = U(x + \Delta x, x)$ into Taylor series in Δx, and keep only the first non-zero term in this expansion. In general, we have

$$U(x + \Delta, x) = U_0(x) + U_1 \cdot \Delta x + U_2(x) \cdot \Delta x^2 + \ldots,$$

where

$$U_0(x) = U(x, x), \quad U_1(x) = \frac{\partial U(x + \Delta x, x)}{\partial(\Delta x)}, \quad U_2(x) = \frac{1}{2} \cdot \frac{\partial^2 U(x + \Delta x, x)}{\partial^2(\Delta x)}. \quad (5.7)$$

Here, by definition of disutility, we get $U_0(x) = U(x, x) = u(x, x) - u(x, x) = 0$. Since the utility is the largest (and thus, disutility is the smallest) when $x' = x$, i.e., when $\Delta x = 0$, the derivative $U_1(x)$ is also equal to 0 – since the derivative of each (differentiable) function is equal to 0 when this function attains its minimum. Thus, the first non-trivial term corresponds to the second derivative:

$$U(x + \Delta x, x) \approx U_2(x) \cdot \Delta x^2,$$

i.e., in other words, that

$$U(\tilde{x}_k, x) \approx U_2(x) \cdot (\tilde{x}_k - x)^2.$$

Substituting this expression into the expression

$$\int_{x_k}^{x_{k+1}} \rho(x) \cdot U(\tilde{x}_k, x) \, dx$$

that needs to be minimized if we want to find the optimal $\widetilde{x}_k$, we conclude that we need to minimize the integral

$$\int_{x_k}^{x_{k+1}} \rho(x) \cdot U_2(x) \cdot (\widetilde{x}_k - x)^2 \, dx. \tag{5.8}$$

This new integral is easy to minimize: if we differentiate this expression with respect to the unknown $\widetilde{x}_k$ and equate the derivative to 0, we conclude that

$$\int_{x_k}^{x_{k+1}} \rho(x) \cdot U_2(x) \cdot (\widetilde{x}_k - x) \, dx = 0,$$

i.e., that

$$\widetilde{x}_k \cdot \int_{x_k}^{x_{k+1}} \rho(x) \cdot U_2(x) \, dx = \int_{x_k}^{x_{k+1}} x \cdot \rho(x) \cdot U_2(x) \, dx,$$

and thus, that

$$\widetilde{x}_k = \frac{\int_{x_k}^{x_{k+1}} x \cdot \rho(x) \cdot U_2(x) \, dx}{\int_{x_k}^{x_{k+1}} \rho(x) \cdot U_2(x) \, dx}. \tag{5.9}$$

This expression can also be simplified if we take into account that the intervals are narrow. Specifically, if we denote the midpoint of the interval $[x_k, x_{k+1}]$ by $\overline{x}_k \stackrel{\text{def}}{=} \dfrac{x_k + x_{k+1}}{2}$, and denote $\Delta x \stackrel{\text{def}}{=} x - \overline{x}_k$, then we have $x = \overline{x}_k + \Delta x$. Expanding the corresponding expressions into Taylor series in terms of a small value Δx and keeping only main terms in this expansion, we get

$$\rho(x) = \rho(\overline{x}_k + \Delta x) = \rho(\overline{x}_k) + \rho'(\overline{x}_k) \cdot \Delta x \approx \rho(\overline{x}_k),$$

where $f'(x)$ denoted the derivative of a function $f(x)$, and

$$U_2(x) = U_2(\overline{x}_k + \Delta x) = U_2(\overline{x}_k) + U_2'(\overline{x}_k) \cdot \Delta x \approx U_2(\overline{x}_k).$$

Substituting these expressions into the formula (5.9), we conclude that

$$\widetilde{x}_k = \frac{\rho(\overline{x}_k) \cdot U_2(\overline{x}_k) \cdot \int_{x_k}^{x_{k+1}} x \, dx}{\rho(\overline{x}_k) \cdot U_2(\overline{x}_k) \cdot \int_{x_k}^{x_{k+1}} dx} = \frac{\int_{x_k}^{x_{k+1}} x \, dx}{\int_{x_k}^{x_{k+1}} dx} = \frac{\frac{1}{2} \cdot (x_{k+1}^2 - x_k^2)}{x_{k+1} - x_k} = \frac{x_{k+1} + x_k}{2} = \overline{x}_k.$$

Substituting this midpoint value $\widetilde{x}_k = \overline{x}_k$ into the integral (5.8) and taking into account that on the k-th interval, we have $\rho(x) \approx \rho(\overline{x}_k)$ and $U_2(x) \approx U_2(\overline{x}_k)$, we conclude that the integral (5.8) takes the form

$$\int_{x_k}^{x_{k+1}} \rho(\overline{x}_k) \cdot U_2(\overline{x}_k) \cdot (\overline{x}_k - x)^2 \, dx = \rho(\overline{x}_k) \cdot U_2(\overline{x}_k) \cdot \int_{x_k}^{x_{k+1}} (\overline{x}_k - x)^2 \, dx. \tag{5.8a}$$

When x goes from x_k to x_{k+1}, the difference $\Delta x = x - \overline{x}_k$ between the value x and the interval's midpoint $\overline{x}_k$ ranges from $-\Delta_k$ to Δ_k, where Δ_k is the interval's half-width:

$$\Delta_k \overset{\text{def}}{=} \frac{x_{k+1} - x_k}{2}.$$

In terms of the new variable Δx, the integral in the right-hand side of (5.8a) has the form

$$\int_{x_k}^{x_{k+1}} (\bar{x}_k - x)^2 \, dx = \int_{-\Delta_k}^{\Delta_k} (\Delta x)^2 \, d(\Delta x) = \frac{2}{3} \cdot \Delta_k^3.$$

Thus, the integral (5.8) takes the form

$$\frac{2}{3} \cdot \rho(\bar{x}_k) \cdot U_2(\bar{x}_k) \cdot \Delta_k^3.$$

The problem (5.5) of selecting the Likert scale thus becomes the problem of minimizing the sum (5.5) of such expressions (5.8), i.e., of the sum

$$\frac{2}{3} \cdot \sum_{k=0}^{n} \rho(\bar{x}_k) \cdot U_2(\bar{x}_k) \cdot \Delta_k^3. \tag{5.10}$$

Here, $\bar{x}_{k+1} = x_{k+1} + \Delta_{k+1} = (\bar{x}_k + \Delta_k) + \Delta_{k+1} \approx \bar{x}_k + 2\Delta_k$, so $\Delta_k = (1/2) \cdot \Delta\bar{x}_k$, where $\Delta\bar{x}_k \overset{\text{def}}{=} \bar{x}_{k+1} - \bar{x}_k$. Thus, (5.10) takes the form

$$\frac{1}{3} \cdot \sum_{k=0}^{n} \rho(\bar{x}_k) \cdot U_2(\bar{x}_k) \cdot \Delta_k^2 \cdot \Delta\bar{x}_k. \tag{5.11}$$

In terms of the membership function, we have $\mu(\bar{x}_k) = k/n$ and $\mu(\bar{x}_{k+1}) = (k+1)/n$. Since the half-width Δ_k is small, we have

$$\frac{1}{n} = \mu(\bar{x}_{k+1}) - \mu(\bar{x}_k) = \mu(\bar{x}_k + 2\Delta_k) - \mu(\bar{x}_k) \approx \mu'(\bar{x}_k) \cdot 2\Delta_k,$$

thus, $\Delta_k \approx \dfrac{1}{2n} \cdot \dfrac{1}{\mu'(\bar{x}_k)}$. Substituting this expression into (5.11), we get the expression $\dfrac{1}{3 \cdot (2n)^2} \cdot I$, where

$$I = \sum_{k=0}^{n} \frac{\rho(\bar{x}_k) \cdot U_2(\bar{x}_k)}{(\mu'(\bar{x}_k))^2} \cdot \Delta\bar{x}_k. \tag{5.12}$$

The expression I is an integral sum, so when $n \to \infty$, this expression tends to the corresponding integral

$$I = \int \frac{\rho(x) \cdot U_2(x)}{(\mu'(x))^2} \, dx. \tag{5.11}$$

Minimizing (5.5) is equivalent to minimizing I. With respect to the derivative $d(x) \overset{\text{def}}{=} \mu'(x)$, we need to minimize the objective function

$$I = \int \frac{\rho(x) \cdot U_2(x)}{d^2(x)} \, dx \tag{5.12}$$

under the constraint that

$$\int_{\underline{X}}^{\overline{X}} d(x)\,dx = \mu(\overline{X}) - \mu(\underline{X}) = 1 - 0 = 1. \tag{5.13}$$

By using the Lagrange multiplier method, we can reduce this constraint optimization problem to the unconstrained problem of minimizing the functional

$$I = \int \frac{\rho(x) \cdot U_2(x)}{d^2(x)}\,dx + \lambda \cdot \int d(x)\,dx, \tag{5.14}$$

for an appropriate Lagrange multiplier λ. Differentiating (5.14) with respect to $d(x)$ and equating the derivative to 0, we conclude that $-2 \cdot \dfrac{\rho(x) \cdot U_2(x)}{d^3(x)} + \lambda = 0$, i.e., that $d(x) = c \cdot (\rho(x) \cdot U_2(x))^{1/3}$ for some constant c. Thus, $\mu(x) = \int_{\underline{X}}^{x} d(t)\,dt = c \cdot \int_{\underline{X}}^{x} (\rho(t) \cdot U_2(t))^{1/3}\,dt$. The constant c must be determined by the condition that $\mu(\overline{X}) = 1$. Thus, we arrive at the following formula (5.15).

Resulting Formula. The membership function $\mu(x)$ obtained by using Likert-scale elicitation is equal to

$$\mu(x) = \frac{\int_{\underline{X}}^{x} (\rho(t) \cdot U_2(t))^{1/3}\,dt}{\int_{\underline{X}}^{\overline{X}} (\rho(t) \cdot U_2(t))^{1/3}\,dt}, \tag{5.15}$$

where $\rho(x)$ is the probability density describing the probabilities of different values of x,

$$U_2(x) \overset{\text{def}}{=} \frac{1}{2} \cdot \frac{\partial^2 U(x + \Delta x, x)}{\partial^2 (\Delta x)},$$

$U(x', x) \overset{\text{def}}{=} u(x, x) - u(x', x)$, and $u(x', x)$ is the utility of using a decision $d(x')$ corresponding to the value x' in the situation in which the actual value is x.

Comment. The above formula only applies to membership functions like "large" whose values monotonically increase with x. It is easy to write a similar formula for membership functions like "small" which decrease with x. For membership functions like "approximately 0" which first increase and then decrease, we need to separately apply these formula to both increasing and decreasing parts.

Conclusion. The resulting membership degrees incorporate both probability and utility information. This fact *explains why fuzzy techniques often work better than probabilistic techniques* – because the probability techniques only take into account the probability of different outcomes.

7   Conclusions and Future Work

While in general, humans behave rationally, there are many known experiments in which humans show seemingly irrational behavior. For example, when a customer is presented with two objects, one somewhat cheaper and another one more expensive and of higher quality, the customer's choice often depends on the presence of the third object, the object that the customer will not select:

- if the third object is cheaper than both two, the customer will usually select the cheaper of the two objects;
- if the third object is more expensive than the both two, the customer will usually select the more expensive of the two objects.

From the rational viewpoint, the selection between the two object should not depend on the presence of other, less favorable objects – but it does!

There are many other examples of such seemingly irrational human behavior. This phenomenon is known since the 1950s, and an explanation for this phenomenon is also well known – such seemingly irrational behavior is caused by the fact that human computational abilities are limited; in this sense, human rationality is *bounded*.

The idea of bounded rationality explains, on the *qualitative* level, why human behavior and decision making are sometimes seemingly irrational. However, until recently, there have been few successful attempts to use this idea to explain *quantitative* aspects of observed human behavior. In this chapter, we show, on four examples, that these quantitative aspects can be explained if we take into account that one of the main consequences of bounded rationality is *granularity*. The main idea behind granularity is that since we cannot process *all* the available information, we only process *part* of it. Because of this, several different data points – differing by the information that we do not process – are treated the same way. In other words, instead of dealing with the original data points, we deal with *granules*, each of which corresponds to several possible data points. For example, if we only use the first binary digit x_1 in the binary expansion of a number $x = 0.x_1x_2\ldots$ from the interval $[0,1]$, this means that, instead of the exact number x, we use two granules corresponding to intervals $[0,0.5)$ (for which $x_1 = 0$) and $[0.5,1]$ (for which $x_1 = 1$).

In this chapter, we have shown, on four examples (including the above customer examples) that granularity indeed explained the observed quantitative aspects of seemingly irrational human behavior. The remaining challenge is to provide a similar explanation for other observed cases of seemingly irrational human behavior and decision making.

Acknowledgements. This work was supported in part by the National Science Foundation grants HRD-0734825 and HRD-1242122 (Cyber-ShARE Center of Excellence) and DUE-0926721.

The authors are thankful to participants of the Joint World Congress of the International Fuzzy Systems Association and Annual Conference of the North American Fuzzy Information Processing Society IFSA/NAFIPS'2013 (Edmonton, Canada, June 24–28, 2013), 4th World Conference on Soft Computing (Berkeley, California, May 25–27, 2014), and 2014 Annual Conference of the North American Fuzzy Information Processing Society NAFIPS'2014 (Boston, Massachusetts, June 24–26, 2014) for valuable discussions, and to the anonymous referees for useful suggestions.

References

1. Ahsanullah, M., Nevzorov, V.B., Shakil, M.: An Introduction to Order Statistics. Atlantis Press, Paris (2013)
2. Arnold, B.C., Balakrishnan, N., Nagaraja, H.N.: A First Course in Order Statistics. Society of Industrial and Applied Mathematics (SIAM), Philadelphia (2008)
3. Bellman, R.E., Zadeh, L.A.: Decision making in a fuzzy environment. Management Science 17(4), B141–B164 (1970)
4. Cohen, M.D., Huber, G., Keeney, R.L., Levis, A.H., Lopes, L.L., Sage, A.P., Sen, S., Whinston, A.B., Winkler, R.L., von Winterfeldt, D., Zadeh, L.A.: Research needs and the phenomena of decision making and operations. IEEE Transactions on Systems, Man, and Cybernetics 15(6), 764–775 (1985)
5. David, H.A., Nagaraja, H.N.: Order Statistics. Wiley, New York (2003)
6. Fishburn, P.C.: Utility Theory for Decision Making. John Wiley & Sons Inc., New York (1969)
7. Fishburn, P.C.: Nonlinear Preference and Utility Theory. The John Hopkins Press, Baltimore (1988)
8. Kahneman, D.: Thinking, Fast and Slow, Farrar, Straus, and Giroux, New York (2011)
9. Keeney, R.L., Raiffa, H.: Decisions with Multiple Objectives. John Wiley and Sons, New York (1976)
10. Klir, G., Yuan, B.: Fuzzy Sets and Fuzzy Logic. Prentice Hall, Upper Saddle River, New Jersey (1995)
11. Kreinovich, V.: Decision making under interval uncertainty (and beyond). In: Guo, P., Pedrycz, W. (eds.) Human-Centric Decision-Making Models for Social Sciences, Springer Verlag. SCI, vol. 502, pp. 163–193. Springer, Heidelberg (2014)
12. Lorkowski, J., Kreinovich, V.: Likert-scale fuzzy uncertainty from a traditional decision making viewpoint: it incorporates both subjective probabilities and utility information. In: Proceedings of the Joint World Congress of the International Fuzzy Systems Association and Annual Conference of the North American Fuzzy Information Processing Society, IFSA/NAFIPS 2013, Edmonton, Canada, Edmonton, Canada, June 24-28, pp. 525–530 (2013)
13. Lorkowski, J., Kreinovich, V.: Fuzzy logic ideas can help in explaining Kahneman and Tversky's empirical decision weights. In: Proceedings of the 4th World Conference on Soft Computing, Berkeley, California, May 25-27, pp. 285–289 (2014)
14. Lorkowski, J., Kreinovich, V.: Interval and symmetry approaches to uncertainty – pioneered by Wiener – help explain seemingly irrational human behavior: a case study. In: Proceedings of the 2014 Annual Conference of the North American Fuzzy Information Processing Society, NAFIPS 2014, Boston, Massachusetts, June 24-26 (2014)
15. Luce, R.D., Raiffa, R.: Games and Decisions: Introduction and Critical Survey. Dover, New York (1989)
16. March, J.: Bounded rationality, ambiguity, and the engineering of choice. The Bell Journal of Economics 9(2), 587–608 (1978)
17. Nguyen, H.T., Kosheleva, O., Kreinovich, V.: Decision making beyond Arrow's "Impossibility Theorem", with the analysis of effects of collusion and mutual attraction. International Journal of Intelligent Systems 24(1), 27–45 (2009)
18. Nguyen, H.T., Kreinovich, V.: Applications of Continuous Mathematics to Computer Science. Kluwer, Dordrecht (1997)
19. Nguyen, H.T., Kreinovich, V., Lea, B.: How to combine probabilistic and fuzzy uncertainties in fuzzy control. In: Proceedings of the Second International Workshop on Industrial Applications of Fuzzy Control and Intelligent Systems, College Station, December 2-4, pp. 117–121 (1992)

20. Nguyen, H.T., Kreinovich, V., Wu, B., Xiang, G.: Computing Statistics under Interval and Fuzzy Uncertainty. SCI, vol. 393. Springer, Heidelberg (2012)
21. Nguyen, H.T., Walker, E.A.: A First Course in Fuzzy Logic. Chapman and Hall/CRC, Boca Raton (2006)
22. Pedrycz, W., Skowron, A., Kreinovich, V. (eds.): Handbook on Granular Computing. Wiley, Chichester (2008)
23. Rabinovich, S.G.: Measurement Errors and Uncertainty: Theory and Practice. Springer, Berlin (2005)
24. Raiffa, H.: Decision Analysis. McGraw-Hill, Columbus (1997)
25. Redelmeier, D., Shafir, E.: Medical decision mading in situations that offer multiple alternatives. Journal of the American Medical Association 273(4), 302–305 (1995)
26. Shafir, E., Simonson, I., Tversky, A.: Reason-based choice. Cognition 49, 11–36 (1993)
27. Sheskin, D.J.: Handbook of Parametric and Nonparametric Statistical Procedures. Chapman & Hall/CRC, Boca Raton (2011)
28. Simonson, I., Rosen, E.: Absolute Value: What Really Influences Customers in the Age of (Nearly) Perfect Information. HarperBusiness, New York (2014)
29. Tversky, A., Simonson, I.: Context-dependent preferences. Management Science 39(10), 1179–1189 (1993)
30. Zadeh, L.A.: Fuzzy sets. Information and Control 8, 338–353 (1965)
31. Zadeh, L.A.: Outline of a new approach to the analysis of complex systems and decision processes. IEEE Transactions on Systems, Man, and Cybernetics 3(1), 28–44 (1973)
32. Zadeh, L.A.: Precisiated natural language-toward a radical enlargement of the role of natural languages in information processing, decision and control. In: Wang, L., Halgamuge, S.K., Yao, X. (eds.) Proceedings of the 1st International Conference on Fuzzy Systems and Knowledge Discovery FSDK 2002: Computational Intelligence for the E-Age, Singapore, November 18-22, vol. 1, pp. 1–3 (2002)
33. Zadeh, L.A.: Computing with words and perceptions – a paradigm shift in computing and decision analysis and machine intelligence. In: Wani, A., Cios, K.J., Hafeez, K. (eds.) Proceedings of the 2003 International Conference on Machine Learning and Applications, ICMLA 2003, Los Angeles, California, June 23-24, pp. 3–5 (2003)
34. Zadeh, L.A.: A new direction in decision analysis-perception-based decisions. In: Ralescu, A.L. (ed.) Proceedings of the Fourteenth Midwest Artificial Intelligence and Cognitive Sciences Conference, MAICS 2003, Cincinnati, Ohio, April 12-13, pp. 1–2 (2003)

A Comprehensive Granular Model for Decision Making with Complex Data

Ying Xie[*], Tom Johnsten, Vijay V. Raghavan,
Ryan G. Benton, and William Bush

Abstract. This chapter describes a comprehensive granular model for decision making with complex data. This granular model first uses information decomposition to form a horizontal set of granules for each of the data instances. Each granule is a partial view of the corresponding data instance; and aggregately all the partial views of that data instance provide a complete representation for the instance. Then, the decision making based on the original data can be divided and distributed to decision making on the collection of each partial view. The decisions made on all partial views will then be aggregated to form a final global decision. Moreover, on each partial view, a sequential M+1 way decision making (a simple extension of Yao's 3-way decision making) can be carried out to reach a local decision. This chapter further categorizes stock price predication problem using the proposed decision model and incorporates the MLVS model for biological sequence classification into the proposed decision model. It is suggested that the proposed model provide a general framework to address the complexity and volume challenges in big data analytics.

Ying Xie
Department of Computer Science,
Kennesaw State University, Kennesaw, Georgia 30144, USA
e-mail: yxie2@kennesaw.edu

Tom Johnsten · William Bush
School of Computing, University of South Alabama, USA
e-mail: tjohnsten@southalabama.edu,
        wmb1321@jagmail.southalabama.edu

Vajay V. Raghavan
The Center of Advanced Computer Studies, University of Louisiana at Lafayette, USA
e-mail: vijay@cacs.louisiana.edu

Ryan G. Benton
Informatics Research Institute, University of Louisiana at Lafayette, USA
e-mail: rbenton@louisiana.edu

[*] Corresponding author.

© Springer International Publishing Switzerland 2015   33
W. Pedrycz and S.-M. Chen (eds.), *Granular Computing and Decision-Making,*
Studies in Big Data 10, DOI: 10.1007/978-3-319-16829-6_2

Keywords: Complex Data, Comprehensive Granular Model, M+1 way decision making, Big Data, Granular Computing.

1   Introduction

Granular Computing [1] focuses on philosophy, methodology and paradigm for problem solving and information processing based on granular structures [3]. By using the concepts like granules, levels, and hierarchies, granular computing promotes structured thinking at the philosophical level and structured problem solving at the practical level [2]. In [4], Y. Y. Yao stated that "the principle of computing, guided by granular structures, is to examine the problem at a finer granulation level with more detailed information when there is a need or benefit for doing so." Furthermore, Y. Y. Yao developed a sequential three-way decision framework based on a hierarchy of multiple levels of information granularity [5]. Decision tree can be viewed as a simple special case of the sequential three-way decision making.

The three-way decision framework assumes a single vertical view of the granular structures, where each layer of information granulation is a complete representation of the original data at a particular coarse level. However, for complex decision making on data that is unstructured or big, it may be challenging or even impossible to form a single hierarchy of information granules for the data in order to carry out the sequential three way decision making. In order to address this issue, this chapter describes a comprehensive granular model for decision making with complex data. This granular model first uses information decomposition to form a horizontal set of granules for each of the data instances. Each granule is a partial view of the corresponding data instance; and aggregately all the partial views of that data instance provide a complete representation for the instance. Then, the decision making based on the original data can be divided and distributed to decision making on the collection of each partial view. The decisions made on all partial views will then be aggregated to form a final global decision. If a partial view is still complex enough, then the decomposition process can be continued on the partial view to form a horizontal set of sub partial views for that partial view. This decomposition process continues until a single hierarchical structure of information granules can be formed for a (sub) partial view. Then, a sequential M+1 way decision making (an extension of 3-way decision making) can be carried out on the collection of each partial view to reach a local decision. Decision forest can be viewed as a simple special case of this comprehensive granular model for decision making.

The proposed comprehensive granular model fits well with parallel/distributed computing frameworks. Each collection of a partial view covers all data instances with respect to that partial view, and is independent from other collections of different partial views. Therefore, local decisions can be made on all collections of partial views in a parallel manner. On each collection of a partial view, the sequential M+1 way decision making process may be implemented as multiple iterations of MapReduce processes. Therefore, the proposed granular model can be viewed as a general solution framework for big unstructured data. This chapter will further categorize stock price predication problem and biological sequence classification problem using the proposed decision model.

2 A Generalized M+1 Way Decision

In [5], Yao described a three way decision model for 2-state classification problems. Compared with traditional two way (acceptance/rejection) decision model, the three way decision model maps a data instance for decision to one of the three disjoint regions, POS, NEG, and BND. Decision of "acceptance" is made on those objects mapped to the POS region; decision of "rejection" is made on those mapped to the NEG region; and "noncommitment" is assigned to those mapped to BND region. A noncommitment assignment suggests that more information of the corresponding data instance is needed in order to accept or reject that object. The essential ideas of three way decision making has been widely used in real-life decision making in different domains, such as medical decision-making, social judgment theory, and hypothesis testing, and peering review processes [6]. In order to apply this idea to multi-class classification problem, Yao's Three Way Decision Model can be naturally extended to a $M + 1$ Way Decision Model. Formally, the $M + 1$ Way Decision Model can be described as follows.

Given a set of data instances U with N elements $\{u_1, u_2, ..., u_N\}$ and a set of class labels C with M elements $\{c_1, c_2, ..., c_M\}$, the $M + 1$ Way decision model divides U into $M + 1$ disjoint regions $\{r_{c_1}, r_{c_2}, ..., r_{c_M}, r_{c_{M+1}}\}$, such that if a data instance $u_i \in r_{c_j}$ (where $1 \leq j \leq M$), then the class label c_j is assigned to x_i; if $u_i \in r_{c_{M+1}}$, a "noncommitment" is assigned to u_i.

In [6], Yao uses a Two-Poset based evaluation to segment U into the acceptance region and the rejection region. We follow the exact same method to extend the Two-Poset based evaluation to a M-Poset based evaluation for $M + 1$ way decision model.

Let $L_C = \left\{ \left(l_{r_{c_i}}, \preccurlyeq_{r_{c_i}} \right) \middle| 1 \leq i \leq M \right\}$ be M posets, and $V = \{ v_{r_{c_i}} : U \to l_{r_{c_i}} | 1 \leq i \leq M\}$ be M evaluation functions. Given $x \in U$ and $1 \leq i \leq M$, $v_{r_{c_i}}(x)$ returns an acceptance value of x to r_{c_i}. For two objects $x, y \in U$ and $1 \leq i \leq M$, if $v_{r_{c_i}}(x) \preccurlyeq_{r_{c_i}} v_{r_{c_i}}(y)$, then x is less acceptable than y to r_{c_i}. Further let $1 \leq i \leq M$ and $l_{r_{c_i}}^+ \subseteq l_{r_{c_i}}$ be the set of designated values of acceptance to r_{c_i}. Then, given an object $x \in U$, we have $x \in r_{c_i}$ if and only if $v_{r_{c_i}}(x) \in l_{r_{c_i}}^+$. In other words, we can define r_{c_i} ($1 \leq i \leq M$) as follows: $r_{c_i}(1 \leq i \leq M) = \{x \in U | v_{r_{c_i}}(x) \in l_{r_{c_i}}^+\}$. Then we can further define $r_{c_{M+1}}$ as follows: $r_{c_{M+1}} = \{x \in U | \forall (1 \leq i \leq M) \to v_{r_{c_i}}(x) \notin l_{r_{c_i}}^+\}$.

3 A Generalized Sequential M+1 Way Decision Algorithm

We further generalize Yao's Sequential Three way decision algorithm [5] to Sequential $M + 1$ Way Decision Algorithm. Assume for each data instance $x \in U$, there exists $n + 1$ levels of granular description of x. By following the same notation used in [5], the $n + 1$ level of granular description of x can be represented as follows:

$$Dec_0(x) \preccurlyeq Dec_1(x) \preccurlyeq \cdots \preccurlyeq Dec_n(x)$$

where the relation $\preccurlyeq$ denotes a "finer than" relationship. In order to make a decision on x, the Sequential $M + 1$ Way Decision Algorithm first uses $Dec_n(x)$ as the representation for x. A decision is made on x by assigning the class label c_i to x, if $\exists (1 \leq i \leq M) \rightarrow v_{r_{c_i}}\big(Dec_n(x)\big) \in l^+_{r_{c_i}}$; otherwise, we continue to use $Dec_{n-1}(x)$ as the representation for x. If we are still unable to make decision on x even reaching the level 0 of granular description of x, we have two options. Option 1: assign "noncommitment" as the label to x as the final decision on x. This option indicates that no decision can be made at required confidence level on x based upon all information available on x. Option 2: assign the class label that corresponds to the largest $v_{r_{c_i}}\big(Dec_n(x)\big)$ $(1 \leq i \leq M)$. Option 2 can be applied to the situations where a data instance has to be categorized to one of the M categories.

4   Decision Making on Complex Data

The sequential M+1 way decision algorithm assumes a single vertical view of the granular structures, where each layer of information granulation is a complete representation of the original data at a particular coarse level. However, for complex decision making on data that is unstructured or big, it may be challenging or even impossible to form a single hierarchy of information granules for the data in order to carry out the sequential $M + 1$ way decision algorithm. In order to address this issue, we propose a comprehensive granular model for decision making with complex data. This granular model first uses information decomposition to form a horizontal set of granules for each of the data instances. Each granule is a partial view of the corresponding data instance; and aggregately all the partial views of that data instance provide a complete representation for the instance. Then, the decision making based on the original data can be divided and distributed to decision making on the collection of each partial view. The decisions made on all partial views will then be aggregated to form a final global decision. More formally, assume that for each data instance $x \in U$, we decompose x into a set of $l \geq 1$ partial views $PV_x = \{x_1, x_2, \ldots, x_l\}$, where $PV_x[i] = x_i$. This decomposition could be lossless or lossy. For lossless decomposition, PV_x contains the same amount of information as what x contains; i.e., there exists an operator op, such that $x = agg_{op}\{x_i | x_i \in PV_x\}$. For lossy decomposition, PV_x contains less information than x. Furthermore, for each of the partial view of x, we assume there exists $n + 1$ levels of granular description of the partial view; i.e., for the ith partial view of x, $PV_x[i]$, we have

$$Dec_0(PV_x[i]) \preccurlyeq Dec_1(PV_x[i]) \preccurlyeq \cdots \preccurlyeq Dec_n(PV_x[i])$$

Overall, the representation of the set of complex data instances can be illustrated in figure 1.

If a partial view is still complex enough, then the decomposition process can be continued on the partial view to form a horizontal set of sub partial views for that partial view. This decomposition process continues until a single hierarchical structure of information granules can be formed for a (sub) partial view. For simplicity of description, we will not further illustrate this situation.

Based on this decomposition on each data instance in U, the set of all data instances U is decomposed into l sets U_1, $U_2, \ldots, U_i, \ldots, U_l$, where $U_i = \{PV_{u_j}[i] | u_j \in U\}$, i.e., the set of the ith partial views from all data instances in U. The decomposition on U can be illustrated in figure 2.

Then the overall decision process on U can be decomposed to a decision process on each U_i. In other words, given a data instance $u_j \in U$, we will have l completely independent decision process that can be carried out by l distributed computing processes. The final decision on u_j will be an ensemble of the results of the l independent decision processes. One possible ensemble approach can be described as follows.

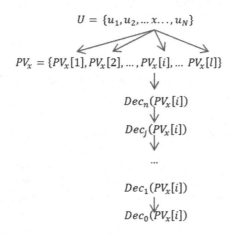

$$U = \{u_1, u_2, \ldots x \ldots, u_N\}$$

$$PV_x = \{PV_x[1], PV_x[2], \ldots, PV_x[i], \ldots PV_x[l]\}$$

$$Dec_n(PV_x[i])$$

$$Dec_j(PV_x[i])$$

$$\ldots$$

$$Dec_1(PV_x[i])$$

$$Dec_0(PV_x[i])$$

Fig. 1 An Illustration of the Proposed Model for Representing a Complex Data Set

$$U \Rightarrow \begin{cases} U_1 = \{PV_{u_1}[1], PV_{u_2}[1], \ldots, PV_{u_N}[1]\} \\ U_2 = \{PV_{u_1}[2], PV_{u_2}[2], \ldots, PV_{u_N}[2] \\ \ldots \\ U_i = \{PV_{u_1}[i], PV_{u_2}[i], \ldots, PV_{u_N}[i] \\ \ldots \\ U_l = \{PV_{u_1}[l], PV_{u_2}[l], \ldots, PV_{u_N}[l] \end{cases}$$

Fig. 2 An Illustration of Decomposing a Data Set U into l sets of Partial Views

Given a data instance $u_j \in U$, assume a class label c_k $(1 \leq k \leq M)$ is assigned to u_j on $Dec_p\left(PV_{u_j}[s]\right)$ in U_s, where $\left(PV_{u_j}[s]\right)$ is the sth $(1 \leq s \leq l)$

partial view of u_j and $Dec_p\left(PV_{u_j}[s]\right)$ is the granular representation of this partial view $PV_{u_j}[s]$ at p $(0 \leq p \leq n)$ level. Then the weight of this decision can be represented as $\theta_{j,s,p,k}(0 < \theta_{j,s,p,k} \leq 1)$, where $\theta_{j,s,p,k} = f(j,s,p,k)$ can be evaluated by a function that reflects the confidence of the decision that assigns c_k to u_j based on U_s at the granularity level p. For instance, a possible expression of this weight function can be $\theta_{j,s,p,k} = \omega_s * (p + 1)/(n + 1)$, where $\omega_s(0 < \omega_s \leq 1)$ reflects the relative significance of using U_s in the overall decision making; $\frac{p+1}{n+1}$ (where p is the granularity level of data representation at which the decision is made and $0, 1, ..., n$ are the available granularity levels from finest to coarsest) suggests that reaching decision at a courser level justifies a stronger differential power of this partial view for u_j. Therefore, the overall weight for assigning c_k to u_j can be calculated as $\sum_{1 \leq s \leq l} \theta_{j,s,p(s),k}$, where $p(s)$ represents the coarsest granularity level at which c_k is assigned to u_j using U_s. The class label with the highest overall weight will be the final class label that is assigned to u_j.

5   Categorizing Stock Price Prediction Using the Proposed Decision Model

As is well known, stock price prediction is a broad yet very challenging research area. In this section, we will categorize one particular stock price prediction task by using the proposed decision model. This prediction task can be described as follows. Given the historical price movement of a stock or index, such as SPY, in a particular time frame (which can be weekly, daily, hourly, 5 minutes, and so on), predict the price movement for the next time unit (next week, next day, next hour, or next 5 minutes). If we visualize the price movement at a historical time unit using a bar as shown in figure 3, the task is to predict what will be the upcoming bar.

We now use the proposed decision model to categorize this prediction task. We first annotate each historical bar by using the bar that immediately follows it. For instance, bar1 in figure 3 can be annotated as Lower Open Lower Close (LOLC), given that its following bar has a lower open and lower close compared to its own close price; bar2 can be annotated as Higher Open Higher Close (HOHC), given that its following bar has a higher open and higher close (HOHC) compared to its own close price; bar3 can be annotated as Lower Open Higher Close (LOHC), given that its following bar has a lower open and higher close compared to its own close price; and bar4 can be annotated as higher open and lower close (HOLC), given that its following bar has a higher open and lower close compared to its own close price. Therefore, we identify a set of 4 class labels {LOLC, HOHC, LOHC, HOLC}. Now, the task is to predict the class label for the current bar which is the last bar shown in figure 3. The class label assigned to the current bar is the predication of the price movement for the upcoming time unit.

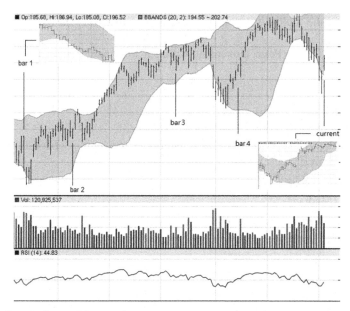

■ Op:195.68, HI:196.94, Lo:195.08, Cl:196.52   ■ BBANDS (20, 2): 194.55 ~ 202.74

bar 1

bar3

bar 4

current

bar 2

■ Vol: 120,825,537

■ RSI (14): 44.83

Fig. 3 A Sample Chart of Stock Price Movement (copy from www.barchart.com)

In order to predict the class label for the current bar, we need to identify features that can be used to describe each bar. This is where complexity comes into the picture, given that each historical bar could be associated with numerous factors that may indicate or correlate with the stock price movement at next time unit. Moreover, those factors may be heterogeneous in nature, formats, and scales; and some features are complex data by themselves. Therefore, it is very challenging for any typical machine learning algorithm to utilize all these features together. However, by using our proposed decision model, each of those features or a combination of a group of features can serve as a partial view for a given bar, and all partial views deliver comprehensive information for that bar. A partial view can be certain technique analysis (TA) features, sentimental features, or fundamental analysis (FA) features. Therefore, the proposed decision model provides a framework for a comprehensive analysis of stock price movement.

Without loss of generality, we only consider some technique analysis features for an illustration. These features include price movement within a time unit, price movement over a period of time, volume, RSI, and moving average. Based on these features, we form two partial views for each bar. The first partial view, denoted as $PV[1]$, is the combination of price movement within a time unit, volume, RSI, and moving average; the second partial view, denoted as $PV[2]$, is the combination of price movement over a period of time, volume, and RSI. As can be seen, $PV[1]$ is for predicting next price movement based on price movement at the current time unit; whereas $PV[2]$ is for predicting next price movement based on price movement cross multiple time units. Furthermore, for each partial view, we form multi-level descriptions with different granularities.

For instance, the multi-level granular descriptions of $PV[1]$ on bar 1 are shown below:

$Dec_2(PV_{bar1}[1])$: Price movement within a time unit – {Open High, Close Low},

   Volume – {High}

   RSI – {Median}

   Moving average – {above moving average}

$Dec_1(PV_{bar1}[1])$: Price movement within a time unit – {High-Low, Low-Low, Open-Low, Close-Low},

   Volume – {High}

   RSI – {Median}

   Moving average – {above moving average}

$Dec_0(PV_{bar1}[1])$: Price movement within a time unit – {time series of price movement within this time unit*},

   Volume – {High}

   RSI – {Median}

   Moving average – {above moving average}

The time series of price movement within the time unit for bar 1 and the current bare are illustrated in figure 3.

Based on the multi-level granular descriptions of $PV[1]$ for each bar, a 4+1 Way Sequential Decision Algorithm can be applied in order to make a decision on the current bar. The algorithm first compare $Dec_2(PV_{current}[1])$ with each $Dec_2(PV_{bar\_x}[1])$, where bar_x is a historical bar. This comparison finds all historical bars that match the current bar on Volume, RSI, moving average, and price movement values. If at least $\alpha\%$ of all matched bars share the same class label, then decision can be reached at this level on this partial view; otherwise; the 4+1 Way Sequential Decision Algorithm goes down to the finer level of granularity to look for a decision. That is, the algorithm further compares $Dec_1(PV_{current}[1])$ with each $Dec_1(PV_{bar\_y}[1])$, where bar_y is one of the matched historical bars found by the comparison at the previous granularity level. For this comparison, we find the K nearest neighbors of the current bar by calculating the similarity on the price movement within the time unit between the current bar and each of the bars selected as a match at the previous granularity level. If at least $\alpha\%$ of the K nearest neighbors share the same class label, then decision can be reached at this level on this partial view. Otherwise, the algorithm further goes down to the finest level of granular descriptions to look for a decision by comparing $Dec_0(PV_{current}[1])$ with each $Dec_0(PV_{bar\_y}[1])$, where bar_y is one of the matched historical bars found by the comparison at the coarsest granularity level. The computing at this finest granular level is most intensive, since each comparison between $Dec_0(PV_{current}[1])$ and $Dec_0(PV_{bar\_y}[1])$

requires computing similarity between two time series by using techniques like Dynamic Time Warping (DTW) [7]. The 4+1 Way Sequential Decision Algorithm reduces the requirement of this intensive computing by 1) trying to make decision at a coarser level of granular representation; and 2) reducing the number of times of executing intensive computing by taking advantage of the results that have been generated at a coarser level.

Similarly, the multi-level granular descriptions of $PV[2]$ on bar 1 are shown below

$Dec_2(PV_{bar\_1}[2])$: Price movement for 2 most recent time units – {Close High, Close Low}*,

bar_1 itself closed low, the previous bar closed high

Volume – {High}

RSI – {Median}

$Dec_1(PV_{bar\_1}[2])$: Price movement for 3 most recent time units – {Close High, Close High, Close Low},

Volume – {High}

RSI – {Median}

$Dec_0(PV_{bar\_1}[2])$: Price movement for 4 most recent time units – {Close Low, Close High, Close High, Close Low},

Volume – {High}

RSI – {Median}

Again, a 4+1 Way Sequential Decision Algorithm can be carried out on $PV[2]$ in order to make a prediction on the current bar. Finally, decision made on each partial view can be assembled to form a final decision.

6   Incorporating the MLVS Model for Biological Sequence Classification into the Proposed Decision Model

Bioinformatics methods are increasingly important for biomedical research, as advances in high throughput sequencing have drastically reduced the cost per base for genomic sequencing. For the first time, we are in an era where our ability to sequence genomes is quickly outstripping our capacity to analyze them in a useful manner. Developing novel methods to detect and analyze features of interest within biological sequences is a critical step in fully utilizing the wealth of information made available by advanced sequencing techniques [8]. A significant challenge in analyzing such data is the extraction and representation of significant features from the data. To address this challenge, we recently developed a model, called Multi-Layered Vector Spaces (MLVS), for representing biological sequences for the purpose of characterization and classification [10, 11]. Experiments show that MLVS-based classifiers are able to outperform or perform on par with existing methods for classifying biological sequences [10, 12].

The MLVS model is based on the idea of mapping biological sequences into a set of vectors. In general, each vector contains the location of h-step ordered pairs of symbols, where a symbol is an element of the alphabet from which the sequence is constructed and h represents the number of spaces between two symbols. If all ordered pairs made up of consecutive symbols of the alphabet form 1-step pairs, U_1, then allowing multiple spaces between the elements of the ordered pair generates a set of m-step pairs, $U_1, U_2, ..., U_k$, forming a multi-layered space. The original sequence can thus be conceptually viewed as the union of all such ordered pairs stratified at k distinct layers. The mapping of biological sequences into such a vector space has the potential to bring out subtle local patterns that may be overlooked by existing methods. We now present a formal description of the MLVS model and illustrate how it can be successfully incorporated into the proposed granular decision making model.

A sequence S of finite length $|S|$ defined over a finite alphabet Σ is viewed to have a multi-layered structure made up of a set of m-step ordered pairs (i, j), with $(i, j) \in \Sigma$, denoted by $U_{h|(i,j)}$, where $1 \leq h \leq k$. The parameter h stands for the number of spaces between the elements of the pair downstream in the flow (left to right) of the sequence, and k is the maximum admissible value of h. Ordered pairs made up of consecutive elements of the sequence are said to form the family of 1-step (one-step) pairs, $U_{1|(i,j)}$. Allowing multiple spaces between the elements of an ordered pair generates a multitude of m-step pairs (families) $U_1, U_2, ..., U_k$, creating a multi-layered k-clustering C_k made up of sets $U_{h|(i,j)}$, $m = 1, 2, ..., k$ as follows: $C_k = \cup_m \cup_{(i,j)} U_{h|(i,j)}$. The upper bound for parameter k is $|S| - 1$. The binding factor between the elements of a particular set $U_{h|(i,j)}$ is the step size h, common for all ordered pairs making up the family. The total number of ordered pairs that can be drawn from the alphabet is $|\Sigma|^2$. A sequence S can be viewed as the union of all such ordered pairs at k distinct layers. The following example demonstrates how the said structures are built.

Table 1 Sample Sequence

1	2	3	4	5	6	7	8	9	10
g	c	t	g	g	g	c	t	c	a
11	12	13	14	15	16	17	18	19	20
g	c	t	a	a	t	g	a	g	c

Example-1: Given the alphabet $\Sigma = \{a, c, g, t\}$, with $|\Sigma| = 4$, $|\Sigma|^2 = 16$, and the following biological sequence S defined over Σ, with $|S| = 20$: $S = [g, c, t, g, g, g, c, t, c, a, g, c, t, a, a, t, g, a, g, c]$.

Table-1 shows the step locations of the elements making up the sequence. The following are sample h-step pairs: 1-step ordered pairs for (g, c) are located at step locations $[1,2], [6,7], [11,12]$, and $[19,20]$; 1-step ordered pairs for (g, g) are located at step locations $[4,5]$, and $[5,6]$; 2-step ordered pairs for (g, t) are located at step locations $[1,3], [6,8]$, and $[11,13]$; 4-step ordered pairs for (c, g) are located at step locations $[2,6]$, and $[7,11]$. For a selected value of h and a given ordered pair $(i, j) \in \Sigma$, the sequence of anchor positions is taken as forming the scalar components of an q-dimensional feature vector $V_{h|(i,j)}$ associated with the ordered pair (i, j). The union of such vectors for all ordered pairs (for a given h) forms a vector cluster $\check{Z}_h$ at step size h, $\check{Z}_h = \cup_{(i,j)} V_{h|(i,j)}$, providing a single-step representation for the sequence.

The union of vector clusters $\check{Z}_h$ provides a multi-layered feature vector space $\check{Z}_k = \cup_m \cup_{(i,j)} V_{h|(i,j)}$, one layer for each value of h, for the original sequence. The grand vector space $\check{Z}_k$ provides the option of controlling the accuracy and resolution of the solution space by selecting m, the step size for ordered pairs, and q, the dimensionality of the vectors $V_{h|(i,j)}$ in an appropriate manner. Vector $V_{h|(i,j)}$, functioning as a feature vector in this paper, represents the sequential positions of the leading anchor elements of ordered pairs throughout the entire sequence.

Feature vectors for each h-step ordered pair can be structured in at least two different ways. One approach is to simply record the step (spatial index) locations of anchor positions as Boolean values $(1,0)$. This approach is suitable for collections of equal length sequences. An alternative approach is to partition a sequence into q equal segments and record the number of anchor positions that fall into each segment. The number of segments q will determine the dimension of the vectors thus formed. The size of q can be adjusted to meet restrictions or expectations on resolution and accuracy. This approach has the advantage of mapping sequences of unequal length into fixed length feature vectors.

Using the alphabet and sequence from the previous example (Table 1), the following are sample feature vectors for a select group of ordered m-step pairs: Anchor positions of 1-step ordered pairs for (g, c) are located at step (index) locations $[1,6,11,19]$; vector $V_{1|(g,c)}$, is represented by the Boolean vector $< 1,0,0,0,0,1,0,0,0,0,1,0,0,0,0,0,0,0,1,0 >$ if step anchor locations are used directly as vector components. If we instead partition the sequence into 4 equal segments ($q = 4$), the vector $V_{1|(gc)}$, is represented by the 4D vector $< 1,1,1,1 >$ with vector components representing the number of anchor elements in each segment; anchor positions of 1-step ordered pairs for (g, g) are located at step (index) locations $< 4,5 >$; vector $V_{1|(g,g)}$ is represented by the Boolean vector $< 0,0,0,1,1,0,0,0,0,0,0,0,0,0,0,0,0,0,0,0 >$ or by the 4D vector $< 2,0,0,0 >$; anchor positions of 2-step ordered pairs for (g, t) are located at step (index) locations $[1,6,11]$; vector $V_{2|(g,t)}$ is represented by the Boolean vector $< 1,0,0,0,0,1,0,0,0,0,1,0,0,0,0,0,0,0,0,0 >$ or by the 4D vector $< 1,1,1,0 >$.

The MLVS model, with its correspondingly very large vector space, is an excellent domain in which to apply the proposed granular model for decision making. There are numerous ways in which to define partial views in the context of the MLVS model. These include the construction of partial views based on individual ordered pairs specified across one or more step sizes, and partial views based on combinations of multiple ordered pairs specified across one or more step sizes. To illustrate, consider the classification of protein sequences. Such sequences are typically defined in terms of an alphabet consisting of twenty amino acids. Thus, a given protein sequence can be transformed into four hundred MLVS feature vectors in which individual vectors correspond to a specific ordered pair / step size combination. Each vector can be regarded as a partial view and collectively all vectors for a given sequence x represent the set of partial views, $PV_x = \{x_1, x_2, ..., x_{400}\}$. A classification of the protein sequences can be obtained based on the output of the four hundred independent decision making processes defined by the partial views. As described in Section 4, each decision making process would start with analyzing MLVS vectors represented at a coarse level of granularity (i.e. low dimension vectors) and, if necessary, repeat the decision making process using vectors represented at finer levels of granularity (i.e. higher dimensional vectors). The results of the individual decision processes are subsequently combined using an ensemble approach to obtain an overall classification of the protein sequences.

The benefits using the proposed granular model for classifying MLVS feature vectors are twofold. First, classification accuracy may improve as a result of processing data at different levels of granularity by minimizing the curse of dimensionality phenomena. This benefit is particularly significant when analyzing high dimensional MLVS vectors. Second, the process of classifying MLVS vectors can be easily distributed across multiple computing processors and therefore provide a reduction in processing time.

7   Conclusion and Future Work

The major challenges of data analytics comes from the complexity and volume of the data. In this chapter, we propose a comprehensive granular model for decision making in order to tackle both challenges. There are different types of data complexity. One is that the structure of a data instance is complex. For this type of complexity, this chapter suggests that a complex data instance is first decomposed to multiple partial views, each of which has simpler structures. Then the proposed decision model can be applied to those partial views for decision making. Another type of data complexity is that each data instance is associated with multiple heterogeneous features with different natures, formats, and scales. For this type of data, we can first categorize those features into different partial views, based on which the proposed granular model can be applied for decision making. This chapter uses protein sequence classification and stock price movement predication to illustrate how to apply the proposed decision model for both types of complexity.

For complex data with large volume, the proposed decision model first distributes the overall decision making process onto different partial views. On each partial view, the decision making starts with coarsest granular descriptions, which typically requires much less intensive computation. For those data instances where more intensive computation is needed at a finer level of granular representation, often the number of data instances that need to be involved in intensive computation can be reduced by using the results generated at some previous coarser level as filters. We plan to conduct large-scale experimental studies on various types of complex data, including stock data and protein sequence data, to further refine the proposed decision model. Another future research work is to implement a computational framework that supports the proposed decision model on big data computing platforms such as Spark [13, 14].

References

1. Bargiela, A., Pedrycz, W. (eds.): Human-Centric Information Processing Through Granular Modeling. Springer, Berlin (2009)
2. Yao, Y.Y.: Perspectives of Granular Computing. In: Proceedings of the 2005 IEEE International Conference on Granular Computing, pp. 85–90 (2005)
3. Yao, Y.Y.: Granular Computing: Past, Present, and Future. In: Proceedings of the 2008 IEEE International Conference on Granular Computing, pp. 80–85 (2008)
4. Yao, Y.Y.: Granular Computing: Basic Issues and Possible Solutions. In: Proceedings of the 5th Joint Conference on Information Sciences, vol. 1, pp. 186–189 (2000)
5. Yao, Y.Y.: Granular Computing and Sequential Three-Way Decisions. In: Lingras, P., Wolski, M., Cornelis, C., Mitra, S., Wasilewski, P. (eds.) RSKT 2013. LNCS, vol. 8171, pp. 16–27. Springer, Heidelberg (2013)
6. Yao, Y.Y.: An Outline of a Theory of Three-Way Decisions. In: Yao, J., Yang, Y., Słowiński, R., Greco, S., Li, H., Mitra, S., Polkowski, L. (eds.) RSCTC 2012. LNCS, vol. 7413, pp. 1–17. Springer, Heidelberg (2012)
7. Keogh, E.J., Pazzani, M.J.: Scaling up dynamic time warping to massive datasets. In: Żytkow, J.M., Rauch, J. (eds.) PKDD 1999. LNCS (LNAI), vol. 1704, pp. 1–11. Springer, Heidelberg (1999)
8. Lee, S.J., Jeong, S.J.: Trading Strategies based on Pattern Recognition in Stock Futures Market using Dynamic Time Warping Algorithm. Journal of Convergence Information Technology 7(10), 185–196 (2012)
9. Desai, N., Antonopoulos, D., Gilbert, J., Glass, E., Meyer, F.: From genomics to meta-genomics. Current Opinion in Biotechnology 23(1), 72–76 (2012)
10. Akkoç, C., Johnsten, T., Benton, R.: Multi-layered Vector Spaces for Classifying and Analyzing Biological Sequences. In: Proceedings of International Conference on Bioinformatics and Computational Biology, pp. 160–166 (2011)
11. Raghavan, V.V., Benton, R.G., Johnsten, T., Xie, Y.: Representations for Large-scale Sequence Data Mining: A Tale of Two Vector Space Models. In: Ciucci, D., Inuiguchi, M., Yao, Y., Ślęzak, D., Wang, G. (eds.) RSFDGrC 2013. LNCS, vol. 8170, pp. 15–25. Springer, Heidelberg (2013)

12. Johnsten, T., Fain, L.A., Fain, L.E., Benton, R., Butler, E., Pannell, L., Tan, M.: Exploiting Multi-Layered Vector Spaces for Signal Peptide Detection. International Journal of Data Mining and Bioinformatics (to appear)
13. Zaharia, M., Chowdhury, M., Franklin, M.J., Shenker, S., Stoica, I.: Spark: Cluster Computing with Working Sets. In: Proceedings of the 2nd USENIC Conference on Hot Topics in Cloud Computing, pp. 10–16 (2010)
14. https://spark.apache.org

Granularity in Economic Decision Making: An Interdisciplinary Review

Shu-Heng Chen and Ye-Rong Du

Abstract. In this article, we attempt to provide a review of the idea of granularity in economic decision making. The review will cover the perspectives from different disciplines, including psychology, cognitive science, complex science, and behavioral and experimental economics. Milestones along this road will be reviewed and discussed, such as Barry Schwartz's paradox of choice, George Miller's magic number seven, Gerd Gingerenzer's fast and frugal heuristics, and Richard Thaler's nudges. Recent findings from human-subject experiments on the effects of granularity on decision making will also be reviewed, accompanied by various learning models frequently used in agent-based computational economics, such as reinforcement learning and evolutionary computation. These reviews are purported to advance our thinking on the long-ignored granularity in economics and the subsequent implications for public policy-making, such as retirement plans. It, of course, remains to be examined whether the good use of the idea of granularity can enhance the quality of decision making.

Keywords: Granularity, Paradox of Choice, Chunks, Modularity, Heuristics, Nudges, Reinforcement Learning, Evolutionary Computation.

1 Motivation and Background

While the idea of granularity is already rooted in Lofti Zadeh's earlier work on fuzzy sets and fuzzy logic, it is his article, " Fuzzy Sets and Information Granularity," [53] that gives a formal notion of granularity. This notion serves as a foundation for the later development in computing with words and granular computing. However, Zadeh himself notices that the technical notion of information granulation employed

Shu-Heng Chen · Ye-Rong Du
AI-ECON Research Center, Department of Economics,
National Chengchi University, Taiwan
e-mail: {chen.shuheng,yerong.du}@gmail.com

ⓒ Springer International Publishing Switzerland 2015
W. Pedrycz and S.-M. Chen (eds.), *Granular Computing and Decision-Making*
Studies in Big Data 10, DOI: 10.1007/978-3-319-16829-6_3

in [53] is "in a *strict* and somewhat *narrower* sense..." (Ibid, p.3; Italics added). Very slightly he did also mention its *broad* sense.

> Taken in its *broad* sense, the concept of information granularity occurs under various guises in a wide variety of fields. In particular, it bears a close relation to the concept of aggregation in economics;... (Ibid, p. 3; Italics added)

Being economists or social scientists, our interest in granularity may not be so much related to its strict and narrow sense of granularity; instead, what interests us is its broad sense, and a more general notion given by Zadeh is as follows:

> There are many situations, however, in which the finiteness of the resolving power of measuring or information gathering devices cannot be dealt with through an appeal to *continuity*. In such case, the information may be said to be *granular* in the sense that the data points within in a granule have to be dealt with *as a whole rather than individually*. (Ibid, p.3; Italics added)

Zadeh's idea is novel and fundamental. He actually pointed out that the elementary unit of information processing is not a number (in the real space), neither a set of real numbers, but a symbol, a concept, a feeling, a linguistic variable, a module, or a chunk, *a whole rather than individuals*. For whatever other possible names, he called it a *granule*. This granule stands at a higher level over its constituents and has a command over them and can manipulate them.[1] We consider the granule as a more general concept than the fuzzy set. Although the fuzzy set is a way to deal with one specific form of granule, namely, linguistic variables, not all granules are linguistic variables and hence not of all them are fuzzy. For example, signs or symbols studied in semiotics can be another type of granule, but they may have a precise definition or meaning and are not fuzzy.

In this article, we shall argue that granularity, in its broad sense, bears a close relation not just to the concept of aggregation in economics, but more to decision making and policy-making in economics. In a nutshell, granules are what make our decisions simple and efficient, having been coined the "fast and frugal heuristics" by Gerd Gigerenzer [21]. Their formation, development, and evolution are what enables human agents to effectively deal with the complex environment surrounding them.

With this in mind, we shall provide a comprehensive review of the literature which all points to the significance of granules as elementary units of information processing. These include Barry Schwartz's paradox of choice [44], George Miller's magic number 'seven' or chunking [35, 2, 45], and Herbert Simon's hierarchial modularity [4]. The use of granules in decision marking has far-reaching implications, as demonstrated in recent studies on the behavioral foundations of public policies, such as Gerd Gigerenzer's fast and frugal heuristics [21] and Richard Thaler and Cass Sunstein's 'nudges' [48]. We also show that our understanding of human-subject experiments can be dramatically different by using or not using the idea of

---

[1] A typical example is the fuzzy set with its membership function. Through the membership function, the constituents of the fuzzy set are *coordinated* in such a way that the set, as a whole, can be presented.

granularity in modeling their behavior [12]. In fact, granularity has been largely ignored in experimental studies involving adaptive artificial agents. Many algorithms for the designs of adaptive artificial agents, such as reinforcement learning or evolutionary computation, when applied to mimic or to replicate human-subject behavior, are often ignorant of the idea of granularity. We, however, demonstrate some exceptions in agent-based economic models which do take granularity into account when designing their artificial agents.

The rest of the paper is organized as follows. Section 2 reviews the theoretical foundations of granules. Section 3 shows their significance in regard to individuals' decision making or institutions' policy-making. Section 4 reviews the use of the idea of granularity in recent economic models of learning and adaptation either when applied to human subjects in the context of laboratory experiments or when applied to artificial agents in the context of agent-based computational economics. Section 5 gives the concluding remarks.

2 Social Science Theory of Granularity

2.1 Granularity of the Set of Alternatives

The three essential pillars of microeconomics are the utility function (objective function), the set of alternatives, and, finally, choice-making. With the publication of their magnum opus *"Theory of Games and Economic Behavior"* in 1944, John von Neumann and Oskar Morgenstern introduced to economists a paradigm to explicitly structure the three pillars, which is known as the expected utility maximization paradigm. This paradigm has repeatedly dominated the mainstream economics already for half a century. However, the reality of this paradigm has been constantly questioned in economics; accordingly, the structure of the three pillars has been incessantly given different conceptualizations. Among the three, maybe the least addressed one is the *set of alternatives* or, more specifically, the *cardinality* or the *granularity* of the set. While Barry Schwartz's influential book *"The Paradox of Choice"* [44] has already raised the possibility that a proper choice mechanism may not exist when the set of alternatives is too large or too fine, the granularity issue is still largely ignored in economic models of decision making.

2.1.1 The Choice Overload Hypothesis

The paradox of choice originates from a series of human-subject experiments which address the behavior related to choice conflicts, choice aversion or choice deferral. Obviously, in this situation, the subject is not well motivated to make a choice and, instead, prefers an indefinitely longer procrastination or simply not to make a choice. In the literature, the paradox of choice is formally known as the *choice overload hypothesis*. The hypothesis basically says that "an increase in the number of options to choose from may lead to adverse consequences such as a decrease in the motivation to choose or the satisfaction with the finally chosen option" ([43], p.73). The choice overload hypothesis was first proposed by Sheena Iyengar and Mark Lepper [31], and they also tested this hypothesis with a series of three experiments.

In their series of choice experiments, Iyengar and Lepper distinguished the designs with psychologically manageable numbers of choices (limited-choice condition), say, six, from the designs with psychologically excessive numbers of choices (extensive-choice condition), say, twenty-four. In their famous jam promotion experiment, different numbers of jam jars were displayed in two separate places (tables) in a supermarket, one with six different types of jam and one with twenty four different types of jam. They found that while the 24-jam table was able to attract more shoppers than the 6-jam one, it did not successfully beef up their purchasing willingness. In fact, only 3% of the shoppers at the 24-jam table subsequently purchased a jar of jam, whereas 30% of the shoppers at the 6-jam table did that.

In their two additional experiments, this 'more is less' result was also confirmed. In one case, students who were offered more topics (30) to write their essays for an extra credit did not show a higher interest to do so than students who were given fewer topics (6); and for those who actually did so did not perform better as opposed to that of the 6-choice group, in terms of the quality of the essay. In the other case, subjects were either given 6 or 30 different chocolates to choose. It was found that subjects with 30 choices might initially be more cheerful with this large assortment, but the choice process turned out to be difficult and frustrating and the result was that they were often not satisfied and felt regretful.

Since the freedom to choose is a cornerstone of any democratic society, the choice overload hypothesis does lead to an upheaval among academics and the public, which prompts more follow-up studies. Scheibehenne, Greifeneder, and Todd [43] provided a meta-analytic review of 29 articles (published and unpolished) with 50 experiments, from 2000 to 2009, involving 5,036 subjects. Among these 50 experiments, the minimum number for the limited-choice condition was 3, whereas the maximum number for the extensive-choice condition was 300. Using a random effecte model, they found that the results were mixed, neither supporting the choice overload hypothesis ("more is less") nor its opposite ("more is better").

In their meta-regression analysis, they, however, tended to suggest that the experimental results may be sensitive to some control variables pertinent to the design. Among them, maybe the most important one is the *experience effect*. If the subject is very experienced with the choice problem presented to them, for example, living in a town for years and having sufficient time to know all the restaurants around, he/she may not have a hard time choosing a place for lunch. In this situation, more choice can be better. In fact, this control variable was carefully fine-tuned when Iyengar and Lepper [31] initialized this line of research.

> In addition, to provide a clear test of the choice overload hypothesis, several additional methodological considerations seemed important. On the one hand, to minimize the likelihood of simple preference matching, *care was taken to select contexts in which most participants would not already have strong specific preferences.* (Ibid, p. 996; Italics added)

This may justify the use of exotic products in testing the choice overload hypothesis. What underlies the experience effect or the familiarity effect is the information regarding each alternative and the mechanism used to process the supplied information. The latter is further related to how the information is presented to the subject, i.e., the structure of the information, to which we now turn.

2.1.2 Characteristic Analysis

While making a choice, subjects may have to ask how this alternative is different from others. If their difference noticeably lies in one dimension, for example, 100 baskets containing different numbers of peanuts, then the choice overload issue may not happen because consumers can at least identify exactly what they want, for example, the basket with the maximum number of peanuts. Nonetheless, each alternative may have a number of attributes and they may differ in each attribute. This may cause the information required to distinguish them overwhelming and make a selection hard. Of course, issues can become simpler if these attributes are not presented in a wide flat, but can be endowed with a hierarchical structure.

In consumer theory, Kelvin Lancaster pioneers a different approach to the choice problem, called *characteristic analysis* [29]. In characteristic analysis, commodities (for example, different brands of toothpaste) are characterized by their attributes (characteristics) and the density (quantity or the quality) of those attributes. When presented with a set of alternatives, consumers search for the commodity which is closest to their desired attributes after taking into account the price they are required to pay. Lancaster further assumed that there is a hierarchical structure of these characteristics [30]. From an information processing viewpoint, this hierarchical structure enables decision makers to have a sequential decision process to deal with complex choice problems.

> A decision process in which a choice involving a *restricted number of parameters* is made, after which a further choice is made from *another restricted set of parameters*, and so on down the sequence, is necessarily hierarchical unless it is purely random. The ordering of the hierarchy determines which set of parameters is considered *first*, *second*, and on through the sequence. (Ibid, p.50; Italics added)

The implication of the quotation above is that the assortment structure is an important control variable while testing the choice overload hypothesis. Findings related to this observation are summarized well in [43]. For example, Mogilner, Rudnick, and Iyengar [36] found that an increase in the number of alternatives decreased satisfaction only if the alternatives were not displayed in categories.

To sum up, the paradox of choice, by and large, may exist only as a transition process as a short-term phenomenon.[2] Although humans are limited in their cognitive capacity, they can learn, adapt and hence cope with complex decision problems by developing decision heuristics to simplify hard choice problems. This adaptation process is, in fact, a granulation process. This granulation process can happen for both suppliers and consumers. For the suppliers, when a large number of options

---

[2] See [24] for some related discussions.

are displayed, these options will be ordered, categorized, grouped, and be given a hierarchical structure. For the consumers, they can take advantage of the given hierarchical structure or search with their own heuristics, for example, the elimination strategy, to sequentially reduce the search space and locate what they really want.

While the granulation process in a hierarchical manner is a way to escape from the paradox choice, how is the degree of granularity at each level determined? Why intuitively must it have a *coarse* division instead of a fine division at each level, as Lancaster has pointed out "*a restricted number of parameters*"? We shall address this question in the next section.

2.2 Cognitive Foundation of Granulation

2.2.1 Chunking and Magic Numbers

In this section, we try to examine the psychological foundation of granulation in the hierarchical form. A classic work which one cannot afford to miss is Miller's famous number *seven* [35]. Miller (1956) [35] is a celebrated contribution to psychology in the discussion of *short memory capacity* or *working memory capacity*. In this regard, it is about the number of items that an individual can discriminate or is about the capacity to remember information over very short periods of time, say, seconds. Based on a few experiments that he reviewed, Miller concluded that most people can correctly recall about 7 ± 2 items. This is the origin of the magic number seven.

For the purpose of this chapter, the significance of this work [35] is three-fold. First, it shows that through the granulation process a human can increase his memory span. In other words, granulation can be understood as a psychological process to enhance humans' capability to deal with a complex environment characterized by a large amount of information. This immediately brings us back to the early discussion of humans' capability to deal with the paradox of choice (Section 2.1). Second, it shows that the granulation process proceeds in a hierarchical manner. Third, while without being given an exact definition, the linguistic variables, as we shall see below, play a pivotal role in the granulation process. Hence, even though we argue in the very beginning of the chapter that a fuzzy set is only a special form of granulation, due to the heavy reliance on linguistic terms in the granulation process, the fuzzy set is clearly indispensable to the development of a general theory of granular computing.

Without rephrasing what he actually said and hence not distorting what he actually meant, we shall use two quotations directly from Miller's article to point out these connections. About the granulation process per se, he said the following.

> In order to speak more precisely, therefore, we must recognize the importance of grouping or organizing the input sequence into units or chunks.... In the jargon of communication theory, this process would be called *recoding*. The input is given in a code that contains *many chunks with few bits per chunk*. The operator recodes the input into another code that contains *fewer chunks with more bits per chunk*. There are many ways to do this recoding, but probably the simplest is to group the input events, apply *a new name* to the group, and *then remember the new name rather than the original input events*. ([35], p.93; Italics added.)

Chunking is probably the most influential idea we learned from Miller's studies [35]. The quotation above makes a distinction between *items* and *chunks*. With this distinction the granulation process can be regarded as a transition process from *many chunks with few bits per chunk* (items) to *fewer chunks with more bits per chunk* (chunks). This transition process is simply a process of *information compression*. While Miller's study was conducted in the middle of the 1950s, almost a decade earlier than the advent of *algorithmic information theory*, independently founded by Andrey Kolmogorov, Ray Solomonoff, and Gregory Chaitin in the mid-1960s, the idea of information compression as formations of chunks is already in the paper, as we quoted above. This helps clarify the subsequent discussions on *what exactly the magic number is*.

In their recent article, Fabien Mathy and Jacob Feldman [34] reconcile two versions of the magic number using a notion of *Kolmogorov complexity* and incompressibility. The two versions refer to the original seven (7 ± 2) and the later version of four (4 ± 1) [18]. Mathy and Feldman [34] assert that four is the true capacity of short-term memory in *maximally compressed units*, while Miller's magic number seven refers to the length of an uncompressed sequence. This number, seven or four, gives us a cognitive reason for granulation. To use our limited cognitive capacity more efficiently in order to increase our memory span, we tend to harness individuals as granules (compress items into chunks), and as a maximum we are able to have three to five chunks *at a level*.[3] These magic numbers correspond well to the choice overload hypothesis, which seems to indicate that if options are not arranged into categories (not compressed into groups), then, when the number of options increases beyond a threshold (the magic number), our motivation to make a choice or the satisfaction resulting from the option chosen will decrease.

2.2.2   Hierarchical Structure of Granules

As to the hierarchical form of granulation, Miller emphatically involved the ideas of what is currently known as *encapsulation*, which will also be discussed in Section 2.3.

> In my opinion the most customary kind of recoding that we do all the time is to translate into a *verbal code*. When there is a story or an argument or an idea that we want to remember, we usually try to rephrase it "in our own words." When we witness some event we want to remember, we make a verbal description of the event and then remember our *verbalization*. Upon recall we recreate by *secondary elaboration the details* that seem consistent with the particular verbal recoding we happen to have made. (Ibid, p. 95; Italics added)

Based on the quotation above, *the details in secondary elaboration* seem to indicate what are *inside* the chunks. While Miller did not make the hierarchical or recursive structure of chunks explicit, the subsequent interpretations of Miller's work do notice these branches. For example, Baddeley raises the following question [2].

---

[3] As we shall see below, there will be a hierarchical structure of these granules.

The situation is further complicated by the possibility of setting up hierarchical structures of chunks. If seven chunks can be held, *can each one be divided into seven subchunks?* Presumably not, because that would suggest that one can hold 49 chunks. Perhaps the number seven, itself, comes from chunking. ...My own view is that it is unlikely that the limit is set purely by the number of chunks, independent of such factors as *the degree to which material within each chunk is integrated as a result, for example, of prior learning.* (Ibid, p. 355; Italics added)

Although Baddeley [2] did not give a clear answer with regard to the hierarchical structure of chunks, he did correctly point out that it would be hard to count the number of chunks independently of *prior learning*. In the following section, we shall introduce another notion related to granularity from the complex system perspective. There we shall see the relationship between learning and the evolution of the hierarchical granular system.

2.3   Hierarchical Modularity

Our discussion in the previous section indicates that what matters for the working memory capacity may not solely just be a number, 7 or 4. The question which we should really ask is: *what is inside the chunks?*. According to algorithmic information theory, chunks are compressed messages like a program with *minimum description length* [34]. How are these compressions actually made? Do they rely on some other existing programs or building blocks to facilitate the compression? If so, where do these building blocks come from? Has learning anything to do with them? This series of questions leads us to a highly influential concept in complex systems, namely, *modularity*. In this section, we shall first briefly review Herbert Simon's original work on *modularity* [46], and we shall then use LISP programming and genetic programming to demonstrate the learning process as a development of a hierarchical modular structure.

2.3.1   Modularity

This section is inspired by Herbert Simon's work on *near decomposability* or *modularity* [46]. Modularity refers to a structural relationship between a system as a whole and the constituent components which can function as independent entities. The interactions of the elements within the same constituent component are strong; however, the interactions of elements across different constituent components are weak, but not zero. The latter property is known as *near decomposability*. The chunk or granule, as a collection of items that have strong associations with one another but have much weaker associations with other chunks concurrently in use, is a typical near decomposable system. As Simon has argued, near decomposability is a key to harnessing a possibly unbounded complex system.

Simon [46] was probably one of the most influential pioneers inspiring many follow-up works in various scientific disciplines [4]. In addition to near decomposability, Simon viewed *hierarchy* as a general principle of complex structures. He advocated the use of a hierarchical measure – the number of successive levels of hierarchical structuring in a system – to define and measure complexity; furthermore, he argued that hierarchy emerges almost inevitably through a wide variety of evolutionary processes, for the simple reason that hierarchical structures are *stable*.

In addition to the depth of a hierarchy, Simon also noticed the span or the width of a hierarchy at each level. He defined the *span of a system* as the number of subsystems into which it was partitioned. Although he did exemplify some hierarchies with large or even indefinitely large spans, the so-called *flat hierarchies*, his attention was mainly drawn to the hierarchies of *moderate* span. While he did not give an exact range for a moderate span, it is our conjecture here that the magic number discussed in Section 2.2.1 can pinpoint a reasonable niche.

2.3.2 LISP

One example which is useful for us to think about the connection between the hierarchies of moderate span and granules or chunks is the *symbolic system*. In a symbolic system, the elementary units are alphabets (symbols). Using the grammar applied to the system, one can, syntactically correctly, generate words, sentences, books, and volumes. Each of these generated objects can be a granule or a chunk at different levels of a hierarchy of moderate span. Each of them, in an encapsulated form, may have some degree of independence, and can be reused as a chunk of other hierarchies.

In computer science, this is basically what *formal language theory* is about. Some computer languages clearly demonstrate this hierarchical structure, for example, LISP.[4] Each LISP program, regardless of its size, as a whole, is a *list*.[5] However, it may have other (sub)lists as its constituents, and each of them is also an

---

[4] LISP stands for List Processing, which is a high-level computer language invented by John McCarthy (1927-2011) in 1958 at MIT as a formalism for reasoning about the use of certain kinds of logical expressions, called recursion equations. This language is strongly motivated as a practical implementation of the λ calculus or the recursive function theory developed in the 1930s by Alonzo Church (1903-1995) and Alan Turing (1912-1954). See [1] for details.

[5] LISP S–expressions consist of either *atoms* or *lists*. Atoms are either members of a *terminal set*, that comprise the data (e.g., constants and variables) to be used in the computer programs, or they are members of a *function set* that consists of a number of prespecified functions or operators that are capable of processing any data value from the terminal set *and* any data value that results from the application of any function or operator in the function set. Lists are collections of atoms or lists, grouped within parentheses. In the LISP language, everything is expressed in terms of operators operating on some operands. The operator appears as the left-most element in the parentheses and is followed by its operands and a closing (right) parenthesis. For example, the S-expression $(+ X \ 3)$ consists of three atoms: from the left-most to right-most they are the function "+", the variable X and the constant 3. As another example, $(\times X \ (- Y \ 3 \ ) \ )$ consists of two atoms and a list. The two atoms are the function "$\times$" and the variable "X," which is then followed by the list $(- Y \ 3 \ )$.

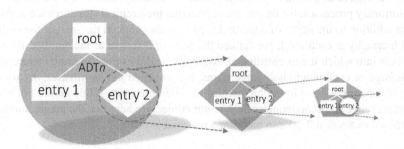

Fig. 1 Hierarchy of Chunks and Automatically Defined Terminals

This figure exemplifies an example of a hierarchy of chunks (granules or modules) through automaticaly defined terminals. Each chunk (granule or module), as represented by a LISP parse tree, has a span of two (two terminals, entry 1 and entry 2). However, each entry itself is also a chunk and has a span of two. This *self-similarity* can recursively go on and on for many levels.

independently implementable program. The macro list can been encapsulated and reused as a subroutine by other programs. This is what John Koza called the *automatically defined functions* (ADFs) [28], or alternatively, automatically defined terminals (ADT) [17] (see Figure 1 for a demonstration). Each automatically defined function or terminal can contribute to information compression as we mentioned in Section 2.2.1, i.e., they help reduce the otherwise more lengthy messages (codes) into smaller ones. Hence, they are chunks used to construct other chunks (see Figure 2 for an illustration).

By presenting all these chunks together, we can see that there is an underlying timeline as indexed by the generation number in Figure 2. To obtain more hierarchical chunks, some low-level chunks have to be developed first, such as ADT8, ADT9, ADT10, and ADT12 (Figure 2). In other words, there is a time order connecting these chunks, for example, ADT12 precedes ADT18 and ADT19. This timeline realized into the real world is, in effect, what from George Miller to Alan Baddeley has been called *learning*. This shows why Miller's magic number seven is not that absolute, because, at most, it only gives us the size of the span at each level. What really matters is the depth of the hierarchy, which is a consequence of learning.

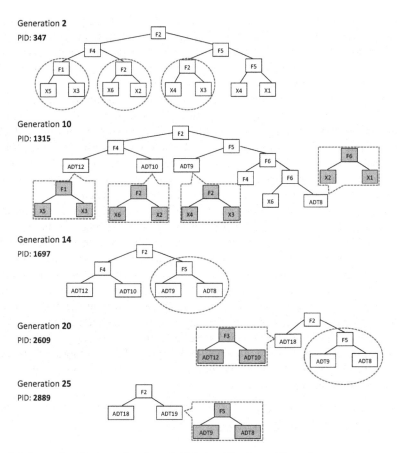

Fig. 2 Information Compression through Existing Chunks and Learning

This figure demonstrates a development process of a hierarchy of chunks (granules, modules). Here, some simpler automatically defined terminals (in terms of depth) are developed first, and they are encapsulated and are used as chunks (building blocks) to form more complex hierarchy. However, by using the existing chunks, the later derived hierarchies, no matter how complex they are, have a depth of no more than two.

3   Policy-Oriented Applications of Granularity

3.1   Granular Decision Making

The previous section, based on psychology, information theory, and complex science, has suggested a behavioral foundation of, as we shall coin in this section, *granular decision making*. The granular decision making refers to a hierarchical decision-making procedure which organizes an entire decision problem into several levels (layers). At each level, only a few granules are presented and need to be looked at; "coarser" granules are arranged at the higher level of the hierarchies, and "finer" granules are arranged at the lower level of the hierarchies. The decision maker starts at the first level (the topmost level) of the decision hierarchy, and stops at the level at which a decision has been made.

While we may have introduced a neologism, by no means do we want to claim its novelty. In fact, some familiar hierarchical decision frameworks have already existed, for example, the *analytical hierarchy process*.[6] However, this hierarchical decision framework is not necessarily applied in a way that is consistent with what is discussed in Section 2. For example, the number of granules required to be given at each level is not constrained by the 'magic number', and a decision cannot be made until all the information in the hierarchy has been processed.

This decision process may sound more systematic, comprehensive, and rigorous, but the fundamental question is: when can we trust the decision made through this process, and when should we better trust our gut feeling? Needless to say, the resolution of the issue is beyond the scope of the paper. However, we would like to point out that there are other hierarchical structures which are consistent with the essence of the granular decision making. Their existence and prevalence can been reviewed from two aspects. First, from the individual viewpoint (Section 3.2), we want to show how the granular decision making has been substantiated into practices to solve many of our daily life problems. Second, from the institutional viewpoint (Section 3.3), we want to show how public policy can become more effective or welfare enhancing if the *choice architecture* can be well taken care of in light of the granular decision-making framework.

3.2   Fast and Frugal Heuristics

The fast and frugal heuristics is pioneered and advocated by the German psychologist and behavioral economist, Gerd Gigerenzer. Over the years, he and the research that he had led at the Max Planck Institute have extensively studied how people actually make decisions when they are presented with complex, vague, and ill-defined problems. It is found that many heuristics, while they may look simple, can effectively solve problems in a fast and frugal manner, and even the results are sometimes better than those of deliberated complex models [23]. In the literature, there is a glossary of such heuristics available, such as the recognition heuristic, fluency heuristic, take-the-best heuristic, and one-good-reason heuristic.[7]

To give a highlight of these heuristics and their relationship with the granular decision making, one needs to know that Gigerenzer considered himself to be a behavioral economist following the legacy of Herbert Simon. Under the influence of Simon's notion of bounded rationality, behavioral economists characterize each

---

[6] Due to the space limitations, we do not intend to give a review of the large body on literature in the analytic hierarchy process. The interested reader is referred to [40, 25].

[7] For a review of these heuristics and many more, the interested reader is referred to [22].

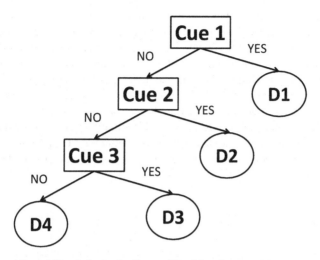

Fig. 3 Fast and Frugal Heuristics in the Form of Decision Trees

decision process with three main stays, namely, a *search rule*, a *stopping rule*, and a *decision rule*. A simple but abstract example is the *alphabet heuristic*.

To use the alphabet heuristic, each alternative is presumably characterized by a 'name' (a series of letters). By means of this heuristic, we search through the name of alternatives letter by letter in reading direction, and assess each letter's position in the alphabet (the search rule). If the letters in the first position differ among the alternatives, we only keep those whose letters appear the earliest in the alphabet, and remove the rest (the decision rule). If there is more than one left, we move to the second position, and then ditto the above procedure until there is only one left (the stopping rule). Then only the remaining one is chosen (the decision rule).

The alphabet heuristic has long been known as *lexicographic ordering* in economics. Its lexicographic structure is generally shared by the aforementioned heuristics, and is consistent with the granular decision making in the following regards. First, it has a hierarchical structure, first letter, second letter, and so on. Second, each letter actually has only two granules (a span of two), "the leading position in the alphabetic order" or "the positions behind it". Third, to make a decision, it is not necessary to go through all levels (or to the last level), as long as a decision can be made at an earlier stage. Fourth, learning plays a role in the formation of this hierarchy. In the case of the alphabet heuristic, one still needs to decide the alphabetical order: should it be 'ABC' or 'CBA'. Psychologists call these *cues*. In principle, one should prioritize those cues which are more informative or discriminating; for some decision problems this order of cues can be determined easily, but for some decision problems this order requires experience and learning.

The other hierarchical structure which is frequently used in machine learning, but less in psychology or behavioral economics, is *decision trees*. In fact, the fast-and-frugal heuristics reviewed above can be regarded as a special kind of decision tree, a binary decision tree as shown in Figure 3. In structure, it is much simpler than the general decision trees. It has only two nodes at each level, and at least one of the two is a terminal node. Because of this characteristic, each cue will appear only once in the tree. However, for a general decision tree, the number of nodes at each level varies, depending on the number of attribute values (not necessarily binary), and each cue can appear more than once as long as it does not repeat itself in the same subtree emanating from itself. Finally, the general decision trees are normally built using statistical algorithms, such as the entropy maximization algorithm, in the data mining context. Hence, learning in this situation is more formal and data-oriented; less depends on personal experiences or memory. Despite these technical differences, general decision trees, by and large, are also a kind of granular decision making.

3.3   Nudges

Social science research reveals that as the choices become more numerous and/or vary on more dimensions, people are more likely to adopt simplifying strategies. The implications for choice architecture are related. As alternatives become more numerous and more complex, choice architects have more to think about and more work to do, and are much more likely to influence choices (for better or for worse). ([48], p. 95)

In Section 2.1.2, we have already mentioned that if options offered to decision makers can be well structured, then choice overload can be alleviated. An implication of this is then the way in which we present the options may affect what is chosen or not chosen. This lesson has already become known as the *framing effect* [49] for three decades. It is about two essentially identical choices provided to subjects, but one is presented in a *positive* frame, say, a bonus, and the other is presented in in a *negative frame*, say, a penalty. Simply by these different "phrases" people will be led to make a different choice. In a sense, the attention drawn to the choice-architecture effect can be regarded as a more extensive awareness of this framing effect.

Before we proceed further, it will be useful to mention one recent influential book in behavioral economics, entitled *Nudges*, authored by Richard Thaler and Cass Sunstein [48]. In a nutshell, what a *nudge* does is to make a *default* option available, and to use it to lessen the choice burden of people and to enable them to make a good choice. Some studies have shown that, by properly designing and incorporating a default option, we can have a choice architecture that can nudge people towards good decisions on spending, saving, health care, borrowing patterns, and organ donations. One of the most illustrative examples of the default option is to make enrollment in 401(k) plans *automatic* for new employees with a form for *opting out* so that if they do not wish to save they would have to register the desire to opt out of the plans. This *opt-out* system with a reasonable default saving plan is likely to result in greater retirement savings than an *opt-in* system [3].

What is the relevance of granularity to defaults? The answer is that, based on what we discuss in Section 2, *defaults are a kind of granule*. Depending on how a default is articulated, it can mean many different things with fine details. Using Miller's term (Section 2.2.1), it is a *name* or a verbal expression used for the information compression purpose. Once after a name is given we do not look further into the details being encapsulated until it is necessary to do so. Therefore, defaults share the psychological, cognitive, information processing, and complexity nature of granules (chunks, modules). The contribution of this chapter is to make their connection with granules clear.

The purpose of this chapter is not to provide an economic analysis of defaults. However, as Madrian [33] has pointed out, the economic studies on defaults is still very limited. Obviously, the significance of defaults in public policy should not be overrated. They are not panaceas. First, to make them work, they have to be properly designed. In our early example, we can have 'pop-in' or 'pop-out' as the default, but the latter (the automatic enrollment) is more effective than the former in increasing the participation rate of the retirement saving plan. Second, the function of a default or defaults should be evaluated in the hierarchy in which they are placed. Choosing a default may not mean the end of the story. For example, after automatic enrollment in the saving plan, the policy-makers can still offer few, but not many, options for employees to choose from.[8] In this situation, the default option can be applied again, which may refer to an endorsed fund. Hence, the design should be continued through the branch emanating from a default node. These additional layers certainly may take better care of the heterogeneous needs of employees than just the one-layer simple default.[9] Hence, from the granularity point of view, good policy-making does not rest upon a single default granule, but a granular hierarchy in which different defaults may be present at different levels.[10]

4   Granularity in Economic Modelling

Based on Duffy [19] and Chen [10], the two major learning or adaptive algorithms used in agent-based modeling are reinforcement learning and evolutionary computation (genetic algorithms and genetic programming). While both of these algorithms have been modified and extended into different forms, little attention and effort have been directed toward the granularity issue involved in the operation of the algorithms.

---

[8] The consideration is mainly to avoid the paradox of choice. Iyengar et al. [26] show that expanding the number of funds offered for the 401(k) contribution plan will lower the participation rate, which drops from 75 to 60 percent when the number of funds offered increases from only 2 to 59 options.

[9] Caroll et al. [8] provide a similar discussion. They actually suggest applying an active choice instead of defaults when there is substantial heterogeneity in consumer preferences.

[10] In practice, Amazon, NetFlix and many of the like, have shown themselves to be the masters of choice architecture as their entire websites are geared to enabling consumers to make desirable choices and continue transacting with them. As to the tools available for choice architects, the interested reader is referred to [27].

4.1 Reinforcement Learning

Let us first look at reinforcement learning. From [39], to [20], and further to [5], we have already extended the simple reinforcement learning algorithm to its generalized version by taking into account various cognitive or mental considerations, such as memory, consciousness, and reasoning. Nonetheless, the most fundamental unit is the choice or the action being reinforced, and hence the set of alternatives is the starting point for any version of reinforcement learning. From the granulation viewpoint, a fundamental question is: to make reinforcement learning applicable to understand a certain class of adaptive behavior, is there a granulation constraint that is required to be satisfied with, for example, the size of the set of alternatives? Although this question has been asked in a few studies, we believe that the granularity issue has never been addressed explicitly in any application of reinforcement learning known to us.

4.1.1 Beauty Contest Game

We shall illustrate this point by referring to some applications of reinforcement learning models to human-subject experiments. The specific one considered by us is the *beauty contest game*, which is also known as the *guessing game*. Players in the *beauty contest game* compete with each other to win a prize by selecting a number between $[0, 100]$. The prize is given to the player whose guessed number is closest to the *target number*, which is calculated by averaging all guesses that are then post-multiplied by a factor p, say, $p = 2/3$. With this parameter, the game is called *the p-beauty contest game*. This game implicitly requires each player to form his/her expectations of other players' expectations. If other players are doing the same thing, the game then suffers from the familiar *infinite regress problem*. Under the homogeneous rational expectations hypothesis, a Nash equilibrium will be reached where everyone chooses an equilibrium of *zero*, which is the result of 50 (the middle point between 0 and 100) post-multiplied by p^∞. However, the resultant beauty contest experiments have demonstrated great deviations from this game-theoretic prediction [37, 38]. The application of reinforcement learning to the repeated beauty contest game was initiated by [5]. In their study, reinforcement learning was applied to a set of 101 alternatives.

In other words, each number that the subject can guess is an independent choice, and by the operation of reinforcement learning in the end the strategies demonstrating larger actual or simulated effects will be chosen more frequently. However, Camerer and Ho [5] found a lack of ability in reinforcement learning for modeling guessing dynamics in the beauty contest game, compared to other noncooperative games which are surveyed in their article. This result was also replicated in [12]. One of the remedies suggested in [5] is to consider the learning when players sophisticatedly realize that other players are learning as well.

Sophistication is a central concept in the beauty contest game for producing level-k reasoning and it has been put into practice in [6]. Chen and Du [12] also considered level-k reasoning during the learning process, yet the motivation and operation is different. In order to explicitly represent the granulation constraint imposed in learning dynamics, they suppose that the basic granules used for learning are only the six level-k rules instead of the 101 numbers. The learning model with these granules seems to be more descriptive than the one [5] which uses individual numbers directly. This result may lead to a fundamental question concerning the applicability of reinforcement learning to the situation when a large number of many possible choices are presented.

4.1.2 Coordination Game

The granularity issue can also be found in the application of the reinforcement learning model to the experimental behavior in the coordination game. The analytical framework was initiated by Van Huyck and his colleagues [51]. Let s_i^j be the action j of individual i and $s_i^j \in S = [0,1]$. The game is defined by the following payoff function $\pi(\cdot)$,

$$\pi(s_i, s_{-i}) = 0.5 - |s_i - \omega M(s)(1 - M(s)|\quad (1)$$

where s_{-i} denotes the actions of other players, $s = (s_i, s_{-i})$ is one action combination, $M(s)$ is the median of s and $\omega \in (1,4]$. There are two strict Nash equilibria, $(s, M) = (0,0)$ and $(1 - 1/\omega, 1 - 1/\omega)$.

Van Huyck and his colleagues conducted the human-subject experiments for this game [50]. The action space is discretized into a finite set containing 101 actions such that $S = \{0, 0.01, 0.02, ..., 0.99, 1\}$. Subjects in the group of five enter each session playing repeatedly for a total of T periods, where $T = 40, 70$ or 75. This experimental data offers an opportunity for researchers, not only Van Huyck and his colleagues, to study the learning behavior by fitting various kinds of learning models [16, 42, 50]. Reinforcement learning serves as a common algorithm fitted in all of those studies.

Some authors have well recognized that the size of the set of alternatives might be too large to be true, and they further suggested that the *similarity* among strategies should be used to simplify the decision problem [16, 42]. They assume that the strategies are ordered numbers and $s^j < s^k$ if $j < k$. The common similarity function being considered in those studies is the *Bartlett similarity function*, which supposes that the degree of similarity decreases linearly as the distance between the chosen strategy s^k and other strategies $s^j, \forall j \neq k$, increases. The similarity function is defined as follows:

$$f(s^j, s^k; h) = \begin{cases} 1 - |j - k|/h & \text{if } |j - k| \leq h, \\ 0 & \text{otherwise.} \end{cases}\quad (2)$$

where h determines the $h - 1$ unchosen strategies on either side of the chosen strategy. The updating rule for the propensity or attraction of strategy j, $R^j(t)$, is governed by

$$R^j(t) = rR^j(t-1) + f(s^j, s^k; h)\pi_k(t) \tag{3}$$

where $\pi_k(t)$ is the payoff of chosen strategy k in round t.

The incorporation of strategy similarity into the operation of reinforcement learning grasps the idea that the neighborhood of actions or strategies (within a granule) should be dealt with *as a whole rather than individually*. On the one hand, we might argue that what makes a granule is not explicitly defined here. On the other hand, if we could consider each chosen *case* as a centroid of a granule, the number of granules might grow throughout the action space as the game is played repeatedly. In the extreme case, the number of granules will be equal to the number of strategies bringing in another granularity issue needing to be solved.

4.1.3   Market

Previous mentioned applications of reinforcement learning, including the guessing game and the coordination game, are both examples of strategic games which are simpler. Market participants usually encounter a more complex learning task. Even though the rules and settings can be controlled and simplified in the laboratory, the time spent for instruction is usually longer and the practice session is always required to insure their comprehension of the *experimental markets*. It is also more difficult to identify what or to characterize how the subjects learned. Give those complexity , it seems that the issue of granularity, even though not been mentioned explicitly, was naturally and inevitably considered when applying reinforcement learning to model the adaptive process in experimental markets.

Chen and Hsieh [13] apply reinforcement learning to an order book-driven experimental prediction market. They assume that the mental representation of the decision problem has been simplified to *the choice of the intensity of limit order submission*. It is found that the intensity of limit order submission is positively correlated with the profit earned. They further assume that a ternary choice problem to reduce the possible dimensions of learning. In particular, the reinforcement mechanism takes effect on updating the propensity of three alternatives including increase the use of limit orders, decrease it, or keep it unchanged. In fact, it assumes what really matter is not a precise degree of intensity but a rough class with a coarse granule.

The second example is an application of generalized reinforcement learning to the adaptive process in *bilateral call market* [7]. In this market, a buyer and a seller are randomly matched each other and privately informed about their values or costs of the goods. They submit bid or ask anonymously and a trade is made at the midpoint when the buyer's bid is higher than the seller's ask. Subjects need to learn how to submit a desirable bid/ask conditional on his private value/cost on hand. For a buyer i, the strategy s_i being considered is the combination of the assigned private value v_i, $v_i \in [0, 200]$, and his submitted bid b_i, which will be determined by the choice of markdown ratio from 0 to 1. In order to implement reinforcement learning model, Camerer, Hsia, and Ho [7] discretize v_i into 10 equal intervals and 16 evenly spaced markdown ratios from 0 to 75%. We might consider this discretization as necessary although the resulting strategy space with 160 alternatives seems to be far from the range of the 'magic number' (Section 2.2.1).

In addition to discretization, Camerer and his colleagues introduced a parameter τ into the generalized reinforcement learning model to characterize the *similarity-based generalization of learning*, which shares the common motivation with other work introduced in Section 4.1.2 [16, 42]. The τs are different for each strategy and are sensitive to the distance between the strategy being considered and the chosen strategy. The definition of τ is given as follows:

$$\tau = e^{-\psi|v-v_i(t)|-\omega|b-b_i(t)|} \tag{4}$$

where $v_i(t)$ is the realized private value and $b_i(t)$ is the chosen bid of buyer i at time t. For the chosen strategy $s_i(t) = (v_i(t), b_i(t))$, τ is equal to its maximum 1 because $v - v_i(t)$ and $b - b_i(t)$ are equal to zero. On the other hand, for the non-chosen ones, τ can be close to zero depending on the magnitude of ψ and ω. The influence of parameter τ is so overwhelming that it can turn on/off the updating process of strategy attraction[11]:

$$A_i(t) = \frac{\phi^\tau N(t-1)A_i(t-1) + \tau\pi_i(b_i, v_i, v_{-i}(t))}{\phi^\tau(1-\kappa)^\tau N(t-1) + \tau} \tag{5}$$

Notice that when $\tau = 0$, $A_i(t)$ is equal to $A_i(t-1)$ indicating all adaptive mechanisms/parameters take no effect.

We have discussed and commented the incorporation of strategy similarity into the operation of reinforcement learning in Section 4.1.2. We would like to make a final remark on it by first quoting a paragraph from [7].

> Similarity-based learning is also arguably a "cognitively economical" heuristic because scarce attention is allocated where it is likely to be most useful - namely, in the vicinity of the current valuation and bid. (Ibid, p. 257)

In other words, subjects may focus only on the chosen strategy due to limited attention, and the other unchosen strategies are reinforced by the "spillover effect" generated from an automatic, unconscious process of having less requirement of cognitive load. This attention-based similarity learning further alleviate the dimensionality problem *when updating the attraction levels*. However, the *further calculation of choice probability* is still based on the strategy attractions of 160 finer granules (after disctretization), which implicitly assumes that subjects, in each round of their choice making, are still required to distinguish the attraction of each of this overwhelming number of alternations. Hence, we arrive the same conclusion that we have indicated in Section 4.1.2: this similarity device alone is not sufficient enough to solve the granularity issue.

---

[11] In order not to distract the attention of the readers, more details of the updating rule refer to [7].

4.2   Evolutionary Computation

Evolutionary computation involves *populations* of programs, strings of symbols, or candidate solutions. In agent-based economic models, sometimes the entire population is used to represent a whole society in a one-to-one mapping, known as *population evolutionary computation*; sometimes, the whole population is used to represent a single individual, known as *individual evolutionary computation*. Economists have frequently used social learning and individual learning to distinguish the former from the latter in their agent-based modeling [52]. The granularity issue does not appear in the former case where each individual agent is only associated with one single strategy and hence no choice. However, it may occur in the case of individual learning where each agent is endowed with a population of strategies and each time upon making a decision or taking an action he/she needs to choose one from them. Will they encounter the issue of the paradox of choice when the population size gets big? Will they need a proper choice architecture to deal with the large population of options? Will they need a search strategy to limit their search efforts. This issue, to the best of our knowledge, has never been raised.

In early days when evolutionary computation had only just been introduced to economics and applied to build agent-based economic models, the idea of 'more is less' had not even been formally proposed. Barry Schwartz's book was available only after 2003, and before 2003 economists were not fully aware of the issue of choice overload. On the other hand, one may argue that the choice overload problem does not apply because in most applications alternatives differ in a single dimension only, namely, fitness or performance. In this case, like the early example which we mention, one only needs to search for the basket with the maximum number of peanuts. Without too many additional attributes, making comparisons among these alternatives is straightforward. Hence, even though some economists may still be concerned with the parameter of population size, they are motivated by different reason [32].

After the middle of the 2000s, some economists started to notice the connection between cognitive psychology and population size, and even Miller's magic number has also been cited as a reference to determine the population size [9]. A research agenda developed along this research line is to use agent-based modeling to address the effect of cognitive capacity on earnings performance under a competitive market environment, and population size is used as a proxy variable for cognitive capacity [14, 15]. Even up to this step, whether or not our artificial agents are overloaded with too many choices is not a concern.

Exceptions do, however, exist. In a follow-up study of the Santa Fe artificial stock market, Tay and Linn [47] state:

> [S]ome might question whether it is reasonable to assume that *traders are capable of handling a large number of rules* for the mapping of market states into expectations, each with numerous conditions, ... We show this by allowing agents the ability to *compress information into a few fuzzy notions* which they can in turn process and analyze with fuzzy logic. (Ibid, p. 322; Italics added.)

Indeed, the above quotation echoes well with the discussion in Section 2.2.1 as well as with the quotation which we cite from Lofti Zadeh at the beginning of the chapter. It makes little sense of the large set of alternatives when they are presented in continuous fashion by rational numbers. It would be more sensible to add a *name* to some of them and treat them *as a whole* (as a granule). In this way, the effective strategy space can be substantially reduced. What Tay and Linn did was to use linguistic variables to group the strategies in such a way that one will not overburden their decisions with just number-crunching trivialities.

A similar approach has also been adopted by Chen and Chie [11] in an agent-based lottery market. In this study, they want to model agents' decision rules regarding lottery participation, and the decision rule depends on the lottery market condition characterized by the Jackpot size. Like Tay and Linn [47], they did not directly use rational numbers to define the market condition; instead, the fuzzy sets and linguistic variables were applied to give different states of the market. The Sugeno style of fuzzy rules was then applied to form the lotto participation rules.

5 Concluding Remarks

Generally speaking, the concept of granularity has been largely ignored by economists in their models of decision marking. However, thanks to the recent interdisciplinary research joining economists and psychologists, which enables us to realize the possible implications of Shannon information capacity for the human mind and to cast doubt on the fundamental assumption of economics: more options can only do good, or at least no harm. This gives us a psychological foundation of the granule (from fuzzy mathematics), chunk (from cognitive psychology), or module (from complex science) as an elementary unit of information processing and decision making. Moreover, through learning and constant information compression, these elementary units are accumulated and arrayed in a hierarchical form so that our memory capacity can be less constrained. Experimental studies with human subjects also find that subjects' learning behavior can be well captured by the reinforcement learning model if the size of the set of alternatives is restricted to a small number, say, 6, and not 101 [12]. In other words, when agents are presented with a large number of individual options, they group them as granules and mentally work with these granules (as a whole) rather than with the constituent individuals.

This chapter just serves as the beginning of a new research direction for both experimental economics and agent-based computational economics. As for the former, it prompts us to design laboratory experiments or field studies to address the effect of the number of options on choice behavior, with or without the choice architecture. The role of defaults and their possible forms can be closely scrutinized in these experiments. As for the latter, it motivates us to apply the granular information processing scheme to modeling artificial agents. In this case, the default at the very top of the hierarchy may mean the *status quo heuristic*.

The *status quo heuristic* has been well studied in psychology and behavioral economics [41]. It means that in many choice situations, people value the *status quo*, and will forego the opportunity to switch to an alternative unless it is really necessary to do so. The status quo heuristic has been applied in agent-based models in different forms, the threshold width in the threshold model, the intensity of choice in the stochastic choice model, and even the steady-state replacement used in evolutionary computation. To some extent, all try to give a weight for the inertial tendency. However, few have explicitly acknowledged the underpinning psychological force.

Finally, given the hierarchical structure of granules (or granular information processing), it is important to take a further look at the neologism. If granules have to be named verbally, then the evolution of 'names' (still using Miller's term) should give us a footprint of a hierarchy from its primitives to the holistic one. For example, in considering 'happiness' as a granule, what inside this is granule may differ over time; at one time, it can mean homeland security, and at another time, it may mean finding a real job. Therefore, it opens a new navigation in the literature.

Acknowledgements. The authors are grateful to Ying-Fang Kao, Bin-Tzong Chie, and two anonymous referees for their suggestions made to the paper. All remaining errors are the authors' sole responsibility. The Ministry of Science and Technology grant MOST 103-2410-H-004-009-MY3 is gratefully acknowledged.

References

1. Abelson, H., Sussman, G.: Structure and Interpretation of Computer Programs, 2nd edn. MIT Press (1996)
2. Baddeley, A.: The magical number seven: Still magic after all these years? Psychological Review 101(2), 353–356 (1994)
3. Beshears, J., Choi, J., Laibson, D., Madrian, B.: The importance of default options for retirement saving outcomes: Evidence from the United States. In: Kay, S., Sinha, T. (eds.) Lessons from Pension Reform in the Americas, pp. 59–87. Oxford University Press, New York (2008)
4. Callebaut, W., Rasskin-Gutman, D. (eds.): Modularity: Understanding the Development and Evolution of Natural Complex Systems. MIT Press, Cambridge (2005)
5. Camerer, C., Ho, T.-H.: Experienced-weighted attraction learning in normal form games. Econometrica 67(4), 827–874 (1999)
6. Camerer, C., Ho, T.-H., Chong, J.-K.: Sophisticated experience-weighted attraction learning and strategic teaching in repeated games. Journal of Economic Theory 104(1), 137–188 (2002)
7. Camerer, C., Hsia, D., Ho, T.-H.: EWA learning in bilateral call markets. In: Rapoport, A., Zwick, R. (eds.) Experimental Business Research, pp. 255–284 (2002)
8. Carroll, G., Choi, J., Laibson, D., Madrian, B., Metrick, A.: Optimal defaults and active decisions: Theory and evidence from 401(k) saving. Quarterly Journal of Economics 124, 163–1674 (2009)
9. Casari, M.: Can genetic algorithms explain experimental anomalies? An application to common property resources. Computational Economics 24, 257–275 (2004)

10. Chen, S.-H.: Reasoning-based artificial agents in agent-based computational economics. In: Nakamatsu, K., Jain, L. (eds.) Handbook on Reasoning-Based Intelligent Systems, pp. 575–602. World Scientific (2013)

11. Chen, S.-H., Chie, B.-T.: Lottery markets design, micro-structure, and macro-behavior: An ACE approach. Journal of Economic Behavior and Organization 67(2), 463–480 (2008)

12. Chen, S.-H., Du, Y.-R.: Heterogeneity in experience-weighted attraction learning and its relation to cognitive ability. Working paper, AI-ECON Research Center, National Chengchi University (2014)

13. Chen, S.-H., Heieh, Y.-L.: Reinforcement Learning in Experimental Asset Markets. Eastern Economic Journal 37, 109–133 (2011)

14. Chen, S.-H., Tai, C.-C.: The agent-based double auction markets: 15 years on. In: Takadama, K., Cioffi-Revilla, C., Deffuant, G. (eds.) Simulating Interacting Agents and Social Phenomena, pp. 119–136. Springer (2010)

15. Chen, S.-H., Yu, T.: Agents learned, but do we? Knowledge discovery using the agent-based double auction markets. Frontiers of Electrical and Electronic Engineering (FEE) in China 6(1), 159–170 (2010)

16. Chen, Y., Khoroshilov, Y.: Learning under limited information. Games and Economic Behavior 44, 1–25 (2003)

17. Chie, B.-T., Chen, S.-H.: Non-price competition in a modular economy: An agent-based computational model. Economia Politica: Journal of Analytical and Institutional Economics XXX, 149–175 (2013)

18. Cowan, N.: The magical number 4 in short-term memory: A reconsideration of mental storage capacity. Behavioral and Brain Sciences 24, 87–185 (2001)

19. Duffy, J.: Agent-based models and human subject experiments. In: Tesfatsion, L., Judd, K. (eds.) Handbook of Computational Economics: Agent-based Computational Economics, vol. 2, pp. 949–1011. Elsevier, Oxford (2006)

20. Erev, I., Roth, A.: Predicting how people play games: Reinforcement learning in experimental games with unique, mixed strategy euilibria. American Economic Review 88, 848–881 (1998)

21. Gigerenzer, G.: Gut Feelings: The Intelligence of the Unconsciousness. Penguin Books (2007)

22. Gigerenzer, G., Gaissmaier, W.: Heuristic decision making. Annual Review of Psychology 62, 451–482 (2011)

23. Gigerenzer, G., Todd, P.: Fast and Frugal Heuristics: The Adaptive Toolbox. In: Gigerenzer, G., Todd, P. (eds.) Simple Heuristics That Make Us Smart, pp. 3–34. Oxford University Press, New York (1999)

24. Hutchinson, J.: Is more choice always desirable? Evidence and arguments from leks, food selection, and environmental enrichment. Biological Reviews 80(1), 73–92 (2005)

25. Ishizaka, A., Labib, A.: Review of the main developments in the analytic hierarchy process. Expert Systems with Applications 38, 14336–14345 (2011)

26. Iyengar, S., Huberman, G., Jiang, W.: How much choice is too much? Contributions to 401(k) Retirement Plans. In: Mitchell, O., Utkus, S. (eds.) Pension Design and Structure: Lessons from Behavioral Finance, pp. 83–95. Oxford University Press (2004)

27. Johnson, E., Shu, S., Dellaert, B., Fox, C., Goldstein, D., Haubl, G., Larrick, R., Payne, J., Peters, E.: Beyond nudges: Tools of a choice architecture. Marketing Letters 23, 487–504 (2012)

28. Koza, J.R.: Genetic Programming II–Automatic Discovery of Reusable Programs. The MIT Press, Cambridge (1994)
29. Lancaster, K.: A new approach to consumer theory. Journal of Political Economy 74(2), 132–157 (1966)
30. Lancaster, K.: Hierarchies in goods-characteristics analysis. Advances in Consumer Research 3, 348–352 (1976)
31. Lyengar, S., Lepper, M.: When choice is demotivating: Can one desire too much of a good thing? Journal of Personality and Social Psychology 79(6), 995–1006 (2000)
32. Lux, T., Schornstein, S.: Genetic learning as an explanation of stylized facts of foreign exchange markets. Journal of Mathematical Economics 41(1-2), 169–196 (2005)
33. Madrian, B.: Applying insights from behavioral economics to policy design. Annual Review of Economics 6, 663–688 (2014)
34. Mathy, F., Feldman, J.: What's magic about magic numbers? Chunking and data compression in short-term memory. Cognition 122, 346–362 (2012)
35. Miller, G.: The magical number seven plus or minus two: Some limits on our capacity for processing information. Psychological Review 63, 81–97 (1956)
36. Mogilner, C., Rudnick, T., Iyengar, S.: The mere categorization effect: How the presence of categories increases choosers' perceptions of assortment variety and outcome satisfaction. Journal of Consumer Research 35(2), 202–215 (2008)
37. Nagel, R.: A survey on beauty contest experiments: Bounded rationality and learning. In: Budescu, D., Erev, I., Zwick, R. (eds.) Games and Human Behavior, Essays in Honor of Amnon Rapoport. Lawrence Erlbaum Associates, Inc., New Jersey (1998)
38. Nagel, R.: Experimental beauty contest games: Levels of reasoning and convergence to equilibrium. In: Plott, C.R., Smith, V. (eds.) Handbook of Experimental Economics Results, vol. 1, ch. 45, pp. 391–410. Elsevier (2008)
39. Roth, A., Erev, I.: Learning in extensive-form games: Experimental data and simple dynamic models in the intermediate term. Games and Economic Behavior, Special Issue: Nobel Symposium 8, 164–212 (1995)
40. Saaty, T.: How to make a decision: The analytic hierarchy process. European Journal of Operational Research 48, 9–26 (1990)
41. Samuelson, W., Zeckhauser, R.: Status quo bias in decision making. Journal of Risk and Uncertainty 1, 7–59 (1988)
42. Sarin, R., Vahid, F.: Strategy similarity and coordination. The Economic Journal 114, 506–527 (2004)
43. Scheibehenne, B., Greifeneder, R., Todd, P.: Can there ever be too many options? A meta-analytic review of choice overload. Journal of Consumer Research 37(3), 409–425 (2010)
44. Schwartz, B.: The Paradox of Choice: Why More Is Less. Harper Perennial (2003)
45. Shiffrin, R., Nosofsky, R.: Seven plus or minus two: A commentary on capacity limitations. Psychological Review 101(2), 357–361 (1994)
46. Simon, H.: The architecture of complexity. Proceedings of the American Philosophical Society 106(6), 467–482 (1962)
47. Tay, N., Linn, S.: Fuzzy inductive reasoning, expectation formation and the behavior of security prices. Journal of Economic Dynamics & Control 25, 321–361 (2001)
48. Thaler, R., Sunstein, C.: Nudge: Improving Decisions about Health, Wealth, and Happiness. Penguin Books (2008)

49. Tversky, A., Kahneman, D.: The framing of decisions and the psychology of choice. In: Wright, G. (ed.) Behavioral Decision Making, pp. 24–41. Plenum (1981)

50. Van Huyck, J., Battalio, R., Rankin, F.: Selection dynamics and adaptive behavior without much information. Economic Theory 33, 53–65 (2007)

51. Van Huyck, J., Cook, J., Battalio, R.: Selection dynamics, asymptotic stability, and adaptive behavior. Journal of Political Economy 102, 975–1005 (1994)

52. Vriend, N.: On two types of GA-Learning. In: Chen, S.-H. (ed.) Evolutionary Computation in Economics and Finance, pp. 233–243. Physica-Verlag, Heidelberg (2001)

53. Zadeh, L.: Fuzzy sets and information granularity. In: Gupta, M., Ragade, R.K., Yager, R.R. (eds.) Advances in Fuzzy Set Theory and Applications, pp. 3–18. North-Holland Publishing Company (1979)

Decision Makers' Opinions Changing Attitude-Driven Consensus Model under Linguistic Environment and Its Application in Dynamic MAGDM Problems

Bapi Dutta and Debashree Guha

Abstract. Reaching acceptable agreement among decision makers before selecting the suitable alternatives is an important issue in multi-attribute group decision making (MAGDM) process. The aim of this chapter is to present a flexible consensus method in linguistic contexts that adopts a new advice generation scheme by incorporating decision maker's attitude to achieve agreement at each round of consensus process. Different consensus models for MAGDM problems have been proposed in the literature. However in all of these processes, it is assumed that all the decision makers are equally interested to change their initial opinions. But practically different decision makers may have different levels of confidences in their own opinions and that make their inclinations in changing opinions significantly different to each others'. Moreover, the decision makers who have sufficient agreement levels may not be interested to change their opinions further. This analysis motivates us to develop a new consensus reaching process under linguistic environment wherein decision makers' opinions will be changed according to the opinion changing indices provided by them and, thus, decision makers' moral right to modify their opinions will be preserved. Theoretical foundation of the proposed consensus model is laid down and we further implement the consensus algorithm in a linguistic MAGDM problem under dynamic environment. The main contribution of our work is that sovereignty of each decision maker is under consideration in the process of reaching consensus. Finally, a practical example is presented to illustrate the functioning of the proposed method.

Keywords: Consensus reaching process (CRP), Decision makers' attitude, linguistic 2- tuple, Multi-attribute group decision making, Opinion changing index.

Bapi Dutta · Debashree Guha
Indian Institute of Technology, Patna, 800013, India
e-mail: {bdutta,debashree}@iitp.ac.in, deb1711@yahoo.co.in

© Springer International Publishing Switzerland 2015
W. Pedrycz and S.-M. Chen (eds.), *Granular Computing and Decision-Making*
Studies in Big Data 10, DOI: 10.1007/978-3-319-16829-6_4

1 Introduction

Decision making is an integral part of human beings in day to day activities. With the increasing complexity of socio-economic environment, it has become more difficult for a single decision maker to solve the decision making problem. As a result decision making processes in group settings have been widely studied in literature. Group setting provides an excellent framework to judge the problems from diverse aspects.

Most of the real-world group decision making (GDM) problems involve fuzzy and qualitative aspects and measuring such aspects require decision makers' perceptions which contain subjectivity, imprecision and vagueness. Therefore, it is difficult for the decision makers to provide their preferences against the alternatives by using exact numerical values. Decision makers might feel more comfortable to articulate their preferences using words by means of linguistic labels. In such scenario, a linguistic computational model is required to capture these linguistic terms within a mathematical framework and to facilitate the computation between linguistic terms. Several feasible and effective computational models [8, 9, 17] have been suggested in literature from different perspectives. We will adopt the 2-tuple linguistic computational model [17] based on an ordinal scale, which has been found to be highly useful, due to its simplicity in computation and its capability to avoid information loss during the aggregation of linguistic labels [31].

Finding the solution of GDM problems consists of two processes: consensus process and selection process [16, 19, 20]. The consensus process refers to how to obtain the maximum degree of agreement among decision makers' opinions which are given for the finite set of alternatives. On the other hand, selection process involves finding the best alternative from the finite set of alternatives on the basis of decision makers' opinions. Therefore, it is desirable in GDM process that the best alternative is unanimously accepted by the group of decision makers.

Several consensus models have been proposed in literature. Literally, by consensus we mean unanimous agreement among the group members, who are involved in the decision making process. But practically the chance of reaching such perfect agreement is not always possible. This fact leads the researchers to redefine the term consensus from different view point. In [23], Kacprzyk and Fedrizzi introduced the notion of soft degree of consensus which is measured by indicating how far a group of decision makers is from consensus. Based on the concept of soft consensus, various consensus reaching processes (CRPs) have been designed in the literature [22]. The existing CRPs for solving GDM problems can be classified in two categories: static consensus models [15, 16, 23, 11, 24, 10, 3] and interactive consensus models [19, 20, 6, 21, 37, 13, 14, 39, 34, 32, 5, 26]. The major distinction between these two kind of models is that the former does not provide any scope for the interaction among the decision makers to increase the consensus level. Basically, these existing static consensus models does not incorporate any kind of feedback mechanism to advice decision makers how to change their opinions to achieve a higher level of agreement while the later does.

However, in both kinds of consensus models decision makers' inclinations or attitudes in changing their original opinions have not been considered during reaching consensus. Palomares *et al.* [28] introduced the concept of group's attitude towards consensus and integrates it in CRP. In their process, at the beginning of the CRP, the moderator determines the group's attitude towards consensus. There is no scope for the decision makers to express their attitudes in changing opinions in a natural way [29].

But in real decision situations, different decision makers possess different levels of knowledge and expertise in their problems domains. Thus, they may have different levels of confidence in their respective opinions and these make their inclination in changing opinions significantly different to one another. Therefore, the decision makers' opinion changing inclinations should incorporate in designing consensus reaching algorithm. Moreover, in the existing models [38, 37, 39, 10, 3, 2, 5], to increase the consensus level, all the concerned decision makers' opinions are changed despite of the fact that some of them have already reached at desired agreement levels. To overcome the above mentioned drawbacks, the objective of our study is to design a new consensus algorithm that adopts the concept of opinion changing inclination index to model decision makers' preferences in changing opinions. Based on the opinion changing inclinations of the decision makers, we design a novel feedback mechanism for modifying decision makers' opinions by following two protocols as described in section 3. Finally, the proposed CRP is implemented to solve dynamic multi-attribute decision making (DMAGDM) problems.

The rest of the chapter is organized as follows. In section 2, a brief introduction of 2-tuple linguistic model is presented. Section 3 introduces the new consensus model and presents the consensus algorithm with detailed mathematical framework. Based on the proposed CRP algorithm, an approach for solving DMAGDM problem is developed in section 4. In section 5, a practical example is given to illustrate the proposed approach while section 6 concludes the discussion.

2 Brief Review of 2-Tuple Linguistic Computational Model

Let $S = \{l_0, l_1, ..., l_t\}$ be a linguistic term set with the odd cardinality $t + 1$. Any term $l_i \in S$ denotes a possible value for linguistic variable. The following properties should hold for the term set S [17, 25]:

- the set S should be ordered, *i.e.*, $l_i \geq l_j$ if $i \geq j$;
- negation of any linguistic term $l_i \in S : neg(l_i) = l_j$ such that $j = t - i$;
- the maximum of any two linguistic terms $l_i, l_j \in S : \max(l_i, l_j) = l_i$ if $l_i \geq l_j$.

The cardinality of linguistic term set S must be small enough so as not to impose useless precision to the users and it should be rich enough to allow discrimination of the performances of each criterion in a limited number of grades [18, 7]. In fact, the psychologists recommended the use of 7 ± 2 labels, less than 5 being not sufficiently informative, more than 9 being too much for a proper understanding of their differences [27]. In view of this, a linguistic term set, S with seven labels can be defined as follows:

$$S = \{l_0 = \text{very low (VL)}, l_1 = \text{low (L)}, l_2 = \text{moderately low (ML)}, l_3 = \text{normal (N)},$$
$$l_4 = \text{moderately high (MH)}, l_5 = \text{high (H)}, l_6 = \text{very high (VH)}\}$$

Here, we have adopted 2-tuple linguistic representation model, which was developed by Herrera and Martínez [17, 25] based on the concept of symbolic translation.

Definition 1. Let us assume that $\beta \in [0, h]$ be the result of symbolic aggregation operation on the indices of the labels of linguistic term set $S = \{l_0, l_1, ..., l_h\}$. If $i = round(\beta)$ and $\alpha = \beta - i$, be two values such that $i \in \{0, 1, ..., h\}$ and $\alpha \in [-0.5, 0.5)$, then α is called the symbolic translation.

On the basis of symbolic translation, Herrera and Martinez [17, 25] represented the linguistic information by means of 2-tuple (l_i, α_i) where $l_i \in S$ represents the linguistic label and $\alpha_i \in [-0.5, 0.5)$ denotes the symbolic translation.

The conversion of symbolic aggregation result into equivalent linguistic 2-tuple can be done using following function [12]:

Definition 2. Let $S = \{l_0, l_2, ..., l_h\}$ be linguistic term set and $\beta \in [0, h]$ be the numerical value which is obtained from symbolic aggregation operation on the labels of S, then the 2-tuple that conveys the equivalent information to β is given by the following function,

$$\Delta : [0, h] \to S \times [-0.5, 0.5),$$

$$\Delta(\beta) = (l_i, \alpha)$$

where $i = round(\beta)$ is the usual round operation on label of index, i.e., i is the index of the considered label closest to β, and α is the value of symbolic translation given by

$$\alpha = \begin{cases} \beta - i, & \alpha \in [-0.5, 0.5) \text{ if } i \neq 0, h \\ \beta, & \alpha \in [0, 0.5) \text{ if } i = 0 \\ \beta - h, & \alpha \in [-0.5, 0] \text{ if } i = h \end{cases}$$

Example 1. Assume that $S = \{l_0, l_1, l_2, l_3, l_4, l_5, l_6\}$ represent a linguistic term set and $\beta = 4.2$ is obtained from symbolic aggregation operation. Then from Definition 2, we can convert $\beta = 4.2$ into linguistic 2-tuple $\Delta(4.2) = (l_4, 0.2)$.

Definition 3. Let $S = \{l_0, l_1, ..., l_h\}$ be a linguistic term set. For any linguistic 2-tuple (l_i, α_i), its equivalent numerical value is obtained by the following function:

$$\Delta^{-1} : S \times [-0.5, 0.5) \to [0, h]$$

$$\Delta^{-1}(l_i, \alpha_i) = i + \alpha_i = \beta_i$$

where $\beta_i \in [0, h]$.

Example 2. Assume that $S = \{l_0, l_1, l_2, l_3, l_4, l_5\}$ represents a linguistic term set and $(l_3, -0.3)$ be a linguistic 2-tuple. Based on Definition 3 the equivalent numerical value of $(l_3, -0.3)$ is $\Delta^{-1}(l_3, -0.3) = 3 + (-0.3) = 2.7$.

From Definition 2 and Definition 3, it is noted that any linguistic term can be converted into a linguistic 2-tuple as follows: $l \in S \Rightarrow (l, 0)$.

The ordering of two linguistic 2-tuples (l_m, α_m) and (l_n, α_n) can be done according to lexicographic order as follows:

(1)If $m > n$ then $(l_m, \alpha_m) > (l_m, \alpha_m)$.
(2)If $m = n$ then

(a)$(l_n, \alpha_n) = (l_m, \alpha_m)$, for $\alpha_m = \alpha_n$
(b)$(l_n, \alpha_n) > (l_m, \alpha_m)$, for $\alpha_n > \alpha_m$

Deviation between two linguistic 2-tuples can be measured as follows [33]:

Definition 4. Let (s_i, α_i) and (s_j, α_j) be the two linguistic 2-tuples. Then, the deviation degree between (s_i, α_i) and (s_j, α_j) can be defined as

$$d((s_i, \alpha_i), (s_j, \alpha_j)) = \frac{|\Delta^{-1}(s_i, \alpha_i) - \Delta^{-1}(s_j, \alpha_j)|}{t} \tag{1}$$

3 A Novel Attitude Driven Consensus Model

The aim of this work is to design a CRP by integrating decision makers' attitude in changing their initial opinions. In the following, we describe how to form consensus opinion from individual opinions based on decision makers' opinions changing attitudes.

3.1 *Representation of GDM Problem and Consensus Framework*

In the present study, a typical GDM problem, involving a set of decision makers $E = \{J_1, J_2, ..., J_p\}$ who express their judgments/opinions on a certain problem to reach a common solution, has been considered. Each decision maker provides the linguistic rating of an alternative using the linguistic term set S with odd cardinality $t + 1$. Let $a_k = (o_k, \alpha_k)(k = 1, 2, ..., p)$, where $o_k \in S$ and $\alpha_k \in [-0.5, 0.5)$, be the initial linguistic opinion of the decision maker J_k. In general, decision makers from diverse backgrounds may have different levels of knowledge and expertises in their respective domains that make their assessments regarding the alternatives significantly different form one another. Therefore, it is possible to arise conflict among the decision makers'. To resolve such conflicts, CRP is essential as it leads us to achieve a more accepted solution by a whole group.

Generally, CRP is an iterative process in which decision makers discuss and modify their initial opinions. This process is often co-ordinated by a human figure known as moderator, who is responsible for supervising and guiding the discussion among the decision makers. In general, CRP consists of three main phases as given below [30].

- *consensus measurement:* All the decision makers' opinions are accumulated to measure the current level of agreement among the decision makers in the group via consensus measures.
- *consensus control:* The consensus degree is compared with the predefined consensus threshold. If the desired level of agreement has reached, the group carries out selection process. If the desired agreement level is not achieved, the group continues the discussion.
- *Consensus progress:* A method is followed to improve the agreement among the decision makers. Such method is often called feedback generation/advice generation. In this method, moderator identifies opinions of the decision makers which are farthest from the consensus opinion and advises them to modify their opinions.

From the above discussion, it is clear that feedback generation plays major role in CRP. In the literature [6, 20, 26, 28], different types of strategies have adopted for feedback generation. Here, we propose a feedback generation process by introducing new strategy which is based on decision makers' opinions changing attitudes. The phases of the new CRP are described below.

3.2 Consensus Measure

Once the decision makers have provided their opinions/judgments over the alternatives, we can compute the consensus level of each decision maker by the help of consensus measure. Several consensus measures have been proposed in the literature [15, 4, 24]. Among them most commonly used method is distance measure which is computed by measuring the distance between group and individual opinions where group opinion is taken as ideal state of consensus. Here, we have utilized deviation degree between linguistic 2-tuples, as defined in Eq. (1), as a measure of consensus level. On the other hand, from individual opinions group opinion is formed by utilizing various aggregation operators [1]. In this study, we utilize arithmetic mean to form group opinion from individual opinions. The arithmetic mean of a set of linguistic 2-tuples can be defined as follows:

Definition 5. Let $(o_k, \alpha_k)(k = 1, 2, ...n)$ be the set of n linguistic 2-tuples. The 2-tuple linguistic arithmetic mean of the collection $(o_k, \alpha_k)(k = 1, 2, ...n)$ is defined as

$$(g, \alpha) = \Delta \left(\frac{1}{n} \sum_{k=1}^{p} \Delta^{-1}(o_k, \alpha_k) \right) \qquad (2)$$

On the basis of linguistic deviation degree (Eq. (1)) and arithmetic mean (Eq. (2)), we can define the consensus level of each decision maker at any stage of CRP as follows:

Definition 6. Let $\{a_1, a_2, ... a_p\}$ be the opinions of the decision makers' set $E = \{J_1, J_2, ..., J_p\}$ where $a_k = (o_k, \alpha_k)(k = 1, 2, ..., p)$ and $\alpha_k \in [-0.5, 0.5)$. If (g, α) be the group opinion obtained from individual opinions by utilizing Eq.(2), the consensus level of J_k^{th} expert is given by

$$CL(J_k) = d(a_k, g) = \frac{|\Delta^{-1}(a_k, \alpha_k) - \Delta^{-1}(g, \alpha)|}{t} \quad (k = 1, 2, ..., p) \qquad (3)$$

3.3   Consensus Control

After computing the consensus level of each expert, the moderator checks whether all decision makers have reached desired level of consensus. Let ε be the desired level of consensus. When the moderator finds that $CL(J_k) \leq \varepsilon$ for all $k = 1, 2, ..., p$ the moderator stops the consensus process as acceptable agreement among group and individuals has reached. Otherwise, the moderator activates the feedback process which is described below. .

3.4   Decision Maker's Attitude Based Novel Feedback Generation Process

In aim of substituting moderator's action by feedback mechanism for generating recommendations to the decision makers to increase the consensus level among all the decision makers, a new decision makers' attitude-driven feedback mechanism is proposed in this section.

The feedback process is guided by the consensus level of each decision maker and his/her attitude of changing opinion. Therefore, before generating advice via feedback generation process, the moderator should aware of this fact.

3.4.1   Modeling Decision Maker's Attitude of Changing Opinions

Most of the existing consensus models use consensus measure to guide the feedback mechanism. However, those models do not take into account the decision makers' interest or inclinations in changing their initial opinions by assuming that that all the decision makers are equally inclined to change their opinions. In reality, different decision makers possess different levels of knowledge and expertise in the problems domains. Thus, they may have different levels of confidence in their respective opinions and these make their inclination in changing opinions significantly different to one another. To address this issue, a new feedback mechanism is incorporated in the proposed consensus model, to generate advice according to the decision makers' opinions changing inclinations. Naturally, expert's attitude is intrinsically connected to his/her opinion changing index. The expert with higher (lower) confidence level in his/her opinion provides lower (higher) inclination in changing opinion. If the desired agreement level among all the decision makers is not achieved, the moderator first identifies the decision makers who have not achieved desired level of consensus. Then it seems reasonable to generate advice for modifying the opinion of the expert who has high inclination to change his/her opinion rather than the opinion of the expert with low inclination in changing opinion. As changing of opinions according to the decision makers' inclinations preserve the sovereignty of the decision makers as much as possible, its implementation in consensus process will improve the GDM processes. With this observation at the background, in the following, we describe the new CRP incorporating a new feedback mechanism that replaces and automates the moderator's tasks by computing and sending customized recommendations to the decision makers according to their opinions changing inclinations.

The decision makers' inclinations in changing their original opinions are influenced by several cognitive factors, such as, subjective estimation, knowledge, experience and confidence. Therefore, expressing the opinion changing inclinations, which basically depend on several human perception based factors, by using exact numerical values, is quite difficult. In this light, decision makers may feel more comfortable to express their attitudes of changing opinions in natural language by using linguistic terms, such as, 'very high', 'low' etc. For example if the decision maker is highly interested to change his/her initial opinion, he/she may use the linguistic term, 'very high'. On the basis of this fact, we introduce the concept of opinion changing index which describes the decision makers' attitude in changing opinions. Subsequently, we can model it by selecting a suitable linguistic term from a predefined linguistic term set. For instance, the moderator may ask the decision makers to provide their opinions changing indices by taking any value from the linguistic terms set, say, $C = \{p_0 = \text{very very low}, p_1 = \text{very low}, p_2 = \text{low}, p_3 = \text{moderately}, p_4 = \text{high}, p_5 = \text{very high}, p_6 = \text{very very high}\}$.

3.4.2 Opinion Modifying Protocols and Mathematical Modelling of the Feedback Mechanism

Once the decision makers have provided their opinions changing indices, the moderator initiates the feedback process by identifying the decision makers who are not in desired level of consensus, i.e., the moderator identifies the decision makers not satisfying the condition $CL(J_k) \leq \varepsilon$. Let $K = \{J_{k_1}, J_{k_2}, ..., J_{k_q}\}$ be the set of the decision makers who does not meet the desired consensus level. Now, the moderator being aware of the decision makers' opinions changing indices, generates advice to modify decision makers' opinions based on the following protocols:

- the expert who satisfies the consensus level at initial round will not change his/her opinion throughout the consensus process

- based on the descending order of opinion changing index, expert's opinion will be changed one by one
- change the expert's opinion in such a way such that it deviates minimum from his/her initial opinion and simultaneously satisfies the consensus level

These protocols of modifying decision makers' opinions can be put into mathematical form as follows:

To present model in simplest form, we are going to use following notations now onward: $K = \{J_{k_i} : i = 1, 2..., q\}$ is the set of decision makers who does not meet the desired consensus level initially and $E - K = \{J_{k_i} : i = q+1, ..., n\}$ is the set of decision makers who satisfy the desired consensus level initially. Let the decision maker J_{k_1} has the highest changing inclination index. Let $\bar{a}_{k_1}$ be the decision maker J_{k_1}'s new opinion and g_1 denote the new group opinion. Then,

$$
\begin{aligned}
g_1 &= \Delta \left(\frac{\Delta^{-1}a_1 + \Delta^{-1}a_2 + ... + \Delta^{-1}\bar{a}_{k_1} + ... + \Delta^{-1}a_n}{n} \right) \\
&= \Delta \left(\frac{\beta_1 + \beta_2 + ... + \beta_n + (\bar{\beta}_{k_1} - \beta_{k_1})}{n} \right) \\
&= \Delta \left(\beta_g + \frac{(\bar{\beta}_{k_1} - \beta_{k_1})}{n} \right)
\end{aligned}
\tag{4}
$$

where $\beta_i = \Delta^{-1}a_i$ and $\beta_g = \Delta^{-1}g$. The deviation degree between new group opinion and $J_{k_1}^{th}$ decision maker's new opinion can be calculated by using Eqs. (1) and (4) as follows:

$$
\begin{aligned}
d(g_1, \bar{a}_{k_1}) &= \frac{|\Delta^{-1}g_1 - \Delta^{-1}\bar{a}_{k_1}|}{t} \\
&= \frac{|\beta_g + \frac{(1-n)\bar{\beta}_{k_1} - \beta_{k_1}}{n}|}{t}
\end{aligned}
\tag{5}
$$

According to the protocols, we choose the decision maker J_{k_1}'s new opinion in such a way such that it has least deviation from his/her initial opinion and the new opinion must satisfy the consensus condition. This can be put into the following optimization problem:

$$
\text{minimization } (\bar{\beta}_{k_1} - \beta_{k_1})^2
$$

$$
(P) \qquad \text{s.t.} \quad \begin{cases} d(g_1, \bar{a}_{k_1}) \leq \varepsilon \\ d(g_1, a_i) \leq \varepsilon \text{ for } J_i \in E - K \end{cases}
$$

Using Eqs. (4) and (5), the optimization problem (P) can be transformed into the following problem:

minimization $(\overline{\beta}_{k_1} - \beta_{k_1})^2$

(P1)   s.t. $\begin{cases} (\beta_g - \frac{\beta_{k_1}}{n} - \varepsilon t)\frac{n}{n-1} \leq \overline{\beta}_{k_1} \leq (\beta_g - \frac{\beta_{k_1}}{n} + \varepsilon t)\frac{n}{n-1} \\ (\beta_i + \frac{\beta_{k_1}}{n} - \varepsilon t - \beta_g)n \leq \overline{\beta}_{k_1} \leq (\beta_i + \frac{\beta_{k_1}}{n} + \varepsilon t - \beta_g)n \ J_i \in E - K \end{cases}$

The decision maker J_{k_1}'s new opinion can be obtained by solving the convex optimization problem (P1). Again, we check whether the desired consensus level is achieved or not. If the desired consensus level is not achieved, then the moderator selects the next decision maker who has the next highest opinion changing inclination index from the set K and changes his/her opinions following the same protocols as above. Suppose the decision maker J_{k_2} has the next highest opinion changing inclination index. Let $\overline{\beta}_{k_2}$ be the new opinion of J_{k_2} and g_2 denotes the new group opinion. The problem of finding β_{k_2} can be transformed into the following optimization problem

minimization $(\overline{\beta}_{k_2} - \beta_{k_2})^2$

(P2)   s.t. $\begin{cases} d(g_1, \overline{a}_{k_2}) \leq \varepsilon \\ d(g_1, \overline{a}_{k_1}) \leq \varepsilon \\ d(g_1, a_i) \leq \varepsilon \text{ for } J_i \in E - K \end{cases}$

The model (P2) can be put in the following optimization problem

minimization $(\overline{\beta}_{k_2} - \beta_{k_2})^2$

(P3)   s.t. $\begin{cases} (\beta_{g_1} - \frac{\beta_{k_2}}{n} - \varepsilon t)\frac{n}{n-1} \leq \overline{\beta}_{k_2} \leq (\beta_{g_1} - \frac{\beta_{k_2}}{n} + \varepsilon t)\frac{n}{n-1} \\ (\beta_{k_2} + \frac{\beta_{k_2}}{n} - \varepsilon t - \beta_{g_1})n \leq \overline{\beta}_{k_2} \leq (\beta_{k_1} + \frac{\beta_{k_2}}{n} + \varepsilon t - \beta_{g_1})n \\ (\beta_i + \frac{\beta_{k_2}}{n} - \varepsilon t - \beta_{g_1})n \leq \overline{\beta}_{k_2} \leq (\beta_i + \frac{\beta_{k_2}}{n} + \varepsilon t - \beta_{g_1})n \ J_i \in E - K \end{cases}$

If all the decision makers' opinions are reached in desirable consensus level, then moderator stops the consensus process, otherwise, the process is repeated. Suppose, we repeat the process $h(\leq q)$ times. At h^{th} stage, let $\overline{\beta}_{k_h}$ be the $J_{k_h}^{th}$ decision makers new opinion and g_h is the new group opinion. Then, $\overline{\beta}_{k_h}$ can be obtained by solving the following optimization problem:

minimization $(\overline{\beta}_{k_h} - \beta_{k_h})^2$

(P4)   s.t. $\begin{cases} (\beta_{g_{h-1}} - \frac{\beta_{k_h}}{n} - \varepsilon t)\frac{n}{n-1} \leq \overline{\beta}_{k_h} \leq (\beta_{g_{h-1}} - \frac{\beta_{k_h}}{n} + \varepsilon t)\frac{n}{n-1} \\ (\beta_{k_{h-1}} + \frac{\beta_{k_h}}{n} - \varepsilon t - \beta_{g_{h-1}})n \leq \overline{\beta}_{k_h} \leq (\beta_{k_{h-1}} + \frac{\beta_{k_h}}{n} + \varepsilon t - \beta_{g_{h-1}})n \\ \vdots \\ (\beta_{k_1} + \frac{\beta_{k_h}}{n} - \varepsilon t - \beta_{g_{h-1}})n \leq \overline{\beta}_{k_h} \leq (\beta_{k_1} + \frac{\beta_{k_h}}{n} + \varepsilon t - \beta_{g_{h-1}})n \\ (\beta_i + \frac{\beta_{k_h}}{n} - \varepsilon t - \beta_{g_{h-1}})n \leq \overline{\beta}_{k_h} \leq (\beta_i + \frac{\beta_{k_p}}{n} + \varepsilon t - \beta_{g_{h-1}})n \ J_i \in E - K \end{cases}$

Let

$$a_{k_h}^L = (\beta_{g_{h-1}} - \frac{\beta_{k_h}}{n} - \varepsilon t)\frac{n}{n-1}$$

$$a_{k_h}^R = (\beta_{g_{h-1}} - \frac{\beta_{k_h}}{n} + \varepsilon t)\frac{n}{n-1}$$

$$a_{k_i}^L = (\beta_{k_i} + \frac{\beta_{k_h}}{n} - \varepsilon t - \beta_{g_{h-1}})n \quad J_{k_i} \in K \text{ and } i = 1,2,...,h-1$$

$$a_{k_i}^R = (\beta_{k_i} + \frac{\beta_{k_h}}{n} + \varepsilon t - \beta_{g_{h-1}})n \quad J_{k_i} \in K \text{ and } i = 1,2,...,h-1$$

$$a_i^L = (\beta_i + \frac{\beta_{k_h}}{n} - \varepsilon t - \beta_{g_{h-1}})n \quad J_{k_i} \in E - K$$

$$a_i^R = (\beta_i + \frac{\beta_{k_h}}{n} + \varepsilon t - \beta_{g_{h-1}})n \quad J_{k_i} \in E - K$$

The problem (P4) can be reformulated as:

$$\text{minimization } (\overline{\beta}_{k_h} - \beta_{k_h})^2$$

$$\text{(P5)} \qquad \text{s.t.} \quad \begin{cases} a_{k_h}^L \leq \overline{\beta}_{k_h} \leq a_{k_h}^R \\ a_{k_i}^L \leq \overline{\beta}_{k_p} \leq a_{k_i}^R, \quad i = 1,2,...,p-1 \\ a_i^L \leq \overline{\beta}_{k_p} \leq a_i^R \quad J_i \in D - K \end{cases}$$

Theorem 1. *The model (P5) is equivalent to following optimization problem:*

$$\text{minimization } (\overline{\beta}_{k_h} - \beta_{k_h})^2$$

$$\text{(P6)} \qquad \text{s.t.} \quad \{a^L \leq \overline{\beta}_{k_h} \leq a^R$$

where,

$$a^L = \max\{\max_{k_i \in K} a_{k_i}^L, \max_{k_i \in E-K} a_i^L\} \tag{6}$$

$$a^R = \min\{\min_{k_i \in K} a_{k_i}^R, \min_{k_i \in E-K} a_i^R\} \tag{7}$$

Proof. From Eqs. (6) and (7), it is clear that constraint of (P6) satisfies all the constraints of (P5). Since $a^L \leq \overline{\beta}_{k_h} \leq a^R$ is the intersection of all the constraints of (P5), the feasible regions of (P5) and (P6) are same. As the objective functions of the both models (P5) and (P6), therefore (P6) is equivalent to (P5). Thus both the models (P5) and (P6) have same optimal solution which is unique. □

In the following theorem, the decision maker J_{k_h}'s new opinion has been obtained.

Theorem 2. *The optimal solution of (P6) is given by*

$$\overline{\beta}_{k_h} = \begin{cases} a^L & \text{if } |a^L - \beta_{k_h}| = \min\{|a^L - \beta_{k_h}|, |a^R - \beta_{k_h}|\} \\ a^R & \text{if } |a^R - \beta_{k_h}| = \min\{|a^L - \beta_{k_h}|, |a^R - \beta_{k_h}|\} \end{cases} \tag{8}$$

Proof. First we show that β_{k_h} does not belong to the closed interval $[a^L, a^R]$. Suppose at the contrary $\beta_{k_h} \in [a^l, a^R]$. It implies that minimize $(\overline{\beta}_{k_h} - \beta_{k_h})^2 = 0$, i.e., $\overline{\beta}_{k_h} = \beta_{k_h}$. It contradicts our assumption that J_{k_h} th decision maker's opinion does not satisfy the consensus condition $d(g_{h-1}, a_{k_h}) > \varepsilon$ and his/her opinion is required to be changed to reach consensus. Therefore, $\beta_{k_h} \notin [a^L, a^R]$. Since the objective function is convex, the minimum value of minimize $(\overline{\beta}_{k_h} - \beta_{k_h})^2$ is obtained at the either left end point of the interval $[a^L, a^R]$ or the right end point of the interval $[a^L, a^R]$ according to the minimum distance of the points a^L and a^R from β_{k_h}. Hence, the optimal solution of (P6) is given by Eq. (8).

If we continue this process then it will take at most $|K|$ steps to reach in consensus, where $|K|$ denotes the cardinality of the set K. Because after $|K|$ steps all the opinions will satisfy the specified consensus level. The consensus process has been summarized in the following algorithm.

Consensus Algorithm

Step 1 Input: initial opinions $\{a_1, a_2, ..., a_n\}$, consensus threshold ε, decision makers' opinion changing indices

Step 2 Check: $d(a_i, g) \leq \varepsilon$ for all i. If yes goto **Step 5**, otherwise goto **Step 3**

Step 3 Find the decision makers' set, $K = \{J_{k_1}, J_{k_2}, ..., J_{k_q}\}$ who do not satisfy consensus level, i.e, $d(a_i, g) > \varepsilon$

Step 4 For $c = 1:|K|$

Find the expert, $J_{k_{i_{(c)}}}$ with highest opinion changing index from the decision makers' set $K \setminus \{J_{k_{i_{(1)}}}, J_{k_{i_{(2)}}}, ..., J_{k_{i_{(c-1)}}}\}$ (if $c = 1$, $K \setminus \{J_{k_{i_{(1)}}}, J_{k_{i_{(2)}}}, ...,$ $J_{k_{i_{(c-1)}}}\} = K$);

Compute expert $J_{k_{i_{(c)}}}$'s new opinion by the help of Eq. (6);

Check: $d(a_{k_i}, g_c) \leq \varepsilon$ for the decision makers' set $K \setminus \{J_{k_{i_{(1)}}}, J_{k_{i_{(2)}}}, ..., J_{k_{i_{(c)}}}\}$, where g_c is the group opinion. If satisfy goto **Step 5**, otherwise, continue;

Step 5 End

Remark 1. It is to be noted that consensus process will terminate after a finite numbers of steps subject to the condition that optimization model (P6), which arises in each round has a non-empty feasible region. However, in some cases such condition may be violated for imposing strict conditions by decision makers, such as, decision makers with significant diverge opinions may have very low inclinations in changing opinions. In these situations moderator can either advice the concerning decision makers to review their opinions changing indices or may think of changing the consensus threshold to reach an acceptable agreement among the decision makers.

Remark 2. Sometimes there may be more than one expert with same opinions changing indices and that may bring conflicts to the moderator that who should be selected first to change his/her opinion. In such cases, the moderator selects the expert with highest deviation from the group opinion first and so on. Implementing the consensus algorithm, in the next section, we develop a consensus based approach for solving DMAGDM problems.

4 Application of CRP for Solving Dynamic Multi-attribute Group Decision Making

In this section, the proposed CRP process is implemented to solve consensus based DMAGDM problems with linguistic information.

In many complex real-life decision making problems, the current and past performances of the alternatives are required to take account in the decision making process. This fact leads the decision makers to assess the alternatives over a fixed time period and these kind of problems are termed as dynamic MAGDM problems (DMAGDM) [35]. DMAGDM problems frequently arises in real-life situations, such as, multi-period investment decision making, medical diagnosis, personnel dynamic evaluation, military system efficiency dynamic evaluation etc. So far as to the best of our knowledge, the consensus process in DMAGDM was discussed in one article by Su et al. [32]. They presented an interactive algorithm for intuitionistic fuzzy DMAGDM in which the consensus level of the group preferences is measured by calculating spearman correlation coefficient for both the group ranking and the individual expert's ranking.

In this present study, we implement the proposed CRP process to develop an approach for solving consensus based DMAGDM problems. The proposed approach can be divided into three key phases: consensus phase, aggregation phase and selection phase. In consensus phase, to increase the level of agreement among the decision makers over their opinions/judgments, the proposed consensus reaching algorithm is implemented. Once the desirable agreement is achieved, the aggregation phase is commenced. In aggregation phase, consensus decision information of different periods is aggregated to form the collective opinion of each decision maker and, thereafter, group opinion is formed by aggregating all the decision makers' collective consensus opinions. In selection phase, by computing each alternative's group overall performance, the best alternative(s) is selected. Before describing the proposed method, a detailed mathematical framework of a dynamic MAGDM problem with 2-tuple linguistic information is presented below.

There is a group of p decision makers $\{J_1, J_2, ..., J_p\}$ and a set of m alternatives $X = \{X_1, X_2, ..., X_m\}$. The decision makers' aim is to choose the best alternative among m alternatives depending on n attributes $A = \{A_1, A_2, ..., A_n\}$. Different attributes may have different importance. Let $w = (w_1, w_2, ... w_n)$ be the weight vector of the attributes satisfying the conditions: $w_j \geq 0$ $(j = 1, 2, ..., m)$ and $\sum_{j=1}^{m} w_j = 1$, where w_j is the relative importance of the attribute A_j.

Due to the complexity of the system decision makers' are decided to consider the alternatives' present and past performances. Suppose the decision makers evaluate the alternatives at r different time periods $t_1, t_2,...,t_r$. Different time periods may have different importance. Let $\lambda(t) = (\lambda(t_1), \lambda(t_2), ..., \lambda(t_r))^T$ be the weight vector of the time periods such that $\lambda(t_s) > 0$ and $\sum_{s=1}^{r} \lambda(t_s) = 1$. Various methods for determining the weights of the time periods have been proposed in the literature [35, 36].

The k-th expert, J_k provides his/her rating of an alternative $X_i (1 \leq i \leq m)$ with respect to the attribute A_j at time period t_s as a linguistic 2-tuple $d_{ij}^{(k)}(t_s) = (t_{ij}^{(k)}(t_s), \alpha_{ij}^{(k)}(t_s))$ where $t_{ij}^{(k)}(t_s)$ belongs to the predefined linguistic term set S and $\alpha_{ij}^{(k)}(t_s) \in (-0.5, 0.5]$. The expert J_k's ratings of the alternatives is summarized in 2-tuple linguistic decision matrix $D_k(t_s) = (d_{ij}^{(k)}(t_s))_{m \times n}$ as follows:

$$D_k(t_s) = \begin{matrix} & A_1 & A_2 & \cdots & A_n \\ X_1 & \begin{pmatrix} d_{11}^k(t_s) & d_{12}^k(t_s) & \cdots & d_{1n}^k(t_s) \\ X_2 & d_{21}^k(t_s) & d_{22}^k(t_s) & \cdots & d_{2n}^k(t_s) \\ \vdots & \vdots & \vdots & \cdots & \vdots \\ X_m & d_{m1}^k(t_s) & d_{m1}^k(t_s) & \cdots & d_{mn}^k(t_s) \end{pmatrix} \end{matrix}$$

On the basis of the above decision inputs, an algorithm for solving MAGDM problem is presented here. The steps of the proposed algorithm are as follows:

Step 1   *Formation of consensus opinions*
         The consensus algorithm proposed in section 4 is implemented here to reach a desirable agreement level among the decision makers. Before its implementation, the decision makers are asked to provide their opinion changing inclinations via opinions changing indices by using the linguistic terms set, $C = \{p_0 = $ very very low, $p_1 = $ very low, $p_2 = $ low, $p_3 = $ moderately, $p_4 = $ high, $p_5 = $ very high, $p_6 = $ very very high$\}$. At time period $t_s (1 \leq s \leq r)$, each decision maker provides his/her opinion against the alternatives $X_i (1 \leq i \leq m)$ with respect to the attribute $A_j (1 \leq j \leq n)$ which may be denoted as $\{d_{ij}^1(t_s), d_{ij}^2(t_s), ..., d_{ij}^P(t_s)\}$. There may arise conflict among decision makers' judgments or opinions. Therefore, to reach consensus among decision makers over each time period, we need to resolve conflict among the decision makers' opinions $\{d_{ij}^1(t_s), d_{ij}^2(t_s), ..., d_{ij}^P(t_s)\}$ for all i, j and r by applying the proposed consensus algorithm. Let from decision makers' initial opinions $\{d_{ij}^1(t_s), d_{ij}^2(t_s), ..., d_{ij}^P(t_s)\}$, we form the consensus opinion $\{\bar{d}_{ij}^1(t_s), \bar{d}_{ij}^2(t_s), ..., \bar{d}_{ij}^P(t_s)\}$ by applying the proposed consensus algorithm for all i, j and r. Then the consensus opinions of the decision maker J_k can be summarized in the following decision matrix

$$\bar{D}_k(t_s) = \begin{array}{c} \\ X_1 \\ X_2 \\ \vdots \\ X_m \end{array} \begin{array}{cccc} A_1 & A_2 & \cdots & A_n \\ \left(\bar{d}_{11}^k(t_s) \right. & \bar{d}_{12}^k(t_s) & \cdots & \bar{d}_{1n}^k(t_s) \\ \bar{d}_{21}^k(t_s) & \bar{d}_{22}^k(t_s) & \cdots & \bar{d}_{2n}^k(t_s) \\ \vdots & \vdots & \cdots & \vdots \\ \left.\bar{d}_{m1}^k(t_s) \right. & \bar{d}_{m1}^k(t_s) & \cdots & \bar{d}_{mn}^k(t_s) \end{array} \right)$$

Step 2   *Formation of collective decision matrix, D_k.*
From each decision maker's consensus decision matrices $\bar{D}_k(t_s)(1 \leq s \leq r)$ over the time periods $t_1, t_2, ..., t_r$, the consensus collective decision matrix $\bar{D} = (\bar{d}_{ij}^{(k)})_{m \times n}$ is constructed by utilizing weighted 2-tuple linguistic arithmetic mean operator

$$\bar{d}_{ij}^k = \Delta \left(\sum_{s=1}^{r} \lambda_s \Delta^{-1} \bar{d}_{ij}^k(t_s) \right) \tag{9}$$

where λ_s is the relative importance of the time period t_s.

Step 3   *Construction of consensus group decision matrix.*
Utilize weighted 2-tuple linguistic arithmetic mean operator (Eq. (2)) to aggregate decision makers' consensus collective decision matrices $\bar{D}_k = (\bar{d}_{ij}^k)_{m \times n}(k = 1, 2, ...p)$ into consensus group decision matrix $D = (\bar{d}_{ij})_{m \times n}$. The group decision matrix is presented as follows:

$$D = \begin{array}{c} \\ X_1 \\ X_2 \\ \vdots \\ X_m \end{array} \begin{array}{cccc} A_1 & A_2 & \cdots & A_n \\ \left(\bar{d}_{11} \right. & \bar{d}_{12} & \cdots & \bar{d}_{1n} \\ \bar{d}_{21} & \bar{d}_{22} & \cdots & \bar{d}_{2n} \\ \vdots & \vdots & \cdots & \vdots \\ \left.\bar{d}_{m1} \right. & \bar{d}_{m1} & \cdots & \bar{d}_{mn} \end{array} \right)$$

where,

$$\bar{d}_{ij} = \Delta \left(\frac{1}{p} \sum_{k=1}^{p} \Delta^{-1} \bar{d}_{ij}^k \right) \tag{10}$$

Step 4   *Computation of alternatives overall evaluations.*
From group decision matrix $D = (\bar{d}_{ij})_{m \times n}$, each alternative X_i's $(i = 1, 2, ..., m)$ overall performance $\bar{d}_i(i = 1, 2, ..., m)$ is calculated by utilizing 2-tuple linguistic arithmetic mean operator (Eq. (2)) as follows:

$$\bar{d}_i = \Delta \left(\sum_{j=1}^{n} w_j \Delta^{-1} \bar{d}_{ij} \right) \tag{11}$$

where w_j is the relative importance of the attribute A_j.

Table 1 Linguistic decision matrices of the decision makers at different time periods

time periods	alternatives	E_1				E_2				E_3			
		A_1	A_2	A_3	A_4	A_1	A_2	A_3	A_4	A_1	A_2	A_3	A_4
	X_1	l_3	l_5	l_2	l_4	l_4	l_5	l_2	l_3	l_5	l_3	l_4	l_2
	X_2	l_2	l_1	l_3	l_2	l_3	l_2	l_4	l_3	l_1	l_4	l_2	l_3
t_1	X_3	l_4	l_6	l_3	l_5	l_6	l_5	l_4	l_3	l_3	l_4	l_3	l_5
	X_4	l_5	l_3	l_4	l_2	l_3	l_5	l_3	l_2	l_2	l_4	l_5	l_3
	X_5	l_1	l_3	l_2	l_1	l_2	l_4	l_3	l_1	l_3	l_2	l_1	l_4
	X_1	l_4	l_6	l_3	l_3	l_5	l_4	l_2	l_4	l_2	l_5	l_3	l_2
	X_2	l_1	l_2	l_2	l_3	l_3	l_1	l_3	l_2	l_4	l_3	l_3	l_3
t_2	X_3	l_5	l_4	l_4	l_3	l_6	l_5	l_3	l_1	l_3	l_4	l_2	l_3
	X_4	l_4	l_2	l_3	l_3	l_4	l_6	l_3	l_4	l_5	l_3	l_6	l_4
	X_5	l_0	l_2	l_3	l_2	l_2	l_1	l_4	l_3	l_1	l_4	l_2	l_1
	X_1	l_2	l_4	l_1	l_3	l_3	l_1	l_2	l_4	l_5	l_3	l_4	l_3
	X_2	l_3	l_2	l_1	l_2	l_2	l_4	l_3	l_2	l_1	l_3	l_1	l_0
t_3	X_3	l_3	l_5	l_3	l_4	l_4	l_6	l_4	l_3	l_2	l_4	l_3	l_5
	X_4	l_6	l_3	l_4	l_2	l_4	l_5	l_2	l_6	l_5	l_4	l_5	l_3
	X_5	l_2	l_1	l_2	l_3	l_1	l_2	l_1	l_3	l_0	l_3	l_2	l_2

Step 5   *Ranking of the alternatives.*
Rank the alternatives $X_i (i = 1, 2, ..., m)$, based on their overall performance values d_i $(i = 1, 2, ..., m)$ by using the comparison method of 2-tuple described in Section 2 and choose the best alternative according to their ranking order.

5   A Practical Example

In this section, we present a health system evaluation problem of different states of India to demonstrate the application of the proposed consensus based dynamic multi-attribute group decision making process.

In a real sense, health is a product of and an input in the process of development. When we consider that the purpose of the development is to facilitate people to lead economically productive, healthful and socially satisfying life, the fruits of development are enjoyed by the individual only when he is healthy. Prosperity and happiness of mankind depends, inter-alia on physical, mental and social well being. Therefore, health development is requisite to social and economic development of a country. In view of this, the Health and Family Welfare Department has a number of schemes to cover the under-privileged sections of society and help them with maternity, post and neo-natal healthcare and family planning. Moreover, financial

Table 2 Consensus linguistic decision matrices of the expert E_2 at different time periods

time periods	alternatives	A_1	A_2	A_3	A_4
	X_1	$(l_4,-0.4)$	$(l_5,-0.1)$	$(l_2,0.1)$	$(l_3,0.4)$
	X_2	$(l_2,0)$	$(l_2,-0.2)$	$(l_3,0)$	$(l_2,0.1)$
t_1	X_3	$(l_4,0)$	$(l_6,-0.4)$	$(l_3,0)$	$(l_5,-0.1)$
	X_4	$(l_3,0.4)$	$(l_4,-0.4)$	$(l_4,0)$	$(l_2,0)$
	X_5	$(l_2,-0.4)$	$(l_3,0)$	$(l_2,0)$	$(l_2,-0.4)$
	X_1	$(l_4,0)$	$(l_5,0.4)$	$(l_3,0)$	$(l_3,0)$
	X_2	$(l_2,0.6)$	$(l_2,0)$	$(l_2,0.1)$	$(l_3,0)$
t_2	X_3	$(l_5,0)$	$(l_4,0)$	$(l_3,0.4)$	$(l_3,-0.1)$
	X_4	$(l_4,0)$	$(l_4,-0.4)$	$(l_4,0.6)$	$(l_3,0.1)$
	X_5	$(l_1,-0.4)$	$(l_2,0)$	$(l_3,0)$	$(l_2,0)$
	X_1	$(l_3,-0.2)$	$(l_3,0.2)$	$(l_2,-0.2)$	$(l_3,0)$
	X_2	$(l_2,0.4)$	$(l_3,-0.4)$	$(l_1,0.1)$	$(l_2,-0.1)$
t_3	X_3	$(l_3,0)$	$(l_5,0)$	$(l_3,0)$	$(l_4,0)$
	X_4	$(l_5,0.4)$	$(l_4,-0.4)$	$(l_4,0)$	$(l_4,-0.4)$
	X_5	$(l_1,0.4)$	$(l_2,-0.4)$	$(l_2,0)$	$(l_3,0)$

support is also provided to various states to strengthen the overall health system. To monitor the improvement through current schemes and services, and to make policies for further improvement, Health and Family Welfare Department plans to evaluate health system of different states of the India. The government also conducts survey and collects data related to the health system from the states to check whether through its various services survival rates and quality of life of the people of different states improve over time.

In aim to rank the health systems of five underdeveloped states of India, the Health and Family Welfare Department sets up a committee which consists of three decision makers $\{J_1, J_2, J_3\}$. Such kind ranking of states' health systems is necessary to make state oriented future health policies to improve the health systems in a more comprehend way than the existing central policy system. The five states, denoted as, X_1, X_2, X_3, X_4 and X_5, can be considered as alternative set, i.e., $X = \{X_1, X_2, X_3, X_4, X_5\}$

Four key determinants to judge the quality of health system are as follows:

- health system practices and policies (A_1)
- physical ability to provide quality care (A_2)
- staff competency and motivation (A_3)
- acceptability of health services to community (A_4)

The above four factors $\{A_1, A_2, A_3, A_4\}$ are set as attributes for evaluating the states' performances. Since all the attributes are not equally important, the relative impor-

Table 3 Consensus linguistic decision matrices of the expert E_2 at different time periods

time periods	alternatives	A_1	A_2	A_3	A_4
	X_1	$(l_4,0)$	$(l_5,0)$	$(l_2,0)$	$(l_3,0)$
	X_2	$(l_3,-0.3)$	$(l_2,0)$	$(l_4,-0.3)$	$(l_3,0)$
t_1	X_3	$(l_5,-0.2)$	$(l_5,0)$	$(l_4,-0.1)$	$(l_4,.05)$
	X_4	$(l_3,0)$	$(l_5,-0.3)$	$(l_3,0.3)$	$(l_2,0)$
	X_5	$(l_2,0)$	$(l_4,-0.3)$	$(l_3,-0.3)$	$(l_1,0)$
	X_1	$(l_5,-0.3)$	$(l_4,0.3)$	$(l_2,0.1)$	$(l_4,-0.3)$
	X_2	$(l_3,0)$	$(l_1,0.3)$	$(l_3,0)$	$(l_2,0.05)$
t_2	X_3	$(l_6,-0.3)$	$(l_5,-0.1)$	$(l_3,0)$	$(l_2,.05)$
	X_4	$(l_4,0)$	$(l_4,0.2)$	$(l_5,0)$	$(l_4,0)$
	X_5	$(l_2,-0.3)$	$(l_1,0.3)$	$(l_4,-0.3)$	$(l_3,-0.3)$
	X_1	$(l_3,0)$	$(l_2,0.2)$	$(l_2,0)$	$(l_4,-0.1)$
	X_2	$(l_2,0)$	$(l_4,-0.3)$	$(l_2,-0.05)$	$(l_2,0)$
t_3	X_3	$(l_4,-0.3)$	$(l_6,-0.3)$	$(l_4,-0.1)$	$(l_3,0.3)$
	X_4	$(l_4,0.3)$	$(l_5,-0.3)$	$(l_3,0.2)$	$(l_4,0.2)$
	X_5	$(l_1,0)$	$(l_2,0)$	$(l_1,0.1)$	$(l_3,0)$

tance of the attributes are set as $w = (0.35, 0.2, 0.15, 0.3)$. Based on the collected data of the years 2008, 2010, and 2012, the decision makers are asked to provide their opinions/judgments against state's health system with respect to the aforementioned four attributes in year-wise fashion. We denote the year '2008' as t_1, '2010' as t_2, and '2012' as t_3. As the system is complex, the states' present and past performance are needed to take into consideration for better results.

As most of the attributes are subjective in nature, decision makers prefer to express their opinions in natural language by means of linguistic terms. They decide to provide their preferences by using 2-tuple linguistic information according to the following linguistic terms set:

$S = \{l_0 =$ very low (VL), $l_1 =$ low (L), $l_2 =$ moderately low (ML), $l_3 =$ normal (N),

$l_4 =$ moderately high (MH), $l_5 =$ high (H), $l_6 =$ very high (VH)$\}$

The linguistic assessments of the states' health system against the attributes provided by the decision makers at different time periods are listed in Table 1. The weight vector of the time periods $t_s (1 \leq s \leq 3)$ are set as, $\lambda = (1/6, 2/6, 3/6)^T$. Now, the proposed decision making algorithm is implemented to rank the states' health systems.

Table 4 Consensus linguistic decision matrices of the expert E_3 at different time periods

time periods	alternatives	A_1	A_2	A_3	A_4
t_1	X_1	$(l_5, -0.3)$	$(l_4, 0.05)$	$(l_3, -0.05)$	$(l_2, 0.3)$
	X_2	$(l_2, -0.4)$	$(l_3, -0.2)$	$(l_3, -0.4)$	$(l_3, 0)$
	X_3	$(l_4, -0.2)$	$(l_4, 0.3)$	$(l_3, 0)$	$(l_5, 0)$
	X_4	$(l_2, 0.3)$	$(l_4, 0)$	$(l_4, 0.4)$	$(l_3, -0.1)$
	X_5	$(l_3, -0.3)$	$(l_3, -0.4)$	$(l_2, -0.4)$	$(l_2, 0.2)$
t_2	X_1	$(l_4, -0.4)$	$(l_5, 0)$	$(l_3, 0)$	$(l_3, -0.4)$
	X_2	$(l_4, -0.3)$	$(l_2, 0.4)$	$(l_3, 0)$	$(l_3, 0)$
	X_3	$(l_5, -0.4)$	$(l_4, 0)$	$(l_2, 0.3)$	$(l_3, 0)$
	X_4	$(l_5, -0.1)$	$(l_3, 0)$	$(l_6, -0.3)$	$(l_4, 0)$
	X_5	$(l_1, 0)$	$(l_2, 0.4)$	$(l_3, -0.4)$	$(l_2, -0.4)$
t_3	X_1	$(l_4, -0.2)$	$(l_3, 0)$	$(l_3, -0.2)$	$(l_3, 0)$
	X_2	$(l_1, 0.3)$	$(l_3, 0)$	$(l_1, 0)$	$(l_1, 0.05)$
	X_3	$(l_3, -0.4)$	$(l_5, -0.4)$	$(l_3, 0)$	$(l_4, 0.4)$
	X_4	$(l_5, 0)$	$(l_4, 0)$	$(l_4, 0.2)$	$(l_3, 0)$
	X_5	$(l_0, 0.3)$	$(l_3, -0.3)$	$(l_2, 0)$	$(l_2, 0.1)$

Step 1   We implement the consensus reaching algorithm to make decision makers' opinions, presented in Table 1, closer. Before starting the process, decision makers are asked to provide their opinions changing indices using the linguistic term set: $C = \{p_0 =$ very very low, $p_1 =$ very low, $p_2 =$ low, $p_3 =$ moderately, $p_4 =$ high, $p_5 =$ very high, $p_6 =$ very very high$\}$. Suppose, the expert E_1's opinion changing index is p_2, E_2's opinion changing index is p_6 and E_3's opinion changing index is p_4. The decision makers also set the consensus level as $\varepsilon = 0.1$. To reach the desired consensus among the decision makers $\{E_1, E_2, E_3\}$ over each alternative's rating against the attributes over different time periods, we implement the proposed consensus algorithm and obtain the decision makers' consensus opinions as summarized in Tables: 2-4. For example, to achieve the higher level of agreement among the decision makers $\{E_1, E_2, E_3\}$ over the alternative X_1's ratings $\{l_3, l_4, l_5\}$ (as provided in Table 1), against the attribute A_1 at time period t_1, we apply the proposed consensus reaching algorithm and find the decision makers' consensus opinions as: $\{(l_4, -0.4), (l_4, 0), (l_5, -0.3)\}$. Similarly, we compute the consensus opinions of the decision makers for the rest of the entries of Table 1.

Table 5 Collective consensus linguistic decision matrices of the decision makers

Expert	alternatives	A_1	A_2	A_3	A_4
	X_1	$(l_3, 0.33)$	$(l_4, 0.22)$	$(l_2, 0.25)$	$(l_3, 0.07)$
	X_2	$(l_2, 0.4)$	$(l_2, 0.27)$	$(l_2, -0.25)$	$(l_2, 0.3)$
E_1	X_3	$(l_4, -0.17)$	$(l_5, -0.27)$	$(l_3, 0.13)$	$(l_4, -0.22)$
	X_4	$(l_5, -0.4)$	$(l_4, -0.4)$	$(l_4, 0.2)$	$(l_3, 0.17)$
	X_5	$(l_1, 0.17)$	$(l_2, -0.03)$	$(l_2, 0.33)$	$(l_2, 0.43)$
	X_1	$(l_4, -0.27)$	$(l_3, 0.37)$	$(l_2, 0.03)$	$(l_4, -0.32)$
	X_2	$(l_2, 0.45)$	$(l_3, -0.38)$	$(l_3, -0.41)$	$(l_2, 0.2)$
E_2	X_3	$(l_5, -0.45)$	$(l_5, 0.32)$	$(l_4, -0.4)$	$(l_3, 0.01)$
	X_4	$(l_4, -0.02)$	$(l_5, 0.47)$	$(l_4, -0.18)$	$(l_4, -0.23)$
	X_5	$(l_1, 0.4)$	$(l_2, 0.05)$	$(l_2, 0.23)$	$(l_3, -0.43)$
	X_1	$(l_4, -0.18)$	$(l_4, -0.16)$	$(l_3, -0.11)$	$(l_3, -0.25)$
	X_2	$(l_2, 0.15)$	$(l_3, -0.23)$	$(l_2, -0.07)$	$(l_2, 0.02)$
E_3	X_3	$(l_3, 0.47)$	$(l_4, 0.35)$	$(l_3, -0.23)$	$(l_4, 0.03)$
	X_4	$(l_5, -0.48)$	$(l_4, -0.33)$	$(l_5, -0.27)$	$(l_3, 0.32)$
	X_5	$(l_1, -0.07)$	$(l_3, -0.42)$	$(l_2, 0.13)$	$(l_2, -0.05)$

Table 6 Consensus group linguistic decision matrix

alternatives	A_1	A_2	A_3	A_4
X_1	$(l_4, -0.35)$	$(l_4, -0.19)$	$(l_2, 0.39)$	$(l_3, 0.17)$
X_2	$(l_2, 0.33)$	$(l_3, -0.45)$	$(l_2, 0.09)$	$(l_2, 0.17)$
X_3	$(l_4, -0.05)$	$(l_5, -0.2)$	$(l_3, 0.17)$	$(l_4, -0.39)$
X_4	$(l_4, 0.37)$	$(l_4, -0.07)$	$(l_4, 0.25)$	$(l_3, 0.42)$
X_5	$(l_1, 0.17)$	$(l_2, 0.2)$	$(l_2, 0.23)$	$(l_2, 0.32)$

Step 2   From the consensus decision matrices of the decision makers presented
in Table 2-4, we compute the collective consensus linguistic decision matrices of the decision makers by utilizing Eq. (9) with associated weight
vector $\lambda = (1/6, 2/6, 3/6)^T$ of the time periods $t_s (1 \leq s \leq 3)$. The results
are summarized in Table 5.

Step 3   From consensus collective linguistic decision matrices presented in Table
5, group consensus linguistic decision matrix is formed by using Eq. (10).
The group consensus linguistic decision matrix is presented in Table 6.

Step 4   The consensus group overall performance of the alternatives are derived by utilizing Eq. (11) with given weight vector, $w = (0.35, 0.2, 0.15, 0.3)$ of the attributes as follows: $\bar{d}_1 = (l_3, 0.35)$, $\bar{d}_2 = (l_2, 0.29)$, $\bar{d}_3 = (l_4, -0.01)$, $\bar{d}_4 = (l_4, -0.02)$ and $\bar{d}_5 = (l_2, -0.12)$

Step 5   According to the overall rating values, the ranking order of the alternatives is $X_4 > X_3 > X_1 > X_2 > X_5$. Hence, the health system of the state X_4 performs better in comparison to other states.

6   Conclusions

In this paper, a new consensus building algorithm has been proposed where decision makers' interest in changing their initial opinions are taken into account during reaching consensus by introducing the notion of opinion changing index. The proposed consensus algorithm has been utilized to solve the dynamic MAGDM problem in linguistic environment. The proposed consensus model has the following features:

- the proposed consensus process is automatic despite of the fact that it is controlled by the moderator. As the moderator generates the decision maker's new opinion by solving optimization problem (P6), there is no direct interaction between moderator and decision makers.
- by considering the decision makers' interest in changing their initial opinions in the model, the proposed approach attempt to preserve sovereignty of the decision makers to some extent.

The proposed model has the potentials to be extended in other environments, such as, fuzzy, intuitionistic fuzzy etc. Also, to form the group opinion different aggregation operators can be utilized to enhance the consensus process. In future work, it is also valuable to consider the investigation of advantages, weakness and restrictions to use the proposed advice generation process over the other existing methods. It is also worthy to explore the real-world applications of the proposed consensus process in group decision making.

Acknowledgements. The first author gratefully acknowledges the financial support provided by the Council of Scientific and Industrial Research, New Delhi, India (Award No. 09/1023(007)/2011-EMR-I).

References

1. Beliakov, G., Pradera, A., Calvo, T.: Aggregation Functions: A Guide for Practitioners. Springer, Berlin (2007)
2. Ben-Arieh, D., Chen, Z.F.: Linguistic-labels aggregation and consensus measure for autocratic decision making using group recommendations. IEEE Trans. Syst., Man, Cybern., A Syst. Humans 36, 558–568 (2006)
3. Ben-Arieh, D., Easton, T., Evans, B.: Minimum cost consensus with quadratic cost functions. IEEE Trans. Syst., Man, Cybern. A, Syst., Humans 39, 210–217 (2008)

4. Bryson, N.: Group decision-making and the analytic hierarchy process: Exploring the consensus-relevant information content. Comput. Oper.Res. 23(1), 27–35 (1996)

5. Chen, S.M., Lee, L.W.: Autocratic decision making using group recommendations based on the ILLOWA operator and Likelihood Based Comparison Relations. IEEE Trans. Syst., Man, Cybern., A Syst. Humans 42, 115–129 (2012)

6. Chiclana, F., Mata, F., Martínez, L., Herrera-Viedma, E., Alonso, S.: Integration of a consistency control module with in a consensus decision making model. Int. J. of Uncertain. Fuzziness and Knowl. Based Syst. 16, 35–53 (2008)

7. Cordón, O., Herrera, F., Zwir, I.: Linguistic modeling by hierarchical systems of linguistic rules. IEEE Trans. Fuzzy Syst. 10, 2–20 (2002)

8. Degani, R., Bortolan, G.: The problem of linguistic approximation in clinical decision making. Int. J. Approx. Reason. 2, 143–162 (1988)

9. Delgado, M., Verdegay, J.L., Vila, M.A.: On aggregation operations of linguistic labels. Int. J. Intell. Syst. 8, 351–370 (1993)

10. Dong, Y.C., Xu, Y.F., Li, H.Y., Feng, B.: The OWA-based consensus operator under linguistic representation models using position indexes. Eur. J. Oper. Res. 203, 455–463 (2010)

11. Dong, Y.C., Zhang, G.Q., Hong, W.H., Xu, Y.F.: Consensus models for AHP group decision making under row geometric mean prioritization method. Decis. Support Syst. 49, 281–289 (2010)

12. Dutta, B., Guha, D., Mesiar, R.: A model based on linguistic 2-tuples for dealing with heterogeneous relationship among attributes in multi-expert decision making. IEEE Trans. Fuzzy Syst (2014), doi:10.1109/TFUZZ.2014.2379291

13. Fu, C., Yang, S.L.: The group consensus based evidential reasoning approach for multiple attributive group decision analysis. Eur. J. Oper. Res. 206, 601–608 (2010)

14. Guha, D., Chakraborty, D.: Fuzzy multi attribute group decision making method to achieve consensus under the consideration of degrees of confidence of experts' opinions. Comput. Ind. Eng. 60, 493–504 (2011)

15. Herrera, F., Herrera-Viedma, E., Verdegay, J.L.: A model of consensus in group decision making under linguistic assessments. Fuzzy Sets Syst. 78, 73–87 (1996)

16. Herrera, F., Herrera-Viedma, E., Verdegay, J.L.: A rational consensus model in group decision making using linguistic assessments. Fuzzy Sets Syst. 88, 31–49 (1997)

17. Herrera, F., Martínez, L.: A 2-tuple fuzzy linguistic representation model for computing with words. IEEE Trans. Fuzzy Syst. 8, 746–752 (2000)

18. Herrera, F., Martínez, L.: A model based on linguistic 2-tuples for dealing with multigranular hierarchical linguistic contexts in multi-expert decision-making. IEEE Trans. Syst. Man. Cybern. B. Cybern. 31, 227–343 (2001)

19. Herrera-Viedma, E., Herrera, F., Chiclana, F.: A consensus model for multiperson decision making with different preference structures. IEEE Trans. Syst., Man, Cybern., A Syst. Humans 32, 394–402 (2002)

20. Herrera-Viedma, E., Mata, F., Martnez, L., Chiclana, F.: A consensus support system model for group decision-making problems with multigranular linguistic preference relations. IEEE Trans. Fuzzy Syst. 13, 644–658 (2005)

21. Herrera-Viedma, E., Alonso, S., Chiclana, F., Herrera, F.: A consensus model for group decision making with incomplete fuzzy preference elations. IEEE Trans. Fuzzy Syst. 15, 863–877 (2007)

22. Herrera-Viedma, E., Cabrerizo, F.J., Kacprzyk, J., Pedrycz, W.: A review of soft consensus models in a fuzzy environment. Inform. Fusion 17, 4–13 (2014)

23. Kacprzyk, J., Fedrizzi, M.: A soft measure of consensus in the setting of partial (fuzzy) preferences. Eur. J. Oper. Res. 34, 316–325 (1988)

24. Kacprzyk, J., Fedrizzi, M., Nurmi, H.: Group decision making and consensus under fuzzy preferences and fuzzy majority. Fuzzy Sets Syst 49, 21–31 (1992)
25. Martínez, L., Herrera, F.: An overview on the 2-tuple linguistic model for computing with words in decision making: Extensions, applications and challenges. Inform. Sci. 207, 1–18 (2012)
26. Mata, F., Martínez, L., Herrera-Viedma, E.: An adaptive consensus support model for group decision-making problems in a multigranular fuzzy linguistic context. IEEE Trans. Fuzzy Syst. 17, 279–290 (2009)
27. Miller, G.A.: The magical number seven, plus or minus two: Some limits on our capacity of processing information. Psychol. Rev. 63, 81–97 (1956)
28. Palomares, I., Liu, J., Xu, Y., Martínez, L.: Modelling experts' attitudes in group decision making. Soft Comput. 16(10), 1755–1766 (2012)
29. Palomares, I., Rodriguez, R.M., Martínez, L.: An attitude-driven web consensus support system for heterogeneous group decision making. Expert Syst. Appl. 40(1), 139–149 (2013)
30. Palmoraes, I., Estrella, F.J., Martínez, L., Herrera, F.: Consensus under a fuzzy context: Taxonomy, analysis framework AFRYCA and experimental case of study. Inform. Fusion 20, 252–271 (2014)
31. Rodríguez, R.M., Martínez, L.: An analysis of symbolic linguistic computing models in decision making. Int. J. Gen. Syst. 42, 121–136 (2012)
32. Su, Z., Chen, M., Xia, G., Wang, L.: An interactive method for dynamic intuitionistic fuzzy multi-attribute group decision making. Expert Syst. Appl. 38, 15286–15295 (2011)
33. Wei, G., Zhao, X.: Some dependent aggregation operators with 2-tuple linguistic information and their application to multiple attribute group decision making. Expert Syst. Appl. 39(5), 5881–5886 (2012)
34. Wibowo, S., Deng, H.: Consensus-based decision support for multicriteria group decision making. Comput. Ind. Eng. 66, 625–633 (2013)
35. Xu, Z.S., Yager, R.R.: Dynamic intuitionistic fuzzy multi-attribute decision making. Int. J. Approx. Reason. 48, 246–262 (2008)
36. Xu, Z.S.: On multi-period multi-attribute decision making. Knowledge-Based Systems 21, 164–171 (2008)
37. Xu, Z.S.: An automatic approach to reaching consensus in multiple attribute group decision making. Comput. Ind. Eng. 56, 1369–1374 (2009)
38. Xu, Z.S., Cai, X.Q.: Group consensus algorithms based on preference relations. Inform. Sci. 181, 150–162 (2011)
39. Xu, J., Wu, Z.: A discrete consensus support model for multiple attribute group decision making. Knowl.-Based Syst. 24, 1196–1202 (2011)

Using Computing with Words for Managing Non-cooperative Behaviors in Large Scale Group Decision Making

Francisco José Quesada, Iván Palomares, and Luis Martínez

Abstract. Normally, in group decision making problems, groups are composed by individuals or experts with different goals and points of view. For these reasons, they may adopt distinct behaviors in order to achieve their own aims. Nonetheless, in such problems in general, specially those demanding a certain degree of consensus, each expert should comply with a collaboration contract in order to find a common solution for the decision problem. When decision groups are small, all experts usually attempt to fulfill the collaboration contract. However, nowadays technologies such as social media allow to make consensus-driven decisions with larger groups, in which many experts are involved, hence the possibility that some of them try to break the collaboration contract might be greater. In order to prevent the group solution from being biased by these experts, it is necessary to detect and manage their non-cooperative behaviors in this kind of problems. Recent proposals in the literature suggest managing non-cooperative behavior by reducing the importance of expert opinions. These proposals present drawbacks such as, the inability of an expert to recover his/her importance if behavior improves; and the lack of expert's behavior measures across the time. This chapter introduces a methodology based on fuzzy sets and computing with words, with the aim of identifying and managing those experts whose behavior does not contribute to reach an agreement in consensus reaching processes. Such a methodology is characterized by allowing the importance recovery of experts and taking into account the evolution of their behavior across the time.

Keywords: Group Decision Making, Computing with Words, Fuzzy sets, Consensus Reaching Processes.

F.J. Quesada · L. Martínez
Department of Computer Science, University of Jaén, 23071 Jaén, Spain
e-mail: fqreal@ujaen.es, martin@ujaen.es

I. Palomares
Built Environment Research Institute, School of the Built Environment, University of Ulster, BT37 0QB Newtownabbey, Northern Ireland (United Kingdom)
e-mail: i.palomares-carrascosa@ulster.ac.uk

© Springer International Publishing Switzerland 2015   97
W. Pedrycz and S.-M. Chen (eds.), *Granular Computing and Decision-Making*,
Studies in Big Data 10, DOI: 10.1007/978-3-319-16829-6_5

1   Introduction

In our daily life we can find a myriad of Group Decision Making (GDM) problems, ranging from the choice of a restaurant to have a dinner with friends to the definition of a marketing strategy for a big company. In all of these situations, joining together experiences and knowledge of a group, makes it easier to face complex decision problems and may lead to better decisions. GDM problems are defined as decision situations in which a group of individuals or experts, try to find a common solution to a particular problem made up of a set of alternatives [1, 2]. To do so, experts have to express their opinions over the distinct alternatives that might be a solution to such a problem.

Many real-life GDM problems are often defined under an environment of uncertainty, so that experts should provide the information about their preferences by using an information domain closer to human natural language, which is suitable to deal with such uncertainty [3, 4]. Within Granular Computing [5], there are different approaches to deal with uncertain information, such as fuzzy set theory [6] and the fuzzy linguistic approach [7, 8, 9], which have been some of the most utilized approaches in decision problems under uncertainty [10]. In particular, the Computing with Words (CW) paradigm [4] has been widely considered as a reasoning methodology in decision making problems [11, 12, 13].

Traditionally, the procedure to solve a GDM problem only consists in an alternative selection process, which after gathering experts' opinions and processing them, aims at finding a solution [14, 15, 16]. However, sometimes it is possible that when applying the selection process solely, one or more experts may not feel identified with the decision made and they do not accept it, because they consider that their individual concerns have not been considered sufficiently to reach the solution made. In order to overcome this drawback, Consensus Reaching Processes (CRPs) were introduced as an additional phase in the resolution process for GDM problems [17]. In a CRP, experts try to achieve a high level of agreement before making a decision, by discussing and modifiying their individual preferences, bringing them closer to each other [18]. In the literature, there are many consensus models proposed by different authors to support and guide groups in CRPs conducted in different GDM frameworks [19, 20, 21, 22, 23, 24] attending several criteria [25].

Classically, GDM problems have been carried out by a small number of experts in organizational and enterprise environments. Nevertheless, the appearance of new technological environments and paradigms to make group decisions, such as social networks, e-democracy or group e-marketplaces, have caused that decision problems in which large groups of experts can take part attain greater importance in the last years [24, 26, 27, 28]. In CRPs in which many experts are involved, it may occur that some experts or subgroups of them, seek their own interests rather than the collective interest, which may lead them to break the collaborative contract established amongst participants, in order to achieve a common solution [29]. Therefore, they might not cooperate to bring their opinions closer to the rest of the group [30]. In such situations, it would be convenient to identify and deal with such non-cooperative behaviors of individuals or subgroups, in order to prevent that they deviate the group solution in their favor. This possible deviation of the solution may affect negatively the normal development of the CRP. Currently, there are several approaches that deal with experts who present a non-cooperative behavior in GDM problems [30] and in CRPs [24, 31]. These approaches penalize non-cooperating experts, driving them out of the GDM solution [30], diminishing the importance of their opinions either along the CRP [24] or based on the experts' behavior at the current phase of the CRP [31].

This chapter proposes a novel fuzzy approach based on CW [4] to detect and manage non-cooperative behaviors in CRPs of large scale GDM problems. The CW paradigm facilitates the definition, comprehension and detection of experts' behaviors such as *cooperative*. Additionally to the analysis of experts' current behaviors to manage the manipulation of the CRP performed by experts, this approach also applies a weighting scheme based on hyper-similarity [32] that takes into account the experts behavior across the time. Therefore, the proposed approach provides a mechanism in which cooperative experts outweigh non-cooperative ones in order to achieve a common agreed solution for a GDM problem.

The chapter is set out as follows: Section 2 introduces some preliminaries about CRPs in GDM problems, some related works which deal with non-cooperative behaviors in these problems and the CW paradigm for reasoning processes. Section 3 presents the approach based on CW for managing experts' behaviors in CRPs with large groups and its integration with a consensus model for GDM problems. In Section 4 it is shown an illustrative example which includes a comparison between our proposal and several previous approaches to attempt penalization in CRPs. Finally, in Section 5 some concluding remarks are pointed out.

2 Preliminaries

This section briefly reviews some basic concepts about CRPs in GDM and different works related with the treatment of experts with non-cooperative behaviors in them. Finally, a short conceptual revision of CW, basis of our proposal for managing the non-cooperative behaviors in such processes, is drawn.

2.1 Consensus Reaching Processes in Large Scale GDM

GDM entails the participation of several individuals or experts, who must make a collective decision to find a common solution for a problem. Decision making processes in which several experts take part, who each has his/her own knowledge and experiences, may sometimes lead to better decisions than those made by one expert only [1].

Formally, a GDM problem is composed by [2]:

- The existence of a decision problem to be solved.
- A set $X = \{x_1, ..., x_n\}(n \geqslant 2)$, of *alternatives* or posible solutions to the problem.

- A set $E = \{e_1, ..., e_m\}\,(m \geqslant 2)$, of individuals or *experts*, who express their opinions or preferences over the set of alternatives X.

Usually, experts utilize a preference structure to express their opinions over alternatives. One of the most widely used preference structures in GDM problems under uncertainty is the *fuzzy preference relation* [2, 23, 33]. A fuzzy preference relation P_i associated to expert e_i is defined by a membership function $\mu_{P_i} : X \times X \rightarrow [0,1]$ and it is represented for X finite as an $n \times n$ matrix:

$$P_i = \begin{pmatrix} - & \cdots & p_i^{1n} \\ \vdots & \ddots & \vdots \\ p_i^{n1} & \cdots & - \end{pmatrix}$$

where each assessment $p_i^{lk} = \mu_{P_i}(x_l, x_k)$ represents the preference degree of the alternative x_l over x_k according to the expert e_i, so that $p_i^{lk} > 0.5$ indicates that x_l is preferred over x_k. If $p_i^{lk} < 0.5$ then x_k is preferred over x_l, and $p_i^{lk} = 0.5$ indicates indifference between x_l and x_k.

Classically, the process to find a solution for a GDM problem consists of an alternative selection process, which is composed of two phases [15] (Figure 1).

1. *Aggregation*: In this phase experts' preferences are combined by using an aggegation operator [34].
2. *Exploitation*: Here, a selection criterion [14, 16] is applied to obtain an alternative or a subset of alternatives, as the solution for the GDM problem.

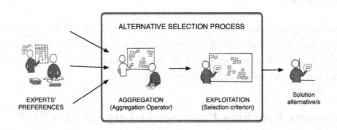

Fig. 1 Selection process in GDM problems

When applying the selection process in a GDM problem solely, it may occur that one or more experts feel that their opinions have not been taken into account sufficiently to reach the solution achieved. This fact could imply that these experts do not feel identified with the solution. There exist some situations in which it is necessary a high agreement level among the participant experts. Therefore, it arises the need of applying a CRP that introduces a new phase in the GDM resolution process with the aim of reaching a high level of agreement between experts before making the decision.

Consensus can be defined as the agreement produced by the mutual consent between all members in a group or several groups [17, 18, 29]. A CRP is a dynamic and iterative process, which is coordinated by a human figure known as *moderator*. The moderator is responsible for supervising and guiding experts over the course of this process [18]. Consensus should be understood as a process in which the final decision may not match the initial position of experts. Thus, experts might change their preferences during several discussion rounds [29]. In the literature, there are many consensus models for a wide variety of GDM frameworks [25]. Figure 2 shows a general CRP scheme followed by many of these models. Its main phases are introduced below [23, 24]:

1. *Gathering Preferences*: Each expert e_i provides his/her preferences on X (e.g. by means of a fuzzy preference relation).
2. *Determine Degree of Consensus*: The moderator calculates the degree of agreement, cr, reached in the group. Such a degree can be a value of the interval $[0,1]$ (where the value 1, indicates a full or unanimous agreement between all experts over all alternatives)[1]. Different consensus measures [25] can be utilized in order to calculate the cr. Such measures are often based on the use of: (i) metrics to calculate degrees of similarity between preferences of experts, and (ii) aggregation operators that obtain the degree of consensus in the group by aggregating similarity values [23, 24].
3. *Consensus Control*: The consensus degree cr is compared with a minimum consensus threshold, $\mu \in [0,1]$. The value of μ is previously established by the group. If $cr > \mu$, then consensus has been reached and after that the group can proceed to the selection phase; otherwise, the CRP must go on with another discussion phase.
4. *Generate Feedback Information*: The moderator calculates the group collective preference, P_c, by aggregating the individual preferences of the experts. On the basis of P_c, he identifies those experts e_i whose assessments p_i^{lk} are farthest to consensus, and advises them how to modify their assessments to increase the consensus degree in the following round (by indicating each expert whether he/she should increase or decrease the value of each assessment). Each recommendation is a triplet with three elements $(e_i, (x_l, x_k), Direction)$, which shows that the expert e_i, should modify the assessment p_i^{lk}, in the direction given by the argument $Direction \in \{Increase, Decrease\}$.

Normally, the consensus process implies the need that experts *accept* to review and modify, in some degree, their opinions on the basis of the recommendations received, with the aim of bringing their opinions closer to the rest of the group. Based on this, it can be assumed that they should accept a priori a collaboration

---

[1] Consensus degrees are normally based on aggregation of similarity values between experts' assessment. Such values belong to the unit interval, hence the resulting agreement values are computed in [0,1].

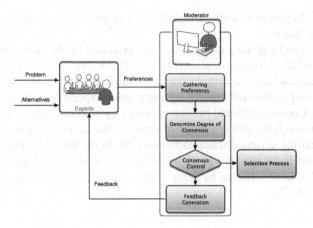

Fig. 2 General CRP scheme

contract [29]. Nevertheless, in large groups, in which normally there are many experts with different aims, it might occur that some experts do not cooperate modifying their assessments as they should do according to the received recommendations, thus breaking the collaboration contract because their own interests outweigh group ones [24]. This chapter considers an expert as cooperative when the group interests outweigh his/her own interests, therefore he/she is willing to change his/her initial opinion. Conversely, an expert is considered as non-cooperative when his/her own goals outweigh group interests, and is not willing to change his/her opinion. This chapter aims at identifying and dealing with the latter type of behaviors. Notwithstanding, there are different proposals, revised in the following subsection, which have already attempted to manage this kind of behaviors. They present important limitations which serve as motivation for our proposal presented in section 3.

2.2   Related Works

In order to solve the shortcoming of experts that break the collaboration contract, trying to strategically deviate the solution for classical GDM problems, Yager proposed in [30] an approach to penalize them. This proposal firstly identifies experts with more drastic opinions (e.g. experts who show a full preference on one alternative and a null preference on the rest of them) as experts with a strategically manipulative pattern of behaviors. After the non-cooperative experts are identified, this approach completely discards the information associated to the preferences of these experts, who are directly excluded from the GDM problem resolution process. This approach might be considered completely drastic, in the sense that it completely eliminates the information associated to the experts whose preferences have been identified as strategically manipulated opinions.

Attempting to solve this issue, Mata et al. [35] extended the previous approach to deal with strategically manipulated preferences in CRPs, assigning experts a weight based not only on drastic opinions, but also on identifying those experts who did not obey the advice received.

More recently, Palomares et al. presented in [24] a methodology, where all expert's opinions have an importance weight $w_i \in [0, 1]$. If an expert's opinion is far from consensus and he/she does not cooperate to bring his/her opinion closer to the group opinion at a given round, his/her weight is decreased. Otherwise, his/her weight keeps invariant. These weights are used to calculate the consensus opinion, given by P_c, in the CRP. In this case, all experts keep taking part during the CRP to some degree, although the opinions of those experts who cooperate, are taken into account to a higher degree than the opinions of experts that do not cooperate. However, if an expert's opinion is penalized at a CRP round, by assigning him/her of a low weight, such a weight cannot be increased at later consensus rounds, even though the expert changes his/her behavior and decides to cooperate again in subsequent discussion rounds [24].

Considering that sometimes, non cooperating in a particular consensus round might be a negotiation strategy, it is necessary that opinions of experts with non-cooperative behavior at some specific rounds only can recover their importance if their behavior is significantly improved afterwards. Palomares et al. proposed in [31], a CW-based methodology to assign experts importance weights depending on their behavior at a given consensus round. Thus, if an expert did not cooperate at a previous consensus round, but he/she cooperates in the current one, his/her opinion may recover the importance previously lost to calculate the group opinion. However, the experts' weights are computed based on their behavior at a given consensus round only, not taking into account the evolution of their behavior since the beginning of the CRP.

2.3 Computing with Words for Reasoning Processes

Human beings use linguistic terms to communicate, explain and understand their surroundings. On the other hand, machines such as computer systems require more complex symbols [36]. One of the most extended proposals to establish an understandable communication gateway between humans and computers, is the CW paradigm [4], which was proposed by Zadeh and it is based on fuzzy sets theory [6]. This methodology gives a framework where the concepts can be modelled by means of fuzzy sets, so that they can be easily understood by both machines and humans.

A key concept in computing with words is the linguistic term. We can define a linguistic term as a word or a phrase in natural language which is used to express the value of an attribute [7, 8, 9]. For instance, let us consider an attribute called *distance*, comprehended as the size of the gap between two places[2]. Some possible linguistic terms to express the value of this attribute could be: "very close", "close",

---

[2] Definition of "Distance" in http://wordnetweb.princeton.edu/

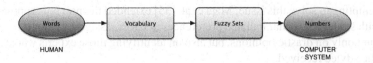

HUMAN   COMPUTER
                                                         SYSTEM

Fig. 3 Paradigm of man-machine understanding. (Taken from [36])

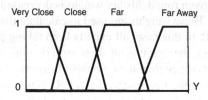

Fig. 4 Different linguistic terms for the attribute *distance*

"far", "far away". Thus, humans can easily understand and reason over their environment with the help of linguistic terms (see Figure 3).

Given the ambiguous and vague nature inherent to the values associated to the linguistic terms, fuzzy sets are a useful tool to formalize the concepts associated to them, thus allowing the comprehension and development of computational processes over these concepts by computers (see Figure 4). If $\mathbf{P}$ is a linguistic term (i.e. "close") of a vocabulary associated with the attribute $\mathbf{A}$ (i.e. *distance*), then $\mathbf{P}$ can be expressed as a fuzzy subset in the domain $Y \subseteq \mathbb{R}$ of $\mathbf{A}$. Given a value $y \in Y$, $\mu_P(y)$ shows the compatibility degree of the value y with the linguistic term $\mathbf{P}$.

The choice of a linguistic term vocabulary to describe the attributes, as well as definition of the associate meaning of these terms is a human task. For instance, humans should facilitate to the computer the linguistic terms that will be used, as well as, theirs semantics given by fuzzy sets.

3 Managing Non-cooperative Behaviors in Large Scale GDM by Using CW and Hyper-Similarity

In order to prevent the bias and manipulation of CRPs carried out for the resolution of large-scale GDM problems, in which experts' own goals and interests are harder to detect, it is necessary to identify and manage the non-cooperative behaviors they might adopt.

This section presents a novel methodology to deal with the shortcomings caused by different patterns of non-cooperative expert behaviors in CRPs carried out for the resolution of large-scale GDM problems. Such a methodology extends the one presented in [31]. It also deals with some of the drawbacks that arose in the approaches presented in [24, 30, 31] (see Section 2.2). More specifically, our approach is characterized by the following features:

- Unlike Yager's approach, in which penalized experts may completely lose their importance, here we take into account experts' preferences across the CRP to a variable degree, depending on their behavior.
- Opinions of non-cooperative experts in an specific round can recover importance if they adopt a more cooperative behavior in the following rounds.
- The overall expert's behavior since the beginning of the discussion process is taken into account.

The methodology is composed by two phases, which will be developed in the following subsections in further detail: (i) a *Cooperativeness Measurement* phase, and (ii) a *Behavior Management* phase. Straightaway, this methodology will be integrated in the general CRP scheme (see Figure 5).

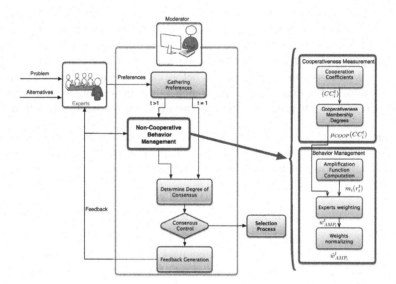

Fig. 5 Integration of the non-cooperative behavior management methodology in the general CRP scheme

3.1 Cooperativeness Measurement Phase

In order to manage experts who present a non-cooperative behavior in large groups, it is firstly necessary to identify them. In this subsection, we define a coefficient that indicates the type of behavior adopted by each expert at a given discussion round. This coefficient is based on the degree of commitment of the collaboration contract amongst experts [29], given by the extent to which they applied changes on preferences based on the feedback received.

Definition 1. Let $\#ADV_i^t$ be the number of advices provided to e_i, advising him/her to modify some of his/her assessments p_i^{lk} at the beginning of the CRP round $t \geq 2$, and let $\#ACP_i^t$ be the number of advice that e_i accepts to modify in accordance with the feedback received. The *Cooperation Coefficient* $CC_i^t \in [0,1]$ of e_i at round t is then defined as follows:

$$CC_i^t = \begin{cases} 1 & \text{if } \#ADV_i^t = 0, \\ \frac{\#ACP_i^t}{\#ADV_i^t} & \text{otherwise.} \end{cases} \tag{1}$$

The value of CC_i^t, represents the degree to which an expert modifies his/her opinions bringing them closer to consensus, as suggested by the advice he/she received. The larger the value of CC_i^t, the more cooperative e_i's behavior regarding this issue. Notice that if an expert does not receive any advice at a given round, this means that all of his/her assessment values are close enough to consensus, therefore we consider that $CC_i^t = 1$ in this case.

The relevance of cooperation varies across the CRP. For example, at the beginning of the CRP, it is usual that experts' opinions might be more distant from each other. Therefore, the level of required cooperation is different than that at the final rounds of the CRP. After several rounds, expert opinions are closer and, consequently, it is necessary to reach consensus before carrying out an excessive number of discussion-rounds. In order to properly model the meaning and the relevance of cooperativeness over the course of the CRP, we use the CW paradigm as follows.

Definition 2. Let $COOP$ be a fuzzy set, associated to the linguistic term *cooperative*. This fuzzy set models the meaning of cooperativeness by means of a semi-trapezoidal increasing membership function $\mu_{COOP}(y)$:

$$\mu_{COOP}(y) = \begin{cases} 0 & \text{if } y < \alpha, \\ \frac{y-\alpha}{\beta-\alpha} & \text{if } \alpha \leq y < \beta, \\ 1 & \text{if } y \geq \beta. \end{cases} \tag{2}$$

being $\alpha, \beta, y \in [0,1], \alpha < \beta$. We then define the *Cooperativeness Membership Degree*, $\mu_{COOP}(CC_i^t)$, which indicates the degree of membership of the cooperation coefficient (CC_i^t) to the linguistic term *cooperative*.

The membership function of $COOP$ may change at each consensus round to become more restrictive with the concept of cooperativeness as the CRP goes on. The following example illustrates this aspect in detail.

Example 1. Consider a fuzzy set $COOP$, whose membership function parameters have the initial values α=0.2 and β=0.5 at the beginning of the CRP. After the fourth consensus round, the value of both parameters will be increased by 0.1 per round, until each one of them reaches a value of 1. Thus, the approach is more restrictive with the behavior of experts as the CRP progresses. Figure 6 illustrates this process.

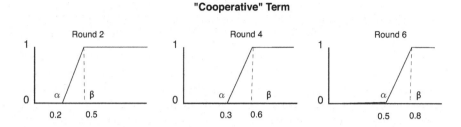

"Cooperative" Term

Fig. 6 Evolution of the fuzzy membership function associated to the linguistic term "cooperative" across the CRP (from the 4th round onwards, values of parameters α, β increase by 0.1 per round)

3.2 Behavior Management Phase

Once it has been applied the cooperativeness measurement phase, the values of $\mu_{COOP}(CC_i^t)$, for each $e_i \in E$, provide an insight of the experts who have non-cooperative behavior at round t. Therefore, we are now in a position to take the necessary actions on these experts in order to prevent the CRP manipulation by them.

Here, it is proposed a flexible approach to manage experts' opinions, so that their importance weights increase or decrease attending to their behavior across the CRP. This approach extends the work presented in [31], therefore it is necessary to firstly introduce the weight computation methodology utilized in such a work.

Proposition 1. *[31] Once it is computed* $\mu_{COOP}(CC_i^t)$, *the weight of* e_i *at round t is calculated as follows:*

$$w_i^t = \begin{cases} 1 & \text{if } t = 1, \\ \mu_{COOP}(CC_i^t) & \text{otherwise.} \end{cases} \tag{3}$$

Such weights are normalized to facilitate computations:

$$\hat{w}_i^t = \frac{w_i^t}{\sum_1^n w_i^t} \tag{4}$$

An example that illustrates this approach to compute each expert's weight is shown below.

Example 2. Let us Suppose a set of three experts, $E = \{e_1, e_2, e_3\}$, who are taking part in a CRP and each expert may adopt one out of these three different behaviors:

- *Full cooperation.* Experts always cooperate along the CRP. This is e_1's behavior.
- *Half cooperation.* In this case, experts with this behavior, cooperate obeying only a half of their feedback advice received. e_2 has this type of behavior.

- *Alternation of null and full cooperation.* Experts with this behavior, alternate null and full cooperation, disobeying all their feedback advice in a specific round and obeying all of them in the following round. e_3 follows this behavior pattern.

At the beginning, all experts' opinions have the same importance because no feedback has been yet generated, therefore it is assigned to them the maximum weight at the first round, $w_i^t = 1$ (see Eq.(3)). Once they receive the feedback advice and modify their assessments, their weights are recalculated according to their behavior by computing the cooperation coefficient, CC_i^t and applying Eq. (3) to calculate $\mu_{COOP}(CC_i^t)$, as explained in the detection phase (Section 3.1). Let us assume that the non-normalized and normalized weights, w_i^t and $\hat{w}_i^t$, of each expert along the CRP, are the ones shown in Table 1.

Table 1 Example of expert opinion weights along the CRP

Rounds \ Experts	e_1^t		e_2^t		e_3^t	
t	w_1^t	$\hat{w}_1^t$	w_2^t	$\hat{w}_2^t$	w_3^t	$\hat{w}_3^t$
1	1	0.33	1	0.33	1	0.33
2	1	0.66	0.5	0.33	0	0
3	1	0.4	0.5	0.2	1	0.4
4	1	0.66	0.5	0.33	0	0
5	1	0.4	0.5	0.2	1	0.4

e_1, presents the most cooperative behavior during all the CRP, and $w_1^t = 1$, $\forall t$, hence he/she is assigned the highest $\hat{w}_i^t$ at each round of the CRP. e_2, has always $w_2^t = 0.5$, therefore his/her opinion weight is smaller than that of e_1. The behavior of e_3, is more variable from one round to another, because he/she alternates full and null cooperation along the rounds of the CRP. Here we can see, how at $t = 2$, e_3 does not cooperate at all, therefore $w_3^t = 0$ and $\hat{w}_3^t = 0$. On the other hand, in the third round, $t = 3$, e_3 decides to cooperate bringing all his/her assessments closer to the rest of the group. Thus, $w_3^t = 1$ and $\hat{w}_3^t = 0.4$. At this point, e_3 has completely recovered the importance of his/her opinion. This pattern of varying behavior is repeated again in the fourth and fifth rounds, as we can observe in Table 1.

This proposal allows that experts can recover the importance of their opinions if they improve their behavior between two consecutive CRP rounds. Nevertheless, it may be possible that after several rounds, some experts attempt to use the weight restriction to bias the GDM problem solution. For example, by comparing the weight values of experts e_1 and e_3 in Example 2, we can see that e_1 has full cooperation across the whole CRP, while e_3 alternates null and full cooperation. This kind of behavior entails that the weight of e_3, in those rounds in which he/she cooperates, has the same influence to calculate the consensus opinion as the weight of e_1, who cooperates constantly throughout all the CRP. At certain rounds of the CRP, e_3 might disobey the advice applying contrary changes as he/she has received, moving his position further from the consensus opinion. At the following round, e_3 obeys the advices and changes all the corresponding assessments accordingly. Acting this way, allows e_3 that after several rounds, his/her opinion is brought closer to his/her initial position again and the consensus opinion moves closer to e_3's position, due to the high importance weight given by his/her cooperative behavior at the previous round, like e_1, even though he/she does not fully cooperate during all the CRP like e_1 does. The effect of e_3's varying behavior is illustrated in Figure 7. This illustration has been made using a graphical monitoring tool to represent a group of experts' preferences [37].

For this reason, it seems necessary to take into account not only each expert's behavior at each discussion round, but also how the overall behavior of an expert evolves along the CRP since its beginning. To do so, we will consider the ideas

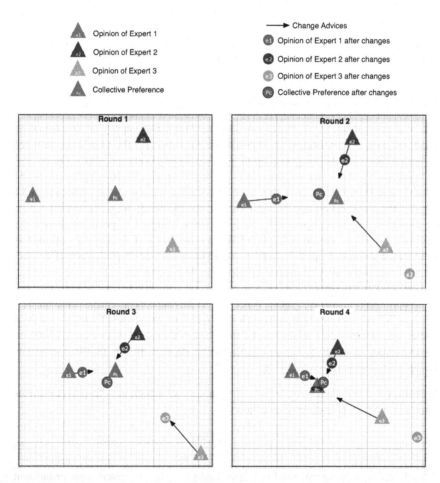

Fig. 7 Example of behavior that attempts to manipulate the CRP, adopted by e_3

expounded by Yager and Petry in [32], where they use hyper-similarity matching to facilitate intuitive decision making.

They suggest that when alternatives are assessed according to certain attributes, those attributes for which an alternative takes an extreme value, should be given a higher importance in the decision making process. For example, for an extremely heavy person, its weight attribute will play an important role in characterizing the weight than in the case of a person with an average weight (i.e. the number of people and the average weight that can be in an elevator). Thus, the focus of their proposal is the effect of these extreme values in amplifiying the effect of the attribute in the characterization of the decision situation.

Definition 3. [32] Let $m_i(S1)$ indicate the amplification effect associated with an attribute A_i for situation S1, denoted as follows:

$$m_i(S1) = f(Dev(A_i(S1)))$$ (5)

Yager and Petry use $Dev(A_i(S1))$ to indicate the deviation of $A_i(S1)$ from normal. The function f:$[0, \infty)$ has the following properties

1. f(0)=1
2. f(a) $\geq$ f(b) if a $>$ b; f is monotonic

In our case, inspired by Yager's penalizing proposal to apply an amplification effect, we aim at reinforcing an expert's weight based on the closeness between his/her preferences and the group opinion. Such an amplification effect will be also taken into account to calculate the expert's opinion weight. Thus, in situations when several experts' opinions have the same importance weight at the current round t (e.g. e_1 and e_3 at round $t = 5$, see Example 2), the opinions of an expert which are closer to the group opinion P_c should receive a higher importance, hence the amplification effect should be higher. On the other hand, if the expert's opinion is far from the collective opinion, he/she should have a lower amplification effect, i.e. the resulting importance of his/her opinion should be smaller.

If the expert's opinion position throughout the consensus rounds is close to the group opinion, it means that either this expert is having a cooperative behavior during the CRP or his/her opinions are close enough to consensus, therefore most of his/her assessments do not need to be modified. Otherwise, if an expert's opinion moves further from the group opinion at some stage, it means that this expert is having a non-cooperative behavior.

Definition 4. Let r_i^t be the rate of e_i's assessments p_i^{lk} which are close to consensus at round t. $r_i^t \in [0, 1]$ is defined as follows:

$$r_i^t = 1 - \frac{\sharp ADV_i^t}{n(n-1)}$$ (6)

Such a rate must be assigned a greater value if e_i has received a lower amount of advice at round t, and vice versa, thus giving a rough insight on how close an expert assessments are to consensus at each round. On the one hand, when $\sharp ADV_i^t$

is smaller, the opinion of e_i might be close to P_c. On the other hand, if the value of $\#ADV_i^t$ is closer to the total number of assessments provided by e_i, $n(n-1)$, then it means that the opinions of e_i are rather far from consensus, hence most of his/her assessments need to be modified. To sum up, the value of r_i^t gives us an insight on the degree of proximity between an expert's opinion and the collective opinion, P_c.

Definition 5. Once r_i^t is computed, and based on Yager's ideas stated above, an amplification function $m(r_i^t)$ is defined as:

$$m(r_i^t) = r_i^t + 1, m(r_i^t) \in [1,2] \tag{7}$$

After the $m(r_i^t)$ calculation, we compute the weights of experts' opinions as:

$$w_{AMP_i}^t = \frac{m(r_i^t)\mu_{COOP}(CC_i^t)}{2}, \tag{8}$$

here it is necessary divide by 2 to bound $w_{AMP_i}^t$ to the unit interval, being $w_{AMP_i}^t \in [0,1]$.

Finally, we re-normalize experts' weights to allow the recovery of opinion importance (see Proposition 1):

$$\hat{w}_{AMP_i}^t = \frac{w_{AMP_i}^t}{\sum_1^n w_{AMP_i}^t} \tag{9}$$

Let us illustrate this weight computation method in the example shown below.

Table 2 Example of expert opinion weights along the CRP after applying amplification function

	e_1^t				e_2^t				e_3^t			
t	μ_{COOP}	$m(r_i^t)$	$w_{AMP_i}^t$	$\hat{w}_{AMP_i}^t$	μ_{COOP}	$m(r_i^t)$	$w_{AMP_i}^t$	$\hat{w}_{AMP_i}^t$	μ_{COOP}	$m(r_i^t)$	$w_{AMP_i}^t$	$\hat{w}_{AMP_i}^t$
1	1	1	1	0.33	1	1	1	0.33	1	1	1	0.33
2	1	1.5	0.75	0.67	0.5	1.5	0.375	0.33	0	1.5	0	0
3	1	1.75	0.875	0.46	0.5	1.70	0.425	0.22	1	1.2	0.6	0.32
4	1	1.85	0.925	0.68	0.5	1.80	0.45	0.32	0	1.4	0	0
5	1	1.95	0.975	0.49	0.5	1.85	0.4625	0.23	1	1.1	0.55	0.28

Example 3. Suppose the same situation as Example 2, but in this case, using our weight management scheme based on Yager's amplification function. Thus, we have three experts who have different behaviors (e_1 cooperates always, e_2 cooperates all the consensus rounds obeying only half of the advices received, and e_3 alternates full and null cooperation along the CRP). Table 2 shows the weights assigned to experts, being in this case computed from $\mu_{COOP}(CC_i^t)^3$ and $m(r_i^t)$ values, thus giving the same importance to the degree of cooperation and the expert opinion position with respect to the consensus opinion.

---

[3] Being $\mu_{COOP}(CC_i^t)$ expressed in the table as μ_{COOP} for the sake of space.

As we can see in Table 2, e_1 has always the highest weight because he/she cooperates over the course of the CRP, so just as the CRP progresses, his/her opinion will be closer to the group's opinion and therefore his/her amplification value m_i^t will be higher. Regarding e_3's behavior, we can appreciate how its weight is smaller as his/her opinion moves further from the collective opinion. For instance, comparing rounds 3 and 5, we can see how its weight decreases in favour of experts that are closer to the group's opinion, hence its strongly variable behavior is now penalized more accordingly. We can see graphically in Figure 8 how the importance of expert opinions evolves across the CRP rounds.

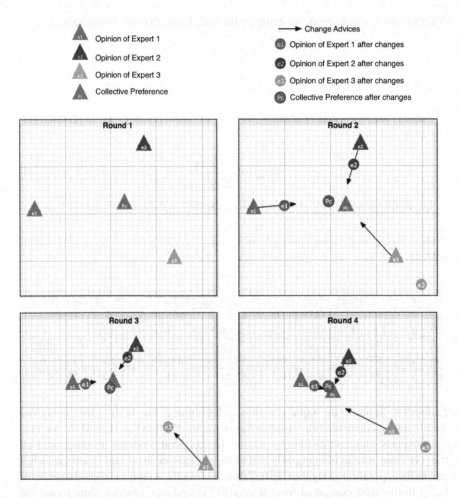

Fig. 8 Example of management non-cooperative behaviors using this proposal

3.3 Integration of the Methodology in the General CRP Scheme

Once it has been defined the cooperativeness measurement and behavior management phases, it is possible to describe in detail the complete scheme of this proposal (see Figure 5).

In the evaluation phase it is computed the cooperation coefficient CC_i^t and the cooperation membership degree $\mu_{COOP}(CC_i^t)$. Once all experts' cooperation coefficients have been computed, we can know the experts' behaviors in the current CRP round, therefore we can start the management phase.

In the management phase, a weight is assigned to each expert depending on his/her behavior. On the one hand it is computed an amplification value which will strengthen or attenuate the expert's importance opinion according to the positions of his/her preferences with respect to the position of the group's opinion. After computing amplification values and taking into account the cooperativeness degrees which have been computed in the detection phase, we can calculate expert's weights $w_{AMP_i}^t$. After that, it is necessary to normalize these weights to allow weight recovery along the consensus rounds. Thus, we obtain the normalized weights of all experts, $\hat{w}_{AMP_i}^t$, which will be used to calculate the collective preference.

In order to integrate the two phases in the general CRP scheme (Figure 2), it is necessary to have, at each round t, the updated expert assessments and change recommendations sent in the feedback at the end of the previous round, $t-1$. For these reasons, this module is firstly applied at the second round of the CRP, because we need the change recommendations of round $t-1$, in this case round 1. Thus, at the first round all the opinions of experts have the same importance because no experts have been assigned any behavior yet. At subsequent consensus rounds, the weights computed will be used to calculate the collective preference, P_c (see Figure 5).

4 Illustrative Example

In this section it is presented an illustrative example which shows the application of the approach presented in this contribution, in order to clarify how it affects in a real CRP for the resolution of a large-scale GDM problem. First of all, the GDM problem and the necessary parameters are introduced and described in detail.

An enterprise committee formed by 30 experts of the different company branches, $E = \{e_1, ..., e_{30}\}$ must reach an agreement about the investment of 100000\$ for the company improvement. There are four possible proposals, $X = \{x_1, x_2, x_3, x_4\}$:

- x_1: TV marketing campaign.
- x_2: Research and development.
- x_3: Replace old production machines.
- x_4: Develop a new corporative software.

Experts express their assessments, p_i^{lk}, over pairs of alternatives, by using fuzzy preference relations. The minimum level of agreement required is $\mu = 0.85$. Regarding the membership function of the fuzzy set $COOP$ to define the meaning of cooperativeness across the CRP, its initial parameter values and the way such values evolve to become more restrictive, we consider the settings previously shown in Example 1 and Figure 6.

Experts present different patterns of behavior, which have been modelled by using a recently developed simulation framework so-called AFRYCA [25]:

- *Cooperative*(experts $e_1 - e_{21}$): These experts always apply all changes suggested on their asssessments, as indicated in the feedback they receive along the whole CRP.
- *Manipulative*(expert e_{22}): this expert alternates null and full cooperation across the CRP in order to manipulate the CRP solution.
- *Non-Cooperative*(experts e_{23}): this expert disobeys always, i.e. he/she does not change his/her assessments at all as it is indicated in the feedback.
- *Undefined*(experts $e_{24} - e_{30}$): the behavior of these experts can change across the CRP, applying or ignoring changes suggested to a variable degree.

This example has been executed on a multi-agent based consensus support system [38] and the results of applying it according to the different approaches will be reviewed in the following subsections. First of all, the results obtained by applying the example on the proposal described in Section 3 are shown. Afterwards, the example will be run by applying proposals [24] and [31]. In section 4.2, a comparison of the results after executing the three proposals is shown. In order to facilitate the results' interpretation, a graphical representation of preferences has been generated for all the cases by using the monitoring tool MENTOR introduced in [37].

4.1 Results Based on the CW and Hyper-Similarity Proposal

Table 3 shows the consensus degree per round and the necessary information to calculate expert's opinion importance, $\hat{w}_{AMP_i}^t$, across the CRP (see Sections 3.1 and 3.2). Notice that this table shows the results associated to a representative set of experts only, over which we can describe what happens attending to the different behaviors previously introduced.

In order to compute $\hat{w}_{AMP_i}^t$ we take into account both the behavior $\mu_{COOP}(CC_i^t)$, and the proximity to the P_c, given by $m(r_i^t)$. For this reason, although several experts have the same behavior, their opinions might have different importance, if their distance to P_c are different.

Once these aspects have been clarified, we start to analyze Table 3.

Experts e_1, e_7 and e_{16} represent experts with *cooperative* behavior. They always accept all their advices, $\sharp ACP_i^t = \sharp ADV_i^t$, so their $\mu_{COOP}(CC_i^t)$ have always the highest value 1. Nevertheless, in order to have the highest $\hat{w}_i^t$ it is also necessary to be close to P_c. e_7 does not receive any advice, $\sharp ADV_7^t = 0, \forall t$, therefore it means that e_7's opinion is very close to P_c, thus he/she has the highest amplification value, $m(r_7^t) = 2$. He/she always has the highest value $\hat{w}_i^t$ in the group. On the other hand, e_{16} is far from P_c, therefore he/she has to change some of his/her assessments, e.g. $\sharp ADV_{16}^2 = 12$,

Table 3 Values of representative experts across the CRP

	t	1	2	3	4	5	6	7
	Conensus Degree	0.6714	0.7153	0.7572	0.7766	0.8099	0.8383	0.8534
	α	0.2	0.2	0.2	0.3	0.4	0.5	
	β	0.5	0.5	0.5	0.6	0.7	0.8	
e_1	$\#ADV_1^t$		6	6	5	4	5	
	$\#ACP_1^t$		6	6	5	4	5	
	$\mu_{COOP}(CC_1^t)$		1	1	1	1	1	
	$m(r_1^t)$		1.5	1.5	1.583	1.666	1.583	
	$w_{AMP_1}^t$		0.75	0.75	0.791	0.833	0.791	
	$\hat{w}_{AMP_1}^t$	0.333	0.397	0.339	0.38	0.424	0.410	
e_7	$\#ADV_7^t$		0	0	0	0	0	
	$\#ACP_7^t$		0	0	0	0	0	
	$\mu_{COOP}(CC_7^t)$		1	1	1	1	1	
	$m(r_7^t)$		2	2	2	2	2	
	$w_{AMP_7}^t$		1	1	1	1	1	
	$\hat{w}_{AMP_7}^t$	0.333	0.529	0.453	0.48	0.509	0.518	
e_{16}	$\#ADV_{16}^t$		12	11	8	7	7	
	$\#ACP_{16}^t$		12	11	8	7	7	
	$\mu_{COOP}(CC_{16}^t)$		1	1	1	1	1	
	$m(r_{16}^t)$		1	1.833	1.333	1.416	1.416	
	$w_{AMP_{16}}^t$		0.5	0.541	0.666	0.708	0.708	
	$\hat{w}_{AMP_{16}}^t$	0.333	0.264	0.245	0.320	0.360	0.367	
e_{22}	$\#ADV_{22}^t$		6	6	6	8	9	
	$\#ACP_{22}^t$		0	6	0	8	0	
	$\mu_{COOP}(CC_{22}^t)$		0	1	0	1	0	
	$m(r_{22}^t)$		1.5	1.5	1.5	1.333	1.25	
	$w_{AMP_{22}}^t$		0	0.75	0	0.666	0	
	$\hat{w}_{AMP_{22}}^t$	0.333	0	0.339	0	0.339	0	
e_{23}	$\#ADV_{23}^t$		6	6	6	6	6	
	$\#ACP_{23}^t$		0	0	0	0	0	
	$\mu_{COOP}(CC_{23}^t)$		0	0	0	0	0	
	$m(r_{23}^t)$		1.5	1.5	1.5	1.5	1.5	
	$w_{AMP_{23}}^t$		0	0	0	0	0	
	$\hat{w}_{AMP_{23}}^t$	0.333	0	0	0	0	0	
e_{24}	$\#ADV_{24}^t$		6	6	6	6	6	
	$\#ACP_{24}^t$		3	4	3	1	3	
	$\mu_{COOP}(CC_{24}^t)$		1	1	0.666	0	0	
	$m(r_{24}^t)$		1.5	1.5	1.5	1.5	1.5	
	$w_{AMP_{24}}^t$		0.75	0.75	0.5	0	0	
	$\hat{w}_{AMP_{24}}^t$	0.333	0.397	0.339	0.24	0	0	

hence although $\mu_{COOP}(CC_{16}^2) = 1$, his/her amplification value is very low, $m(r_{16}^2) = 1$. Nevertheless, given her cooperating behavior pattern across the CRP, $\sharp ADV_{16}^t$ will decrease over time (e.g. $\sharp ADV_{16}^4 = 8, \sharp ADV_{16}^5 = 7$) and his/her opinions will take more importance as the CRP goes on ($w_{16}^2 = 0.5, w_{16}^4 = 0.666, w_{16}^5 = 0.708$).

Manipulative behavior pattern is presented by e_{22}. In this case, the expert alternates null and full cooperation ($\sharp ADV_{22}^1 = 6, \sharp ACP_{22}^2 = 0.\sharp ADV_{22}^3 = 6, \sharp ACP_{22}^4 = 6.$) in order to deviate the collective opinion in his favor, P_c, towards opinion position. Nonetheless, with this kind of behavior, at rounds in which he does not cooperate, his opinion is still far to the P_c and his/her amplification value decreases over time ($m(r_{22}^5) = 1.333, m(r_{22}^6) = 1.25$).

The *non-cooperative* is another type of behavior that can be adopted, in this case by (e_{23}). This expert does not cooperate across the CRP (i.e. $\sharp ADV_{23}^2 = 6, \sharp ACP_{23}^2 = 0$). For this reason, he/she always has the minimum possible value ($\hat{w}_{23}^2 = \hat{w}_{23}^3 = \hat{w}_{23}^4 = \hat{w}_{23}^5 = \hat{w}_{23}^6 = 0$). Thus, his/her opinion is not taken into account to compute the P_c.

Finally, the *undefined* behavior (e_{24} and e_{30}) consists in cooperating and non cooperating in an alternative way across the CRP. For instance, e_{24} only modifies some assessments, (e.g. $\sharp ACP_{24}^4 = 3, \sharp ADV_{24}^4 = 6$). Thus, he/she will have $\mu_{COOP}(CC_{24}^4) = 0.666$ according the α and β values at $t = 4$.

4.2 Performance Analysis

Once we have reviewed the illustrative example data on this contribution's approach, we solve this GDM problem with the other two proposals introduced in [24, 31].

Table 4 Proposals comparison

Non-Cooperative Experts	[24]	[31]	Current Proposal
Participate during all the process	Yes	Yes	Yes
Expert's opinion can recover importance	No	Yes	Yes
Takes into account the behavior across the CRP	No	No	Yes

In [24], it is assigned to experts a weight according to their distance to P_c, being weights values between 0 and 1. In this proposal, experts whose opinion have lost importance can not recover it anymore. On the other hand, [31] allows the importance recovery. In this case, experts with the most cooperative behavior, always have the highest importance, because the weighting of opinions is done attending the cooperation coefficient. It is pointed out again that, in the proposal introduced in this chapter, it is possible for experts to partially recover their importance if their behavior improves. Moreover, the importance of experts' opinions is computed based on the degree of cooperation and the distance to P_c.

In order to analyze our results easily, we have used [37] to extract the associate images to the example application data. Figure 9 shows several representative

Table 5 Weight values of MENTOR comparison

t		1	2	3	4	5	6	7
e_1	Palomares et al.[24]	1.0	0.3137	0.3137	0.3137	0.3137	0.3137	0.3137
	Palomares et al. [31]	0.0333	0.0379	0.0347	0.0380	0.0396		
	Current proposal	0.333	0.397	0.339	0.38	0.424	0.410	
e_7	Palomares et al.[24]	1.0	0.8594	0.7489	0.7489	0.7206	0.7206	0.7206
	Palomares et al. [31]	0.0333	0.0379	0.0347	0.0380	0.0396		
	Current proposal	0.333	0.529	0.453	0.48	0.509	0.518	
e_{16}	Palomares et al.[24]	1.0	0.0041	0.0041	0.0041	0.0041	0.0041	0.0041
	Palomares et al. [31]	0.0333	0.0379	0.0347	0.0380	0.0396		
	Current proposal	0.333	0.264	0.245	0.320	0.360	0.367	
e_{22}	Palomares et al.[24]	1.0	0.0	0.0	0.0	0.0	0.0	0.0
	Palomares et al. [31]	0.0333	0.0	0.0347	0	0.0396		
	Current proposal	0.333	0	0.339	0	0.339	0	
e_{23}	Palomares et al.[24]	1.0	0.2030	0.0	0.0	0.0	0.0	0.0
	Palomares et al. [31]	0.0333	0.0	0.0	0.0	0.0		
	Current proposal	0.333	0	0	0	0	0	
e_{24}	Palomares et al.[24]	1.0	0.2712	0.1172	0.1172	0.1087	0.1087	0.1087
	Palomares et al. [31]	0.0333	0.0379	0.0347	0.0042	0.0352		
	Current proposal	0.333	0.397	0.339	0.24	0	0	

rounds for the different proposals whose general features are compared in Table 4, and the experts' weights computed throughout the CRP are summarized in Table 5.

- Round 1: the results here are similar for the three proposals. We can see the positions of expert opinions and P_c position (P).
- Round 3: it shows how experts are moving their opinions since the beginning of the CRP.
- Consensus achieved: this is the round when experts reach consensus. In [24] consensus is reached at round 9, in [31] it is reached at round 6 and the proposal of this contribution achieves consensus at round 7.

A further analysis of Figure 9 shows in our proposal that:

- Some experts remain far from the consensus opinion due to their non cooperative behavior at several consecutive rounds.
- The position of consensus strongly reinforces the opinions of highly cooperating experts.

We can conclude that using the proposal issued in this chapter the experts' opinions with an overall cooperative behavior throughout the CRP have more importance than in the other two approaches.

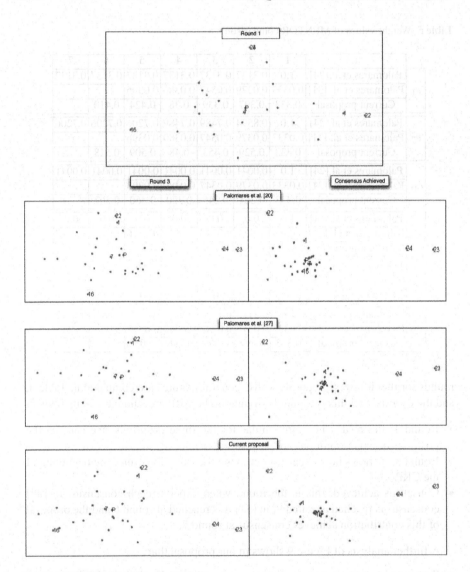

Fig. 9 MENTOR comparison images

5   Concluding Remarks and Future Works

The more experts take part in a CRP for a GDM the more possibility of some of them
try to manipulate the CRP to their own interests. Different approaches in the literature
were issued to address this problem, however none of them consider the change of
behavior of these experts across the CRP. Therefore, in this chapter it has been introduced a new approach to detect and manage non-cooperative behaviors in large-scale
GDM problems that use the CW paradigm in order to facilitate the management of
changing behaviors across the whole consensus reaching process providing a more

flexible and fair negotiation framework. The approach has been integrated in a consensus model and applied to an illustrative example that shows its performance and advantages over other previous approaches to deal with different behaviors of experts.

The proposal presented in this chapter opens the door for future research in decision problems within the field of group recommender systems and social networks among others, because of it will provide a ground to support negotiation processes in large-scale group decision making problems in which biased solutions could be achieved if non-cooperative misbehavior are not properly managed.

Acknowledgment. This research was supported by Research Project TIN-2012-31263 and ERDF.

References

1. Lu, J., Zhang, G., Wu, F.: Multi-Objective Group Decision Making. Imperial College Press (2007)
2. Kacprzyk, J.: Group decision making with a fuzzy linguistic majority. Fuzzy Sets and Systems 18(2), 105–118 (1986)
3. Bellman, R.E., Zadeh, L.A.: Decision-making in a fuzzy environment. Management Science 17(4), 141–164 (1970)
4. Zadeh, L.A.: Fuzzy logic equals computing with words. IEEE Transactions on Fuzzy Systems 4(2), 103–111 (1996)
5. Pedrycz, W.: Granular computing: The emerging paradigm. Journal of Uncertain Systems 1(1), 38–61 (2007)
6. Zadeh, L.A.: Fuzzy sets. Information and Control 8(2), 338–353 (1965)
7. Zadeh, L.A.: The concept of a linguistic variable and its application to approximate reasoning—[i]. Information sciences 8(3), 199–249 (1975)
8. Zadeh, L.A.: The concept of a linguistic variable and its application to approximate reasoning—[ii]. Information sciences 8(4), 301–357 (1975)
9. Zadeh, L.A.: The concept of a linguistic variable and its application to approximate reasoning-[iii]. Information sciences 9(1), 43–80 (1975)
10. Herrera, F., Martínez, L.: A 2-tuple fuzzy linguistic representation model for computing with words. IEEE Transactions on Fuzzy Systems 8(6), 746–752 (2000)
11. Martínez, L., Ruan, D., Herrera, F.: Computing with words in decision support systems: An overview on models and applications. International Journal of Computational Intelligence Systems 3(4), 382–395 (2010)
12. Martínez, L., Herrera, F.: An overview on the 2-tuple linguistic model for computing with words in decision making: Extensions, applications and challenges. Information Sciences 207(1), 1–18 (2012)
13. Rodríguez, R.M., Martínez, L.: An analysis of symbolic linguistic computing models in decision making. International Journal of General Systems 42(1), 121–136 (2013)
14. Herrera, F., Herrera-Viedma, E., Verdegay, J.L.: A sequential selection process in group decision making with a linguistic assessment approach. Information Sciences 85(4), 223–239 (1995)
15. Roubens, M.: Fuzzy sets and decision analysis. Fuzzy sets and Systems 90(2), 199–206 (1997)
16. Orlovsky, S.: Decision-making with a fuzzy preference relation. Fuzzy Sets and Systems 1(3), 155–167 (1978)

17. Butler, C., Rothstein, A.: On Conflict and Consensus: A Handbook on Formal Consensus Decision Making. Takoma Park (2006)
18. Saint, S., Lawson, J.R.: Rules for Reaching Consensus. A Modern Approach to Decision Making. Jossey-Bass (1994)
19. Bryson, N.: Group decision-making and the analytic hierarchy process: exploring the consensus-relevant information content. Computers & Operations Research 23(1), 27–35 (1996)
20. Parreiras, R.O., Ekel, P.Y., Martini, J.S.C., Palhares, R.M.: A flexible consensus scheme for multicriteria group decision making under linguistic assessments. Information Sciences 180(7), 1075–1089 (2010)
21. Wu, Z., Xu, J.: Consensus reaching models of linguistic preference relations based on distance functions. Soft Computing 16(4), 577–589 (2012)
22. Herrera-Viedma, E., Martinez, L., Mata, F., Chiclana, F.: A consensus support system model for group decision-making problems with multigranular linguistic preference relations. IEEE Transactions on Fuzzy Systems 13(5), 644–658 (2005)
23. Mata, F., Martínez, L., Herrera-Viedma, E.: An adaptative consensus support model for group decision-making problems in a multigranular fuzzy linguistic context. IEEE Transactions on Fuzzy Systems 17(2), 279–290 (2009)
24. Palomares, I., Martínez, L., Herrera, F.: A consensus model to detect and manage noncooperative behaviors in large scale group decision making. IEEE Transactions on Fuzzy Systems 22(3), 516–530 (2014)
25. Palomares, I., Estrella, F.J., Martínez, L., Herrera, F.: Consensus under a fuzzy context: Taxonomy, analysis framework afryca and experimental case of study. Information Fusion 20, 252–271 (2014)
26. Zhu, J.J., Di, Q.: Research on large scale group-decision approach based on grey cluster. In: IEEE International Conference on Systems, Man and Cybernetics, SMC 2008, pp. 2361–2366. IEEE (2008)
27. Carvalho, G., Vivacqua, A.S., Souza, J.M., Medeiros, S.P.J.: Lasca: A large scale group decision support system. In: 12th International Conference on Computer Supported Cooperative Work in Design, CSCWD 2008, pp. 289–294. IEEE (2008)
28. Carvalho, G., Vivacqua, A.S., Souza, J.M., Medeiros, S.P.J.: Lasca: a large scale group decision support system. Journal of Universal Computer Science 17(2), 261–275 (2011)
29. Martínez, L., Montero, J.: Challenges for improving consensus reaching process in collective decisions. New Mathematics and Natural Computation 3(2), 203–217 (2007)
30. Yager, R.R.: Penalizing strategic preference manipulation in multi-agent decision making. IEEE Transactions on Fuzzy Systems 9(3), 393–403 (2001)
31. Palomares, I., Quesada, F.J., Martínez, L.: An approach based on computing with words to manage experts behavior in consensus reaching processes with large groups. In: 2014 IEEE World Congress on Computational Intelligence, Beijing, China (2014)
32. Yager, R.R., Petry, F.E.: Intuitive decision-making using hyper similarity matching. In: IFSA World Congress and NAFIPS Annual Meeting (IFSA/NAFIPS), pp. 386–389. IEEE (2013)
33. Herrera-Viedma, E., Herrera, F., Chiclana, F.: A consensus model for multiperson decision making with different preference structures. IEEE Transactions on Systems, Man and Cybernetics, Part A: Systems and Humans 32(3), 394–402 (2002)
34. Beliakov, G., Pradera, A., Calvo, T.: Aggregation functions: A guide for practitioners. Springer Publishing Company, Incorporated (2008)
35. Mata, F., Martínez, L., Martínez, J.C.: Penalizing manipulation strategies in consensus processes. In: Spanish Congress and Fuzzy Logic and Technologies, ESTYLF 2008, pp. 485–491 (2008)

36. Yager, R.R.: Concept representation and database structures in fuzzy social relational networks. IEEE Transactions on Systems, Man and Cybernetics, Part A: Systems and Humans 40(2), 413–419 (2010)
37. Palomares, I., Martínez, L., Herrera, F.: MENTOR: A graphical monitoring tool of preferences evolution in large-scale group decision making. Knowledge-based Systems 58(Spec. Iss. Intelligent Decision Support Making Tools and Techniques: IDSMT), 66–74 (2014)
38. Palomares, I., Sánchez, P.J., Quesada, F.J., Mata, F., Martínez, L.: COMAS: A multi-agent system for performing consensus processes. In: Abraham, A., Corchado, J.M., González, S.R., De Paz Santana, J.F. (eds.) International Symposium on Distributed Computing and Artificial Intelligence. AISC, vol. 91, pp. 125–132. Springer, Heidelberg (2011)

A Type-2 Fuzzy Logic Approach for Multi-Criteria Group Decision Making

Syibrah Naim[*] and Hani Hagras

Abstract. Multi-Criteria Group Decision Making (MCGDM) is a decision tool which is able to find a unique agreement from a group of decision makers (DMs) by evaluating various conflicting criteria. However, most multi-criteria decision making techniques utilizing a group of DMs (MCGDM) do not effectively deal with the large number of possibilities inherent in a domain with a variety of possibilities, different judgments, and ideas on opinions. In recent years, there has been a growing interest in developing MCGDM using type-2 fuzzy systems which provide a framework to handle the encountered uncertainties in decision making models. In addition, fuzzy logic is regarded as an appropriate methodology for decision making systems which are able to simultaneously handle numerical data and linguistic knowledge. In this paper, we will aim to modify the fuzzy logic theories based multi-criteria group decision making models to employ a suite of type-2 fuzzy logic systems in order to provide answers to the problems that are encountered in the real experts' decision. In the proposed framework, we will present a MCGDM method based on interval type-2 fuzzy logic combined with intuitionistic fuzzy evaluation (from intuitionistic fuzzy sets). This combination handles the linguistic uncertainties by the interval type-2 membership function and simultaneously computes the non-membership degree from the intuitionistic evaluation. However, the interval values with hesitation index cannot fully represent the uncertainty distribution associated with the decision makers. Hence, we will present a final component of our framework employing general type-2 fuzzy logic based approach for MCGDM which is more suited for higher levels of

Syibrah Naim
School of Informatics and Applied Mathematics
Universiti Malaysia Terengganu, Kuala Terengganu, 21030, Terengganu, Malaysia
e-mail: syibrah@umt.edu.my

Hani Hagras
School of Computer Science and Electronic Engineering University of Essex, United Kingdom
e-mail: hani@essex.ac.uk

[*] Corresponding author.

© Springer International Publishing Switzerland 2015   123
W. Pedrycz and S.-M. Chen (eds.), *Granular Computing and Decision-Making*,
Studies in Big Data 10, DOI: 10.1007/978-3-319-16829-6_6

uncertainties. In order to optimally find the type-2 fuzzy sets parameters (including interval type-2 and general type-2), we have employed the Big Bang Big Crunch (BB-BC) optimisation method. In order to validate the efficiency of the proposed systems in handling various DMs' behaviour and opinion, we will present comparisons which were performed on two different real world decision making problems. As will be shown in the various experiment sections, we found that the proposed type-2 MCGDM based system better agrees with the users' decision compared to type-1 fuzzy expert system and existing type-1 fuzzy MCDMs including the Fuzzy Logic based TOPSIS (Technique for Order of Preference by Similarity to Ideal Solution). In addition, we will show how the different type-2 fuzzy logic based MCGDM systems compare to each other when increasing the level of uncertainties where the general type-2 MCGDM will outperform the MCGDM based interval type-2 fuzzy logic combined with intuitionistic fuzzy evaluation which will outperform the MCGDM based on interval type-2 fuzzy sets. Hence, this work can be regarded as a step towards producing higher ordered fuzzy logic approach for MCGDM (HFL-MCGDM) which could be applied to complex problems with high uncertainties to produce automated decisions much closer to the group of human experts.

Keywords: MCDM, Type-2 Fuzzy Logic, Intuitionistic Fuzzy Sets, BB-BC.

1   Introduction

Decision Making is the act of choosing between two or more courses of action and a thought process of selecting a logical choice from the available options [1,2]. It is regarded as a result of mental processes leading to a particular selection when surrounded by a number of alternatives, criteria, factors, variables, etc. [1,2]. Every decision making system produces the best and final outcomes after considering available options evaluated by a number of Decision Makers (DMs). According to MakeItRational (2013), the purpose of evaluation is to gain reliable information on the strengths, weaknesses and overall utility of each alternative and criterion [3]. The decision theory helps identify the alternative with the highest expected or ranking value. The produced results can be an action or an opinion.

MCDM is the study of methods and procedures which is concerned with how multiple conflicting criteria can be formally incorporated not only in the management planning process, but in other areas such as medical decision and intelligent systems [4]. MCDM aims to provide a fine selection involving a number of alternatives, criteria, factors or variables. On the other hand, GDM deals with a decision making systems which considers the opinions of a group of experts whose decisions are subject to linguistic uncertainties. In GDM, there is a group of Decision Makers (DMs) try to aggregate all the individual preferences into collective preferences. According to Tindale (2013), groups are seen as superior to individuals as decision making entities when groups can represent a larger and more diverse set of perspectives and constituencies [5]. Thus, they tend to produce fairer decisions by providing "voice" or input from a greater portion of the body for which the decision is made [6,7,8]. Decision making could be considered to include Multi-Criteria Decision Making (MCDM) and Group Decision Making (GDM) which is MCGDM.

This paper aims to modify the higher ordered fuzzy logic theories based-multi-criteria group decision making models to employ a suite of type-2 fuzzy logic systems including interval and general type-2 (generalisations of type-1 fuzzy logic system) [9] in order to provide answers to the problems encountered in the real experts' decision. The presented suite of type-2 fuzzy MCGDMs employ various type-2 fuzzy sets that deal with the various levels of uncertainties. In the proposed framework, we will present a MCGDM method based on interval type-2 fuzzy logic combined with intuitionistic fuzzy evaluation (from intuitionistic fuzzy sets (IFS) [10]). This combination handles the linguistic uncertainties by the interval type-2 fuzzy sets and simultaneously computes the non-membership degree from the intuitionistic evaluation. In addition, the interval type-2 fuzzy values are extended into intuitionistic values to evaluate the hesitation values that is lacking in type-2 fuzzy systems. However, the interval type-2 values with hesitation index (from IFS) cannot fully represent the uncertainty distribution associated with the decision makers. Hence, we will present a general type-2 fuzzy logic based approach for MCGDM which is better suited for higher levels of uncertainties. In order to optimally find the third dimension of type-2 fuzzy sets (including interval type-2 and general type-2), the Big Bang Big Crunch (BB-BC) [11,12] optimisation method is employed, which has low computation overhead and fast convergence.

In order to validate the efficiency of the proposed systems in handling various DMs' behavior and opinion, comparisons were performed on two different real world decision making problems. The first involved employing intelligent decision making systems to select the preferred lighting level during reading where we carried out various experiments in the intelligent apartment (iSpace) located at the University of Essex involving 15 users. The second problem concerned the assessment of the best location for postgraduate study (the evaluation involved 10 candidates who were asked to determine their preferred location of postgraduate study). As shown in the various experiment sections, we found that the proposed type-2 MCGDM based system better agrees with the users' decision compared to type-1 fuzzy expert system and existing type-1 fuzzy MCDM including the Fuzzy Logic based TOPSIS (Technique for Order of Preference by Similarity to Ideal Solution). In addition, it is shown how the different type-2 fuzzy logic based MCGDM systems compare to each other when increasing the level of uncertainties. It will be shown that the general type-2 MCGDM outperform the MCGDM based interval type-2 fuzzy logic combined with intuitionistic fuzzy evaluation which outperform the MCGDM based on interval type-2 fuzzy sets.

This paper is divided into seven parts. In part II, an overview of Fuzzy MCDM is explained. Part III covers the theoretical background for this research. In part IV, the proposed HFL-MCDGM systems (including interval type-2 fuzzy logic with hesitation index and general type-2 fuzzy logic based approach) are demonstrated. The optimal type-2 fuzzy sets (including interval type-2 and general type-2) also have been presented by using Big Bang-Big Crunch (BB-BC) algorithm in part V. Part VI gives all results and part VII concludes this research by highlighting the outcome from this paper and includes the future endeavours.

2   Fuzzy Multi-Criteria Group Decision Making

Nowadays, MCGDM contributes massively to a group decision makers' evaluation. However, the comprehension, analysis and support of the process becomes increasingly difficult due to the ill-structured, dynamic environment and the presence of multiple DMs since each one carries their own viewpoint on the way the problem should be handled and regarding how a decision should to be reached [13]. Thus, the current MCGDM techniques do not effectively deal with the large number of possibilities that cause disagreement amongst different judgments and the variety of ideas and opinions that lead to high uncertainty levels. The less capability of a system to evaluate these uncertainties, the higher the chances that it will generate an uncertain or wrong decision. It is therefore crucial to investigate the available uncertainties and constraints in order to find the ultimate decision output. In order to deal with these challenges in MCGDM, there is a need to employ techniques that can handle the various levels of uncertainties caused by such challenges.

According to Wang (2012), MCGDM aims to find a supreme desirable alternative from a set of feasible alternatives based on the decision information on criteria values provided by a group of decision makers [14]. Therefore, the number of criteria, alternatives and diverse categories of a group decision maker can cause massive ambiguity, hesitation and vagueness. Lately, researchers have realised that the existing techniques lack in their consideration of the uncertainty, impreciseness, ambiguity, hesitancy and conflicts of information. There are many sources of uncertainties facing the decision making methods in the real world-application; we list some of them as follows ([15,16]):

- ✓ Uncertainties and subjectivity of the linguistic term evaluated by the DMs as the meaning of words that are used in the antecedents and consequents linguistic labels can be uncertain as words mean different things to different people.
- ✓ Hesitation and conflict in DMs evaluation in choosing the preferable preferences in inference decision among the criteria and alternatives.
- ✓ Ambiguity in the DMs judgments where the DMs might come from different backgrounds, experiences and level of education.

The uncertainty and vagueness encountered in a decision system can be handled by using appropriate techniques such as hierarchical structures and decision trees (Analytic Hierarchy Process (AHP)), separation measure by Euclidean distances (Technique for Order of Preference by Similarity to Ideal Solution (TOPSIS)) and fuzzy logic theory. In literature, there are many published hybridisation of AHP and TOPSIS. The AHP method can provide solutions through the analysis of quantitative and qualitative decision. In addition, it presented simple solution using hierarchical model [17]. However, the TOPSIS method gives a simple concept and is easy to implement, it's computationally efficient and is easily understood [17]. Both AHP and TOPSIS need each other to support their weaknesses and the concentration is not in handling the uncertainties in group decision making.

Fuzzy logic based MCGDM has both qualities from the AHP and the TOPSIS methods. The MCGDM is structured naturally so it is easy to compute. The matrices allow a number of experts to evaluate the multi-criteria and alternative in matrices. At the same time, it enables DMs to settle hesitation and conflicts through fuzzy rule based. Each DM represents their opinion in rule base form and the MCGDM computes and aggregates their opinions simultaneously. The application of fuzzy logic in a decision system enables the accumulation of the DMs' opinion objectively through the fuzzy rules development. The evaluation and consideration of uncertainties vary according to different applications. In fuzzy logic decision system, the linguistic label is characterised by fuzzy membership function.

There are various kinds of fuzzy sets which could be employed such as type-1 fuzzy sets, intuitionistic fuzzy sets and type-2 fuzzy sets. The developments of fuzzy membership functions (type-1 fuzzy sets, intuitionistic fuzzy sets and type-2 fuzzy sets) in decision systems require a very comprehensive evaluation so as to aggregate the uncertainties. The involvement of a number of DMs or experts such as in multi-criteria decision making system (MCDM) or group decision making (GDM) critically needs an optimised membership function in order to present the DMs opinions. The sketch of the membership function (type-1 fuzzy sets, intuitionistic fuzzy sets or type-2 fuzzy sets) is crucial and it is one of the important parts in the decision making procedures. The membership function is able to accumulate the subjectivity of the linguistic term and the ambiguity of the information among DMs.

Type-1 fuzzy set (widely known as fuzzy sets) and MCDM have been developed along the same lines, although with the help of fuzzy set theory a number of innovations have been made possible; the most important methods are reviewed and a novel approach interdependence in MCDM is introduced [18]. Bellman (1970) was the first to relate fuzzy set theory to decision making problems [19]. After that, several aggregation methods (such as score function, choquet integral, weighted average, Euclidean distance, etc.) based on fuzzy set theory have been proposed to combine the individual opinions on decision making. Aggregation operators have been extensively developed until today. In Hsu (1996), the similarity aggregation method is proposed in order to combine the individual subjective estimates by positive trapezoidal fuzzy numbers [20]. Karsak (2001) introduced a fuzzy decision making procedure utilising a computational elective alternative to rectify some of the difficulties posed by the existing evaluation techniques in financial evaluation of advanced manufacturing system investments under conditions of inflation [21]. Bozdag (2003) and Kahraman (2003) reviewed three different fuzzy multi-attribute methods of group decision making that are used for the justification of computer integrated manufacturing systems [22,23]. They are as follows (i) a fuzzy model of group decision proposed by Blin (1974), (ii) a weighted goals method created by Yager (1978) and (iii) a fuzzy analytic hierarchy process (AHP) [24,25]. fuzzy relations approach are the least complex, while fuzzy AHP has many computational steps and is the most complex among the four [23,24].

In decision making, Intuitionistic evaluation has been extensively applied after Atanassov (1986) proposed the Intuitionistic Fuzzy Sets (IFSs) which is an extension of the complimentary fuzzy sets (from type-1 fuzzy sets theory) [10]. The idea has contributed massively to decision making methods where IFSs evaluate the membership degree (positive), non-membership degree (negative) and intuitionistic index (hesitation margin). IFSs have been shown to provide a suitable framework for dealing with decision making systems involving membership, non-membership and hesitation that showed good results when dealing with conflicting criteria. In the literature, hundreds of decision making approaches have been proposed based on IFSs whereas in Castillo (2007), a new approach for calculating the output of an intuitionistic fuzzy system and fuzzy inference was applied to real world plants [26]. In Lin (2007), a method was proposed to allow the degrees of satisfiability and non-satisfiability of each alternative with respect to a set of criteria [27]. A new decision model under vague information was proposed by Wang (2009) which extended the max–min–max composition of IFS [28]. Intuitionistic fuzzy preference relations have been applied by Gong (2011) for selecting industries with higher meteorological sensitivity [29]. In Chen (2012), multi-criteria fuzzy decision making method employing IFSs was used, which outperformed the method proposed by Ye (2009) [30,31]. Furthermore, in Zhang (2012) an optimisation model was presented to determine attribute weights for MCDM problems with incomplete weight information of the criteria under interval valued IFSs [32]. Finally, in Xu (2012), operational laws, score function and accuracy function were proposed for the intuitionistic fuzzy numbers [33].

Over the past few years, numerous publications have emerged in MCDM using Interval type-2 fuzzy sets. Currently, the applications of interval type-2 fuzzy sets are quickly gaining popularity especially in decision making systems [34,35,36]. Type-2 fuzzy sets are able to evaluate the uncertainty better than type-1 membership functions in certain environments [37]. In Ozen (2003) and (2004), it is shown that type-1 FL is not capable of handling the linguistic uncertainties in terms of the flexibility and consistency in experts' decision making [38,39]. Hasuike (2009) proposed a method of interval type-2 fuzzy portfolio selection while Chen (2010a) and Lee (2008) present a new method for handling fuzzy multiple attribute group decision-making problems [35,36,40]. In Liang (2000), an interval type-2 fuzzy logic system (FLS) was presented for decision making in respect to connection admission control in ATM networks [34]. In Chen (2012), a simple method was proposed to complement the methods presented in Chen (2010a) for fuzzy multiple attribute group decision making based on interval type-2 fuzzy sets [30,36]. In Chen (2010b), a method was proposed to handle evaluating the membership values represented by non-normal interval type-2 fuzzy sets [41]. Chen (2010c) proposed a new TOPSIS method for handling fuzzy multiple attribute group decision making problems based on the ranking values of interval type-2 fuzzy set [42]. The triangular interval type-2 fuzzy set extension of AHP was introduced by Chiao (2011) and in 2012, he presented a trapezoidal interval type-2 fuzzy set (IT2FS) for AHP [43,44]. Recently, Wang (2012) introduced work that investigated group decision making problems in which all the information provided by decision makers (DMs) was expressed as interval type-2 fuzzy decision matrices [14].

Recently several researchers have begun to explore the application of general type-2 fuzzy sets and systems. Wagner 2009a, presented two methods for the automatic design of general type-2 fuzzy sets using data gathered through a survey on the linguistic variables [45]. A series of results presented in Wagner 2010 are related to the different levels of uncertainty handled by the different types of Fuzzy Logic Systems (FLSs) including general type-2 fuzzy logic systems [46]. To the best of our knowledge, general type-2 fuzzy sets have not yet been implemented in decision making area.

The basic Fuzzy MCGDM is explained as follows:

Let A be a set of alternatives, let X be as set of criteria and let D be a set of experts/DMs, where $A = \{l_1, l_2, ..., l_e\}$, $X = \{x_1, x_2, ..., x_n\}$ $D = \{z_1, z_2, ..., z_m\}$, respectively. A MCGDM problem can be concisely expressed in a matrix format as follows:

$$
D_r^k = \begin{array}{c|cccc}
 & X_1 & X_2 & \cdots & X_n \\
\hline
X_1 & X_{11} & X_{12} & \cdots & X_{1n} \\
X_2 & X_{21} & X_{22} & \cdots & X_{2n} \\
\vdots & \vdots & \vdots & & \vdots \\
X_n & X_{n1} & X_{n2} & \cdots & X_{nn}
\end{array} \tag{1}
$$

In what follows, we state the basic approach to fuzzy MCGDM without considering risk attitude and confidence. A general decision making problem with e alternatives $l_r(r = 1, ..., e)$, n criteria $x_t(t = 1, ..., n)$ and m experts or decision makers $z_k(k = 1, ..., m)$ can be concisely expressed as: $D = |l_r|$. Here D refers to a *DM* (where the entry X_{ij} represents the rating of the rule formed by criteria X_i and criteria X_j) where $i = 1, ..., n$ and $j = 1, ..., n$.

The illustrated matrix above represents the process of DMs' evaluation, transformed into the score/membership values in matrices to perform the next processes which are aggregations and accumulations to conclude the optimum decision based on the ranking.

2.1 Review on Fuzzy Logic Based on MCGDM

The utilisation of fuzzy logic and decision making method has become very popular in attempting to resolve the problems of imprecision and uncertainty in medical diagnosis. Pattern recognition methods were the focus of artificial intelligence (AI) application in medical diagnosis until 1974 when Shortliffe published the first rule based approach for therapy advice in infectious diseases [47]. Rule based programs use the "if-then-rules" in chains of deductions so as to reach a conclusion. Research by [48] observe that rule based systems are good for narrow domains of medicine, but most serious diagnostic problems are so broad and complex that straightforward attempts to chain together larger sets of rules encounter major difficulties. Fuzzy logic based decision making also has been proposed to shape the decision in software development [49] and etc. Dalalah (2009) found that multi-criteria decision making technique using fuzzy reasoning can merge quantitative and qualitative factors to handle different groups of actors and opinions of experts [50]. They applied the fuzzy rules as an input to the model in order to calculate the competencies between the alternatives.

The most popular methods in the real world application are TOPSIS and AHP and their hybrids are employed for real world decision problems. There are no better or worse techniques; some techniques better suit a particular decision problem than others do [51]. Uzokaa (2011) study attempts to conduct a case comparison of the fuzzy logic and Analytical Hierarchy Process (AHP) methods in the development of a medical diagnosis system (malaria diagnosis). Their results demonstrate that fuzzy logic diagnoses vary a little bit more strongly with the conventional diagnosis results of the AHP method (one of the most popular MCDM methods) [52]. Recently, Alkhawlani (2011) proposed a system combines fuzzy logic and TOPSIS, a MCDM algorithm, to the problem of Vertical Handover (VHO) decision problems [53].

To the best of our knowledge, none of the methods present an algorithm which could produce results correlated with numbers of DMs. Thus, we believe that fuzzy logic based approach for MCGDM could provide a new dimension to real decision problem which could encounter group of DMs' opinion in order to find the consensus decision. The fuzzy logic rule base allows each DM to express their opinions and our proposed system will accumulate all these opinions by the aggregations which considered type-2 fuzzy membership functions so as to handle the variation among DMs' opinions.

3   Theoretical Backgrounds

In this part, we explain the theoretical fuzzy sets backgrounds related to the study. In the beginning, we briefly elaborate on what fuzzy sets which are also known as type-1 fuzzy sets, fuzzy preference relations, fuzzy logic, intuitionistic fuzzy sets and type-2 fuzzy sets.

Definition 1 [54] Let X be a space of points (object), with a generic element of X denoted by x. Therefore, $X = \{x\}$. A fuzzy set (class) A in X is characterized by a membership (characteristic) function $f_A(x)$ which associates with each points in X a real number in the closed interval $[0,1]$ with the value of $f_A(x)$ at x representing the "grade of membership" of x in A. Specifically, a fuzzy set on a classical set X is defined as follows:

$$A = \{(x, \mu_A(x)) \mid x \in X\} \tag{2}$$

A concept of preference relation is widely used in decision making. The preference score between two elements measured in decision matrices shows the relations between each other.

Definition 2 [55] A linguistic preference relation L on the set X is represented by a linguistic decision matrix $L = \left(l_{ij} \right)_{n \times n} \subset X \times X$ with $l_{ij} \in \overline{S}$, $l_{ij} \oplus l_{ji} = s_o$, $l_{ii} = s_o$, for all $i, j = 1, 2, ..., n$ where l_{ij} denotes the preference degree of the alternative x_i over x_j. In particular, $l_{ij} = s_o$, indicates indifference between x_i and x_j, $l_{ij} > s_o$, indicates that x_i is preferred to x_j, and $l_{ij} < s_o$, indicates that x_j is preferred to x_i.

Fuzzy logic is a generalisation of the conventional set theory as a mathematical way to represent vagueness of parameters [56]. In fuzzy logic the basic idea is that statements are not just 'true' or 'false' since partial truth is also accepted [52]. According to Mendel (1995), the extension of crisp logic to fuzzy logic is made by replacing the bivalent membership functions of crisp logic with fuzzy membership function [57]. It is much closer to human thinking and natural language by provides an effective means of capturing the approximate, inexact nature of the real world [58].

A fuzzy logic system maps crisp inputs into crisp outputs. It contains four components: rules, fuzzifier, inference engine, and defuzzifier [57]. The most important component in fuzzy logic system is the rule base [64]. It is basically a logic-based model that provides the means to model system behavior in the form of a rule base where each rule is made up of antecedents and consequents [64]. The IF-THEN statement "IF u is A, THEN v is B," where $u \in U$ and $v \in V$, has a membership function $\mu_{A \rightarrow B}(x, y)$ where $\mu_{A \rightarrow B}(x, y) \in [0, 1]$. Note that $\mu_{A \rightarrow B}(x, y)$ measures the degree of truth of the implication relation between x and y. Once the rules have been established, a Fuzzy Logic System can be viewed as a mapping from inputs to outputs (the solid path in Figure 1, from "Crisp Inputs" to "Crisp Outputs"), and this mapping can be expressed quantitatively as $y = f(x)$ [57]. The rule structure within fuzzy logic system is the standard Mamdani type fuzzy logic system rule structure employed in type-1 and type-2 fuzzy logic system.

Rules may be provided by experts or can be extracted from numerical data [57]. The fuzzifier maps crisp number into fuzzy sets. This is needed in order to activate rules which are in terms of linguistic variables, which are associated with them [57]. The inference engine of fuzzy logic system maps fuzzy sets into fuzzy sets [57]. It handles the way in which rules are combined [57]. In many applications, crisp numbers must be obtained at the output of a fuzzy logic system. The defuzzifier map output sets into crisp number [57].

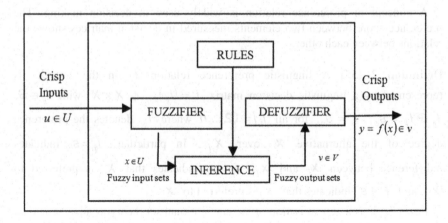

Fig. 1 A Fuzzy Logic System

Twenty years after Zadeh introduced the fuzzy sets, Atanassov (1986) expanded Zadeh's idea by using the concept of dual membership degrees in each of the sets by giving both a degree of membership and a degree of non-membership which is more or less independent of one another whereby the sum of these two grades should not be greater than one [10,59]. This idea, which is a natural generalisation of a standard fuzzy set, seems useful in modeling many real life situations [60]. The idea was derived from the capacity of humans to develop membership functions through their own natural understanding. It can also entail linguistic truth-values about this knowledge.

Definition 3 [61] An intuitionistic fuzzy set (IFS) A on a universe X (shown in Figure 3.1) is defined as an object of the following form:

$$A = \{(x, \mu_A(x), \upsilon_A(x)) \mid x \in X\})\tag{3}$$

Here $\mu_A : X \to [0,1]$ defines the degree of membership and $\upsilon_A : X \to [0,1]$ is the degree of non-membership of the element $x \in X$ in A, and for every $x \in X$, $0 \le \mu_A(x) + \upsilon_A(x) \le 1$. Obviously, each ordinary fuzzy set may be written as $\{(x, \mu_A(x), 1 - \upsilon_A(x)) \mid x \in X\}$.

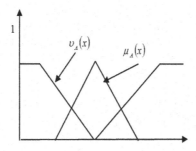

Fig. 2 Intuitionistic Fuzzy Sets with Membership and Non-Membership Function

Recently, the necessity has been stressed about taking into consideration a third parameter $\pi_A(x)$, known as the intuitionistic index or hesitation degree, which arises due to the lack of knowledge or 'personal error' in calculating the distances between two fuzzy sets [62]. In Atanassov (1986), the intuitionistic fuzzy sets are expanded by adding the intuitionistic index (called hesitation) which existed because of the uncertainty of knowledge from a conflicting environment (membership and non-membership) [10]. The hesitation index is defined as follows:

$$\pi_A(x) = 1 - \mu_A(x) - \upsilon_A(x) \qquad (4)$$

with the condition $\mu_A(x) + \upsilon_A(x) + \pi_A(x) = 1$.

Hence, with the introduction of hesitation degree, an intuitionistic fuzzy set A in x may be represented as $A = \{(x, \mu_A(x), \upsilon_A(x), \pi_A(x)) \mid x \in X\}$ with the condition $\mu_A(x) + \upsilon_A(x) + \pi_A(x) = 1$. In a fuzzy set, non-membership value equals 1 - membership value or the sum of membership degree and non-membership value is equal to 1. This is logically true. But in the real world, this may not be true as humans may not express the non-membership value as 1-membership value. Thus, logical negation is not equal to practical negation. This is due to the presence of uncertainty or hesitation or the lack of knowledge in defining the membership function.

This part continues with the explanations of type-2 fuzzy sets theory. Type-2 fuzzy sets theory includes interval type-2 fuzzy sets and general type-2 fuzzy sets.

Type-2 fuzzy systems are based on type-2 fuzzy sets which are an extension of classical type-1 fuzzy sets [63]. Interval type-2 systems have been applied in a large number of applications and have been shown to outperform type-1 fuzzy systems, particularly in fuzzy environment which involved high level of uncertainties [15,63,64,65]. Consider the transition from ordinary sets to fuzzy sets. When we cannot determine the membership of an element in a set as 0 or 1, we use fuzzy sets of type-1. Similarly, when the circumstance is so fuzzy that we have trouble determining the membership grade even as a crisp number in $[0,1]$, we use fuzzy sets of type-2. According to Mendel (2002), the concept of a type-2 fuzzy set was introduced by Zadeh (1975) as an extension of the concept of an ordinary fuzzy set (henceforth called a type-1 fuzzy set) [9,66]. A type-2 fuzzy set is a set in which we also have uncertainty about the membership function, i.e., a type-2 fuzzy set is characterised by a fuzzy membership function whose graded for each element is a fuzzy set $[0,1]$.

Definition 4 [16] A type-2 fuzzy set, denoted $\tilde{A}$, is characterised by a type-2 membership function $\mu_{\tilde{A}}(x, u)$, where $x \in X$ and $u \in J_x \subseteq [0,1]$, i.e.,

$$\tilde{A} = \{((x, u), \mu_{\tilde{A}}(x, u)) \mid \forall x \in X, \forall u \in J_x \subseteq [0,1]\} \qquad (5)$$

in which $0 \le \mu_{\tilde{A}}(x, u) \le 1$. $\tilde{A}$ can also be expressed as

$$\tilde{A} = \int_{x \in X} \int_{u \in J_x} \mu_{\tilde{A}}(x, u)/(x, u) \quad J_x \subseteq [0,1] \qquad (6)$$

where $\iint$ denotes union over all admissible x and u. J_x is called primary membership of x, where $J_x \subseteq [0,1]$ for $\forall x \in X$ [66]. The uncertainty in the primary memberships of a type-2 fuzzy set consists of a bounded region that is called the Footprint of Uncertainty (FOU) [66], which is the aggregation of all primary memberships [66].

According to Hagras (2004), the upper membership function is associated with the upper bound of the footprint of uncertainty $FOU(\tilde{A})$ of a type-2 membership function. The lower membership function is associated with the lower bound of $(\tilde{A})$ [16]. For example, the upper and lower MFs of $\mu_{\tilde{A}}(x)$ are $\bar{\mu}_{\tilde{A}}(x)$ and $\underline{\mu}_{\tilde{A}}(x)$, so that $\mu_{\tilde{A}}(x)$ can be expressed as:

$$\mu_{\tilde{A}}(x) = \int_{u \in [\bar{\mu}_{\tilde{A}}(x), \underline{\mu}_{\tilde{A}}(x)]} 1/u \qquad (7)$$

In interval type-2 fuzzy sets (shown in Figure 3a, Figure 3b), the secondary membership functions are interval sets (as shown in Figure 3b). Since all the memberships in an interval type-1 set are unity, in the sequel, an interval type-1 set is represented just by its domain interval, which can be represented by its left and right end-points as $[l, r]$ [34].

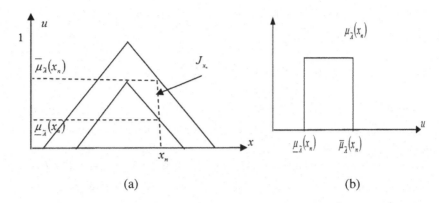

Fig. 3 (a) The Primary Membership, J_{x_n} (b) The Secondary Membership $\mu_{\tilde{A}}(x)$ is an Interval Set

Definition 5 [66] A general type-2 fuzzy set (as shown in Figure 4), denoted $\tilde{A}$, is characterized by a general type-2 fuzzy membership function $\mu_{\tilde{A}}(x, u)$, where $x \in X$ and $\mu \in J_x \subseteq [0, 1]$, i.e.,

$$\tilde{A} = \{((x, u), \mu_{\tilde{A}}(x, u)) | \forall x \in X, \forall u \in J_x \subseteq [0, 1]\} \tag{8}$$

in which $0 \le \mu_{\tilde{A}}(x, u) \le 1$. $\tilde{A}$ can also be expressed as follows:

$$\tilde{A} = \int_{x \in X} \int_{u \in J_x} \mu_{\tilde{A}}(x, u)/(x, u) \quad J_x \subseteq [0, 1] \tag{9}$$

where $\iint$ denotes union over all admissible x and u. J_x is called primary membership of x in $\tilde{A}$ (as shown in Figure 4a), where $J_x \subseteq [0, 1]$ for $\forall x \in X$ [66].

According to Liang (2000), a general type-2 fuzzy set can be thought of as a large collection of embedded type-1 sets each having a weight to associate with it. At each value of x, say $x = x'$, the 2-D plane whose axes are u and $\mu_{\tilde{A}}(x', u)$ is called a vertical slice of $\mu_{\tilde{A}}(x, u)$ [34,66]. A secondary membership function is a vertical slice of $\mu_{\tilde{A}}(x, u)$ [73]. Hence, $\mu_{\tilde{A}}(x', u)$ for $x' \in X$ and $\forall u \in J_{x'} \subseteq [0, 1]$ could be written as [66]:

Definition 6 [66] At each value of x, say $x = x'$, the 2D plane whose axes are u and $\mu_{\tilde{A}}(x', u)$ is called a vertical slice of $\mu_{\tilde{A}}(x, u)$. A secondary membership function is a vertical slice of $\mu_{\tilde{A}}(x, u)$. It is $\mu_{\tilde{A}}(x = x', u)$ for $x = x'$ and $\forall u \in J_x \subseteq [0,1]$. i.e.,

$$\mu_{\tilde{A}}(x = x', u) \equiv \mu_{\tilde{A}}(x') = \int_{u \in J_{x'}} f_{x'}(u)/u, \qquad J_{x'} \subseteq [0,1] \tag{10}$$

in which $0 \le f_{x'}(u) \le 1$. Because $\forall x' \in X$, the prime notation on $\mu_{\tilde{A}}(x')$ is dropped and $\mu_{\tilde{A}}(x)$ is referred to as a secondary membership function [16]; it is a type-1 fuzzy set which is also referred to as a secondary set [see Figure 4b) [66].

Based on the concept of secondary sets, we can reinterpret a type-2 fuzzy set as the union of all secondary sets, i.e., using Equation (7), we can re-express $\tilde{A}$ in a vertical-slice manner as:

$$\tilde{A} = \{(x, \mu_{\tilde{A}}(x)) \mid \forall x \in X\} \tag{11}$$

or, as

$$\tilde{A} = \int_{x \in X} \mu_{\tilde{A}}(x)/x = \int_{x \in X} \left[\int_{u \in X} f_x(u)/u \right], \qquad J_x \subseteq [0,1] \tag{12}$$

Definition 7 [66] The domain of a secondary membership function is called the primary membership of x. In Equation (12), J_x is the primary membership of x, where $J_x \subseteq [0,1]$ for $\forall x \in X$.

Definition 8 [66] The amplitude of a secondary membership function is called a secondary grade. In Equation (12), $f_x(u)$ is a secondary grade; in Equation (8), $\mu_{\tilde{A}}(x', u')(x' \in X, u' \in J_{x'})$ is a secondary grade.

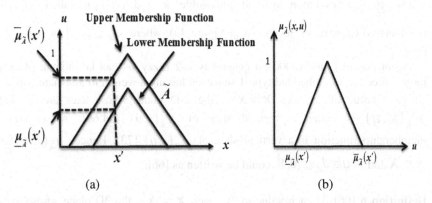

(a)   (b)

Fig. 4 General Type-2 Fuzzy Set (a) The Primary Membership, J_x. (b) The Secondary Membership is a Fuzzy Set

4   The Proposed Higher Ordered Fuzzy Logic Based Approach for Multi-Criteria Group Decision Making (HFL-MCGDM)

In decision making systems, a decision matrix is used to represent the DMs' opinion based on the various alternatives and criteria. The aggregations in decision matrices utilise many kinds of theories and formula such as weighted average, preference relations, entropy, distance measure, score function, arithmetic average, geometric average, ordered weighted average, linguistic quantifier, and others. In order to aggregate the assessment, we utilised fuzzy weighted average and fuzzy preference relations in the proposed system. In the future, HFL-MCGDM can potentially be investigated by different aggregation operators.

This section presents steps to determine the ranking of the outputs. The proposed steps involve the whole process simultaneously in order to determine the fuzzy output. The method combines fuzzy logic systems with MCGDM. This phase includes the fuzzifier process and decision making procedure. The rules and fuzzifier are used in this method while the inference engine is replaced with the MCGDM method and we do not use defuzzification but we rank the output values of the priority weights. The hybrid HFL-MCGDM method (as shown in Figure 5) has been proposed to deal with real expert systems which is very challenging in any real application especially in engineering and medical systems. The efficiency of fuzzy logic systems (FLS) deploying rule bases has been proved in various publications. Thus, we believe this hybridisation is able to cope with the competence of the decision system.

The criteria and alternatives in MCGDM represent the input and output in FLS, respectively. In HFL-MCGDM, we have utilised the fuzzifier and fuzzy logic rule-base to construct decision matrices representing decision makers' opinions and judgments. We modified the pairwise comparison matrices (Equation (1)) by inserting the fuzzy logic rule-base. Generally, each x_{ij} (refer to Equation (1)) to be calculated in the decision matrix above should satisfy the following rule according to the expert opinions:

$$\text{IF } x_i \quad \text{is } \tilde{F}_i \quad \text{and } x_j \quad \text{is } \tilde{F}_j \quad \text{THEN} \quad \tilde{l}_r \qquad (13)$$

The fuzzy logic rule-base then gives the membership values in pairwise comparison matrices. The aggregation parts utilise fuzzy arithmetic averaging operator to compute the membership values in the decision matrix. The normalisations and the priority weights are calculated to determine the ranking for the final outputs.

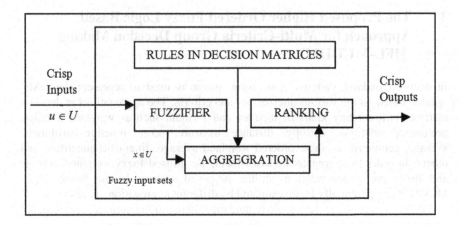

Fig. 5 Higher Ordered Fuzzy Logic in Multi-criteria Group Decision Making

At this point, an important limitation for the previous approaches in MCDM should be highlighted as those approaches lose some of the original decision information in the process of information aggregation and this can cause difficulty to prioritise the given alternatives [55]. By using fuzzy weighted average, the proposed system tries to avoid undistinguished ranking order between the alternatives for the utmost decision, which is proven in [55] by using fuzzy majority.

4.1   The Proposed Interval Type-2 Fuzzy Logic with Hesitation Index Based Approach for MCGDM

Previous research has shown the power of intuitionistic fuzzy sets and interval type-2 fuzzy sets to handle the linguistic uncertainties in many real world applications. In this part, we propose a hybrid method combining interval type-2 fuzzy logic and intuitionistic fuzzy sets to develop the 1st component of the higher ordered Fuzzy Logic based on Multi-Criteria Group Decision Making (HFL-MCGDM) system (based on Figure 5). The intuitionistic evaluation in interval type-2 membership functions was derived from the proposed method which includes seven steps for the aggregation and ranking of the preferred alternatives.

We found that the Footprint of Uncertainty (FOU) - of interval type 2 fuzzy sets -and hesitation index - of intuitionistic fuzzy sets (IFSs) - are able to provide a measure of the uncertainties presented among the various decision makers in the proposed HFL-MCGDM system. The proposed system is more in agreement with the users' decision compared to type-1 and interval type-2 fuzzy logic based systems.

4.1.1   Interval Type-2 Fuzzy Sets with Hesitation Index

According to Deschrijver (2007), Atanassov (1999), and Liu (2007), IFS and Interval-valued Fuzzy Sets (IVFSs) are equivalent [59,61,67]. However, in Liu (2007), it was argued that both types of sets are different [67]. This is because when using interval-valued fuzzy sets, the items $[\mu_1, \mu_2]$ evaluated by experts are the lower and upper approximations. In addition, interval-valued fuzzy sets assigned to each element an interval that approximates the membership degrees values which are not precisely known. However, in IFSs the evaluation is focused on the membership degree and non-membership degree contained in the definition of IFSs. The only constraint is that the sum of the two degrees (membership degree and non-membership degree) does not exceed one because of the hesitation margin $\mu + \upsilon \leq 1$. Both approaches (IFSs and IVFSs) have the virtue of complementing fuzzy sets that are able to model vagueness and the ability to model uncertainty, as well.

According to Atanassov (1999), and Zadeh (1975), this equivalence is only formal even though there are mathematical connections [9,61]. Indeed, a couple $(\mu_A(x), \upsilon_A(x))$ can be mapped bijectively onto an interval. According to Atanassov (1999), the IFSs and interval valued fuzzy sets are equipotent generalisations of the notion of a fuzzy set [67]. Hence we can have (as shown in Figure (6a)),

$$\mu_A(x) = \underline{\mu}_A(x) \tag{14}$$

$$\upsilon_A(x) = 1 - \overline{\mu}_A(x) \tag{15}$$

In Deschrijver (2007), it is claimed that the class of interval-valued fuzzy sets can be embedded in the class of type-2 fuzzy sets because the membership degrees themselves are fuzzy sets in the unit interval [59].

Hence, in this research we present interval type-2 fuzzy sets with intuitionistic evaluation as mentioned in Mendel (2002), where the class of intuitionistic fuzzy sets can be embedded in the class of type-2 fuzzy sets [66]. Thus, we propose a multi-criteria decision making method based on interval type-2 fuzzy logic with hesitation index (from IFSs) as shown in Figure (6b). In Figure 6a, at point x_n, we interpret the interval type-2 membership function for the left and right end-points, $[l, r]$ in intuitionistic value, $[\mu_A(x), 1 - \upsilon_A(x)]$.

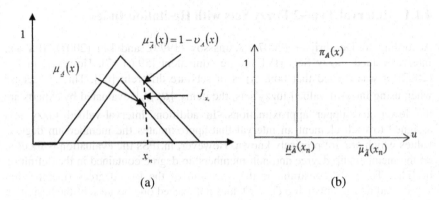

(a)   (b)

Fig. 6 (a) Interval Type-2 Fuzzy Set with Intuitionistic Evaluated at Point X_n Figure (b) Interval Type-2 Fuzzy Set with Hesitation Index from IFS

Similarly, J_{X_n} can represent the hesitation value at X_n where $J_{x_n} = 1 - v_A(x) - \mu_A(x) = \pi_A(x)$ as defined in Equation (16) and Figure (6b).

Thus, the hesitation index employed within an interval type-2 fuzzy set according to Equations (7.1) and (7.2) can be defined as follows:

$$\pi_A(x) = 1 - \underline{\mu}_A(x) - \left(1 - \overline{\mu}_A(x)\right)$$
$$= \overline{\mu}_A(x) - \underline{\mu}_A(x) \tag{16}$$

4.1.2 The Proposed Steps

This section presents the seven steps employed by our proposed system to determine the ranking of the outputs. The proposed steps are used to guide DMs/experts system step by step. Below is the overview of the steps of the proposed fuzzy MCGDM utilised interval type-2 fuzzy logic with hesitation index (from IFSs):

Step 1: Consider a multi-criteria group decision making problem, let $A = \{l_1, l_2, \ldots, l_e\}$ be a discrete set of alternatives (output parameters), $X = \{x_1, x_2, \ldots, x_n\}$ be a set of criteria (input parameters), and $D = \{z_1, z_2, \ldots, z_m\}$ be a set of DMs. The DM $z_k \in D$ provides a judgment based on the rules given and constructs the decision matrix.

Step 2: Let us say we have the input values for each criteria. Then, we utilise the interval type-2 membership function (trapezoidal membership functions was created from the aggregation of DMs' opinion for each linguistic variable (refer to section C) to define the interval membership values, $X_{ij} = \left(\underline{\mu}_{ij}, \overline{\mu}_{ij} \right)$ for each rule defined in the decision matrices and we then identify the fired rule.

Step 3: Then, we use the min operator to compute the firing strength for each rule. This leads to the construction of the fuzzy decision matrices. Based on the DMs/experts $z_k \in D$, we can construct reciprocal decision matrices. Generally, each X_{ij} calculated in the decision matrix above should satisfy the following rule according to the experts' opinions:

$$\text{IF } X_i \text{ is } \tilde{F}_i \text{ and } X_j \text{ is } \tilde{F}_j \quad \text{THEN} \quad \tilde{T}_r \tag{17}$$

$$X^{(r, k)} = \left(x_{ij}^{(r, k)} \right)_{n \times n} \tag{18}$$

In this step, we define Hesitation Index π_{ij} (see Equation (16)) from the interval type-2 values and construct a decision matrix (refer to Equation (1)) where: $x_{ij}^{(r, k)} = \left(\underline{\mu}_{ij}^{(r, k)}, \overline{\mu}_{ij}^{(r, k)}, \pi_{ij}^{(r, k)} \right)$, for interval type-2 FL-MCGDM system we will use only the lower and upper values.

Step 4: Use the fuzzy arithmetic averaging operator to aggregate all $x_{ij}^{(r, k)}$ over the k experts as follows:

$$x_{ij}^{(r)} = \frac{1}{m} \sum_{k=1}^{m} x_{ij}^{(r, k)} \tag{19}$$

Step 5: Use the fuzzy arithmetic averaging operator to aggregate all $x_{ij}^{(r)}$ corresponding to n the criteria:

$$x_t^{(r)} = \frac{1}{n} \sum_{t=1}^{n} x_{ij}^{(r)} \tag{20}$$

Step 6: Find the average of each $x_t^{(r)}$ (where $t = 1, ..., n$), this average is called $x_{tavg}^{(r)}$. Next, normalise the matrix so that each element in the matrix can be written as follows:

$$x_{tnorm}^{(r)} = \frac{x_{tavg}^{(r)}}{\sum_{t=1}^{n} x_{tavg}^{(r)}} \tag{21}$$

Step 7: Find the priority weights, l^r of each alternative as:

$$l^r = \frac{1}{n} \sum_{t=1}^{n} x_{tnorm}^{(r)} \qquad (22)$$

where $l^r > 0$, $r = 1, \ldots, e$, $\sum_{r=1}^{e} l^r = 1$. The priority weighted values allowed the system to provide a ranking and simultaneously choose the highest ranking values which will be the crisp output values.

For complete computation according to the proposed steps, please refer to [68].

4.1.3   Aggregation of Linguistic Variables for each Criterion

All fuzzy sets representing the linguistic labels for each criterion were modelled with trapezoidal type-2 fuzzy membership functions (as shown in Figure 7). Essentially, the linguistic label type-2 fuzzy sets are created by the combination of DMs' opinions (modelled by symmetrical triangular type-1 fuzzy sets as shown in Figure (7)). The minimum, maximum and the average values defined by the intersection of DMs' opinion are demarcated to create the support for each trapezoidal type-2 fuzzy set. For the analysis, by using type-2 and intuitionistic index, we constructed trapezoidal type-2 membership functions and then we evaluated the hesitation index (as we consider both membership degree and non-membership degree) from the interval type-2 in secondary membership values.

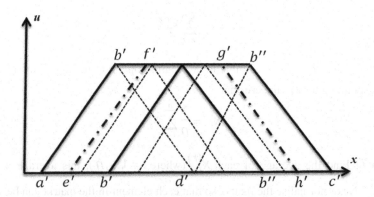

Fig. 7 Type-2 Fuzzy Set from DMs' Opinion (Plotted In Thick Lines) for 'Medium Age' as Generated from 4 DMs' Symmetrical Triangular Type-1 Fuzzy Sets (plotted in thin dashed lines) and the used Type-1 Fuzzy Sets for Comparison in Thick Dashed Line

In the experiments, the aggregations of 15 symmetrical type-1 triangular fuzzy sets were created to build trapezoidal type-1 and type-2 membership functions for the evaluations. Figure 7, shows only an example of having 4 decision makers evaluating the meaning of 'medium age' linguistic variable (4 symmetrical type-1 triangular fuzzy sets - plotted in thin dashed lines as shown in Figure 7- represent the meaning of 'medium' from 4 DMs). Each symmetrical type-1 triangular fuzzy number is defined by three points of (a, b, c). We aggregate $a = \{a_1, \dots, a_m\}$, $b = \{b_1, \dots, b_m\}$ and $c = \{c_1, \dots, c_m\}$ to find the lowest, vertex and upper points according to the number of decision makers, $(k = 1, \dots, m)$. In order to draw the generated type-2 and type-1 fuzzy membership function for each linguistic variable, the following points have to be defined as shown in what follows:

$$a' = \min\{a_1, \dots, a_m\} \tag{23}$$

$$b' = \min\{b_1, \dots, b_m\} \tag{24}$$

$$b'' = \max\{b_1, \dots, b_m\} \tag{25}$$

$$c' = \max\{c_1, \dots, c_m\} \tag{26}$$

$$d' = \frac{b' + b''}{2} \tag{27}$$

Table 1 below shows how we developed the membership functions for type-2 fuzzy membership function and type-1 membership function.

Table 1 Generation of the Aggregated Type-2 and Type-1 Membership Functions

Membership Function	*Aggregations of Each Support*
Upper bound for Type-2	a', b', b'' and c'
Lower bound for Type-2	b', d' and b''
Left support for Type-1	e' (average of a' and b') and f' (average of b' and d')
Right support for Type-1	g' (average of b'' and d') and h' (average of b'' and c')

Different interpretations given by each DM are a major problem in any decision problem. The various interpretations needs have to be measured thoroughly, as they involved a high level of uncertainties especially when the judgements came from a range of different backgrounds (age, origin, sex, level of education, etc.). From the survey we found that DMs' opinion about certain linguistic variables exhibit variation and this allowed us to sketch an aggregation of a trapezoidal fuzzy set from the interval values given by the DMs (interval values were sketched as symmetrical triangle fuzzy sets). Table 2 and 3 show two different opinions from two DMs regarding the variable: very young and young for the criterion Age. DMs' opinion sometimes varying for certain linguistic variables, thus we sketched the membership functions based on the majority and ignore the outliers.

Table 2 The Meaning of Age by Decision Maker One

Linguistic Variable for Age by DM 1	Opinion (in the interval from which suitable for reading application)
Very young	3 years old – 16 years old
Young	16 years old - 27 years old

Table 3 The Meaning of Age by Decision Maker Two

Linguistic Variable for Age by DM 2	Opinion (in the interval forms which suitable for reading application)
Very young	1 years old – 18 years old
Young	18 years old - 25 years old

4.2 The Proposed General Type-2 Fuzzy Logic Based Approach for MCGDM (GFL-MCGDM)

In this paper, we present the 2nd component of HFL-MCGDM system which is a general type-2 fuzzy logic based approach for MCGDM (GFL-MCGDM) which is more suited for higher levels of uncertainties. The correlations value - which shows the agreements between FL-MCGDM and DMs - provided by interval type-2 with hesitation index in FL-MCGDM shows that the proposed system is not practical enough in real decision problems to evaluate the uncertainties. Thus, we extend our research into general type-2 application in FL-MCGDM. General type-2 fuzzy set is well known to be capable to extend the ability to capture more uncertainties in real world application. Even though, a general type-2 fuzzy logic system is too complicated, inferencing and output processing is prohibitive [70]. We found that the implemented strengthened the proposed steps by the cumulative DMs' opinion. The proposed method utilises general type-2 fuzzy sets to evaluate the linguistic uncertainties in the DMs' judgments regarding the linguistic variables. The aggregation operation in the proposed method aggregates the various DMs opinions which allow transforming the disagreements of DMs' opinions into a unique approval. Here, we present results from the proposed system deployment for the assessment of the postgraduate study. The proposed system was able to model the variation in the group decision making process exhibited by the various decision makers' opinion. In addition, the proposed system showed agreement between the proposed method and the real decision outputs from DMs which outperformed the MCGDM systems based on type-1 fuzzy sets, interval type-2 fuzzy sets and interval type-2 with hesitation index.

4.2.1 The Architecture of FL-MCGDM Based on General Type-2 Fuzzy Sets (GFL-MCGDM)

In our previous work, we developed type-1, interval type-2 and interval type-2 fuzzy logic with hesitant index based on IFSs (Intuitionistic Fuzzy Sets) for MCGDM. Figure (8, a, b, c and d) show the secondary membership functions for the type-1, interval type-2, interval type-2 fuzzy sets with hesitant index based on IFSs and general type-2 fuzzy sets, respectively. In this part, we consider, the membership value at a given x' for a general type-2 fuzzy set is a type-1 fuzzy set (refer to Figure 8d) in the third dimension. The representation of a fuzzy set in third dimension (or as secondary membership function) would increase the performance instead by just using interval value and hesitation index.

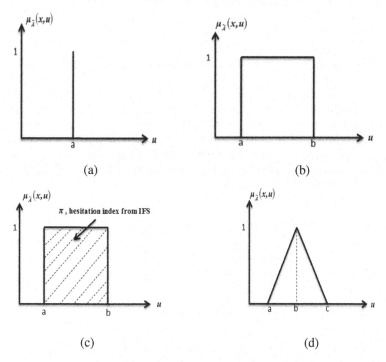

(a)   (b)

(c)   (d)

Fig. 8 View of the Secondary Membership Function in the third dimension; $x - u$ plane (a) Type-1 Fuzzy Set. (b) Interval Type-2 Fuzzy Set. (c) Interval Type-2 Fuzzy Set with Hesitation index (from IFSs). (d) General Type-2 Fuzzy Set

In related works in the literature, we found that when utilising the general type-2 fuzzy sets, the complexity of the system is increased in order to evaluate higher level of uncertainties. However in GFL-MCGDM, we are able to simplify the complexity from various DMs' opinions into a cumulative type-1 fuzzy set to represent the disagreements among the group showing in the example in the following section. The architecture of the proposed system is slightly different from HFL-MCGDM which applies type-1 and type-2 fuzzy membership function

.The aggregation operation in the proposed method aggregates the various DMs' opinions and this allows handling the disagreements of DMs' opinions into a collective approval. The ranking components in the proposed GFL-MCGDM utilise fuzzy arithmetic averaging operators to compute the membership values in decision matrices. The normalisations and the priority weights are calculated to determine the final output.

4.2.2 The General Type-2 Membership Functions

All fuzzy sets representing the linguistic labels for each criterion were modelled in the x-u domain with trapezoidal type-2 fuzzy membership functions (as shown in Figure 9a). Essentially, the linguistic labels type-2 fuzzy sets are created by the combination of DMs' opinions – modelled by symmetrical triangular type-1 fuzzy sets as shown in Figure (9a). The minimum, maximum and the average values defined by the intersection of DMs' opinion are demarcated to create the support for each trapezoidal type-2 fuzzy set.

Thus, the generated general type-2 fuzzy set (in Figure 13a) upper membership function is formed by points a', b', b'' and c' while the lower membership function is formed by points b', d' *and* b''. The type-1 fuzzy sets (which are used when comparing the performance of a type-1 fuzzy based system with the proposed system) consist of the points e' *(average of a' and b')*, f'*(average of b' and d')*, g' *(average of b'' and d')* and h' *(average of b'' and c')* as shown in Figure 13a. The sketch of the general type-2 membership function is the same for interval type-2 membership function. However, the third dimension is represented by a fuzzy set as shown in Figure 9b.

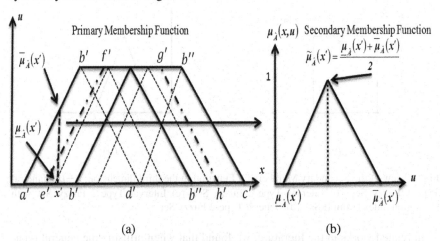

(a)   (b)

Fig. 9 Generation of General Type-2 Fuzzy Set from DMs' opinion (a) Primary Membership Function (plotted in thick lines) as generated from the DMs' type-1 Fuzzy Sets (plotted in thin dashed lines) and the used type-1 Fuzzy Sets for comparison in thick dashed line (b) A Secondary Membership Functions in the Third Dimension

4.2.3 The Proposed Steps

This section presents eight steps so as to determine the ranking of the outputs. This phase involves the fuzzifier process and decision making process. Below is an overview of the steps of the proposed GFL-MCGDM:

Step 1: Consider a multi-criteria group decision making problem, let $A = \{l_1, l_2, ..., l_e\}$ be a discrete set of alternatives (output parameters), $X = \{x_1, x_2, ..., x_n\}$ be a set of criteria (input parameters), and $D = \{z_1, z_2, ..., z_m\}$ be a set of DMs. The DM $z_k \in D$ provides his/her judgment based on the rules given, and constructs the rule-based reciprocal decision matrix.

Step 2: Under the assumption that we have the input values for each criteria, we utilize general type-2 membership function (explained in the previous section and as shown in Figure 9) to define the membership degree for each rule defined in the reciprocal decision matrices and we then identify the rule that is fired.

Each $x_{ij}'^{(r,k)}$ calculated in the decision matrix should satisfy the following rule according to DMs' opinions:

$$\text{IF } x_i \text{ is } \tilde{F}_i \text{ and } x_j \text{ is } \tilde{F}_j \text{ THEN } l_r \tag{28}$$

$$X^{(r,k)} = \left(x_{ij}'^{(r,k)} \right)_{n \times n} \tag{29}$$

Hence, for each $x_{ij}'^{(r,k)}$, we have $\overline{\mu}_{ij}^{(r,k)}, \underline{\mu}_{ij}^{(r,k)}$, for all $i, j = 1, 2, ..., n$.

Step 3: In this step, we define $\tilde{\mu}_{ij}^{(r,k)}$ as follows:

$$\tilde{\mu}_{ij}^{(r,k)} = \frac{\underline{\mu}_{ij}^{(r,k)} + \overline{\mu}_{ij}^{(r,k)}}{2} \tag{30}$$

Hence, for each entry, we have $x_{ij}^{(r,k)} = \left(\underline{\mu}_{ij}^{(r,k)}, \tilde{\mu}_{ij}^{(r,k)}, \overline{\mu}_{ij}^{(r,k)} \right)$, for all $i, j = 1, 2, ..., n$.

It is important to note that in all the operations below all operations on x will be carried on $\underline{\mu}, \tilde{\mu}$ and $\overline{\mu}$ independently in decision matrices.

Step 4: Then, we use the min operator to compute the firing strength for each rule. This leads to the construction of the fuzzy decision matrices. Based on the DMs/experts $z_k \in D$, we can construct reciprocal decision matrices.

Step 5: The general type-2 fuzzy values of each $x_{ij}^{(r,k)}$ are then aggregated. The aggregated set can be determined by $x_{ij}^{(r)} = \left(v_{ij}^r, w_{ij}^r, y_{ij}^r\right)$ for $(k = 1,..., m)$ where,

$$v_{ij}^r = \min_k \left\{\underline{\mu}_{ij}^{(r,k)}\right\} \tag{31}$$

$$w_{ij}^r = \frac{1}{m} \sum_{k=1}^{m} \underline{\mu}_{ij}^{(r,k)} \tag{32}$$

$$y_{ij}^r = \max_k \left\{\overline{\mu}_{ij}^{(r,k)}\right\} \tag{33}$$

Step 6: Use the fuzzy arithmetic averaging operator to aggregate all $x_{ij}^{(r)} = \left(v_{ij}^r, w_{ij}^r, y_{ij}^r\right)$ corresponding to the n criteria.

$$x_t^{(r)} = \frac{1}{n} \sum_{t=1}^{n} x_{ij}^{(r)} \tag{34}$$

Step 7: Find the average of each $x_t^{(r)}$ (where $t = 1,..., n$), this average is called $x_{tavg}^{(r)}$. Next, normalise the matrix so that each element in the matrix can be written as follows:

$$x_{tnorm}^{(r)} = \frac{x_{tavg}^{(r)}}{\sum_{t=1}^{n} x_{tavg}^{(r)}} \tag{35}$$

Step 8: Find the priority weights, l^r of each alternative as:

$$l^r = \frac{1}{n} \sum_{t=1}^{n} x_{tnorm}^{(r)} \tag{36}$$

where $l^r > 0$, $r = 1,..., e$, $\sum_{r=1}^{e} l^r = 1$.

For complete computation according to the proposed steps, please refer to [68 and 70]

5   Big-Bang Big-Crunch (BB-BC) Optimised Type-2 Fuzzy Logic Approach for Multi-Criteria Group Decision Making

The optimisation of fuzzy membership functions is crucially needed in a fuzzy system to find the best parameters in order to achieve the required objective. The performance of a fuzzy logic system is very sensitive to the sketch of the fuzzy set membership functions, the base lengths of the membership functions and the location of their peaks [71]. The type of membership functions varies according to the employed system. The subjectivity involved in interpreting the linguistic variables exists because of the deviation of human interpretation. This problem leads to the complexity of the system by the high level of uncertainties. Thus, in order to sketch subjective membership functions in a very complex system, we crucially need an optimisation algorithm so as to find the best base lengths of the membership functions and the location of their peaks.

In this part (the 3rd component/final of the research) , we propose a type-2 fuzzy logic (including interval type-2 and general type-2) based approach for MCGDM with the optimised membership functions selected by Big-Bang Big-Crunch (BB-BC). The decision method utilises type-2 fuzzy sets to evaluate the linguistic uncertainties within the DMs' judgments regarding the linguistic variables. The aggregation operation in the proposed method aggregates the various DMs opinions which allow handling the disagreements of DMs' opinions into a unique approval. The type-2 FL-MCGDM utilises fuzzy membership functions from BB-BC optimisation to maximise the percentage of correlation between decision system and human decision. The proposed system shows agreement between the system output and decision outputs from DMs as quantified by the Pearson Correlation. In addition, the Pearson correlation values given by the BB-BC based on type-2 FL-MCGDM, outperform the type-2 FL-MCGDM systems based on interval type-2 fuzzy sets, interval type-2 with hesitation index and general type-2 (without BB-BC algorithm). We did not apply the BB-BC for type-1 membership functions is because the purpose of this research to is find the optimum FOU in type-2 fuzy sets based on the type-1 fuzzy sets sketch from DMs' opinion.

The Big Bang–Big Crunch (BB-BC) method is inspired by the theories of the evolution of universe [11,12]. The work in Kumbasar(2011) introduces the Big Bang Big Crunch (BB-BC) theory so as to solve an optimisation problem. In the Big Bang phase, the system generates random points while in the Big Crunch phase, it shrinks those points to a single representative point via a center of mass or minimal cost approach [72] . It is shown that the performance of the new (BB–BC) method outperforms the classical genetic algorithm (GA) for many benchmark test functions [72]. The potentialities of BB-BC are its inherent numerical simplicity, high convergence speed, and easy implementation [11]. Determination of the fuzzy membership functions in a given fuzzy logic system is the key factor for achieving the the best performance [73]. Therefore, the development of fuzzy membership functions in HFL-MCGDM system needs a very comprehensive evaluation to aggregate the uncertainties. The involvement number of DMs or experts such as in multi-criteria decision making system

(MCDGM) or group decision making (GDM) and fuzzy logic system (FLS) critically need an optimised membership function in order to present all their opinions to form the best output. Thus, we believe that BB-BC is potentially able to optimise fuzzy membership function which is one of the most important parts in a decision system. Different parameters of each fuzzy set used in the fuzzy system might produce different outcomes.

5.1   Computation of the Membership Function Using BB-BC Steps

In previous parts, we have shown how we defined the aggregations of DMs opinion to define the type-2 membership function for each of the criteria. Thus, we have found 4 values points (e', f, g' and h') for right and left support (as shown in Figure 10) which show the aggregations of type-1 fuzzy sets. In this analysis, we use parameter V to increase the FOU of type-2 membership function, from 1% to 100% (as shown in Figure 10). The only parameter we have to optimise is V valued which shows the percentage of the uncertainty involved in the system. In order to find the optimal V value, we allow the BB-BC to utilise a random number from 0% to 100%. According to previous parts, we find the optimal V value where the V value for each point for trapezoidal type-2 membership function must be between some points which have been chosen as the maximum area of the trapezoidal shape as shown in Figure 14. The maximum area for left

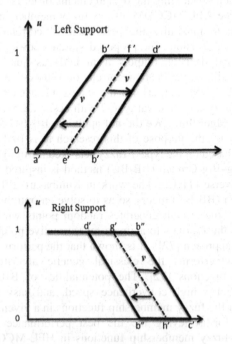

Fig. 10 The Increasing of FOU for Left and Right Support based on v Parameter for BB-BC Optimisation

support should be *a'*, *b'* and *b'*, *d'*, while for the right support should be *b''*, *d'* and *c'*, *b''*. The starting value for parameter *V* to expand the FOU are for the left support: *e'* and *f'*, while for right support: *h'* and *g'* showing the type-1 membership function.

5.2 Maximizing Correlations between Decision Makers and HFL-MCGDM System

The efficiency of the proposed system can be evaluated by the correlation values between the DMs' decision and the output ranking. In this study, Pearson Correlation is used to find the correlation between the DM's decision and the various MCGDM's decisions. Thus, for the proposed type-2 FL-MCGDM based on BB-BC, we use Pearson Correlations as the cost function. The objective of the study is to maximize the Pearson correlation as a Cost Function. The Pearson Correlation which is used to find the correlation between the user's decision and the FL-MCGDM's decision is as follows:

$$\rho_{X,Y} = \frac{COV(X,Y)}{\sigma_X \sigma_Y} = \frac{E[(X - \mu_X)(Y - \mu_Y)]}{\sigma_X \sigma_Y} \tag{37}$$

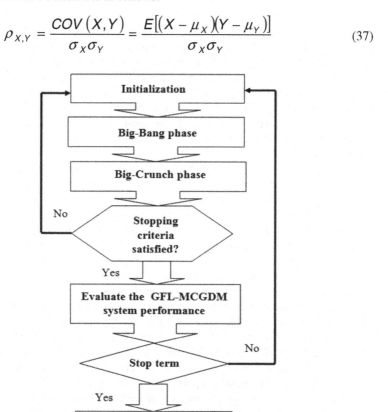

Fig. 11 A Type-2 HFL-MCGDM system based on Optimised Membership Function from BB-BC

According to Figure 11, the optimised membership functions determined by the BB-BC algorithms will be applied in HFL-MCGDM system in order to find the highest correlation value (cost function) among the populations generated by BB-BC, showing the agreement between DMs and the system as decision output. According to Figure 11, the optimisation process starts by initialising the set parameters and gets the fitness function to define the new correlation value, representing the membership function parameter set of values.

6   Experiments

To evaluate the proposed HFL-MCGDM systems, we analyse with two different type of experiments showing different level of uncertainties as follows:

6.1   The Analysis of Postgraduate Survey in Decision Making

In order to evaluate the proposed HFL-MCGDM system, we have conducted a survey among postgraduate students in the Department of Computer Science, University of Essex. Ten participants were chosen and classified as decision makers (DMs) for the group decision making system. In this analysis, DMs are chosen based on their knowledge and experience in choosing a university for their postgraduate study. The group of DMs consists of local students, foreign students and internship students. The system will determine three types of decisions which are foreign postgraduate study, local postgraduate study and internship postgraduate study. Based on the DMs' judgment and assessment, we generate the system.

The purpose of this analysis is to create a decision making system which can determine the preferences of the student for choosing their postgraduate university. For example, in certain circumstances, students might face difficult questions in their lives: "Where should they continue their studies?" Possibilities of choosing the best university (local, foreign or for internships) are crucially considered by almost all students. Each decision needs significant consideration by studying many factors/criteria. The students who have an intention to continue their studies at a postgraduate level usually have to consider certain criteria such as their financial situation, age, the distance from hometown to the university and how many dependants they have to support during their postgraduate study, etc. Three output fuzzy variables were used in this system: study 'Foreign', study 'Local' and study 'Internship'. The developed decision system provides a decision output (foreign, local or internship) based on the input criteria from the students. The fuzzy system for postgraduate decisions was based on four main input variables (criteria); Financial support (cost of living per month to be taken into consideration); Age; preferred living Distance from hometown to the university (travelling time); Dependants (how many (if any) dependants they need to support during their postgraduate study).

6.2   The Analysis of Preferred Lighting Level Selection

Selecting a suitable lighting level for reading is crucial to the overall success of reading comprehension. The light level preferences vary amongst users specifically according to the changing environmental conditions. A reading application considers proper lighting as crucial to the overall success of a reading comprehension. Thus, the preferred and suitable level of lighting design at home, study room, classroom, office or library is very important to be investigated and assessed. In order to assist in determining the decision making strategy, we developed a real-world application where the participants were asked to decide on their preferred level of the ceiling lights as the ambient luminance conditions change when they are reading. The application was deployed at iSpace which is a purpose-built and fully-furnished two-bedroom apartment in the University of Essex, UK. The intelligent space/apartment includes a spacious open plan kitchen and living area, bathroom, a master bedroom and a study room. It has distributed sensors and actuators which are connected in a homogenous manner over the iSpace network by the use of UPnP middleware.

The reading application uses the light sensors which are scattered around the living room area of the iSpace in which the values are aggregated to account for the perceived ambient luminance. The application employs a GUI displayed on the mobile device Apple iPad. By using this interface, the users can interact with the environment and they are able to change the dimmable ceiling light levels depicted on a scale of [0-10] which represents the percentage of the brightness in numeric format ranging between 0 (lights off) and 100 (maximum brightness). For example, by touching the 7th bar of the scale on the iPad, the user can switch on the ceiling light levels to 70%.

In particular, the reading application employs a simplified version of our Fuzzy Task Agent presented by Bilgin (2012) where we limited the operation of the intelligent embedded agent to account for logging the users' ceiling light level preferences and some of the criteria that will be used as inputs to the overall system [74]. As shown earlier, the alternatives for the preferred level of output ceiling lights can be 'very low', 'low', 'medium', 'high' and 'very high'. The various decision criteria may differ depending on the need of the organisation and changing of environment [75]. Consequently, the criteria that may influence the user's preference of the ceiling light levels were chosen to be the time of day, ambient luminance, age of the user, text size used in the document, distance of eyesight from the reading material and the width of the reading material. All these criteria together with the interaction of the user through the GUI (on an Apple iPad) and the alternatives (preferred ceiling light levels) can be visualised in the photos from the experiments and are shown in Figure 12.

These real world experiments were performed on different days with a total number of 15 participants. The users were asked to be seated on the sofa in the living room of the iSpace. There were two dimmable lights positioned above their seats. Next to them, they had access to a range of reading materials including a dictionary, magazine, book, etc. together with a set of boxes varying in volume which the users were required to use on their laps. The different documents served the purpose of having diverse text size and width whereas the different volume of boxes helped to realise the changing distance of the eyesight from the reading material. Moreover, in order to simulate various lighting conditions, the blinds and the curtains within the living room of the iSpace were operated. For example, closing the curtains meant that the time of day was evening, night, etc.

In order to be more practical in the decision making analysis, we designed the embedded agent to log some of the criteria such as the time of day and the light sensor value in numeric format as the rest of the criteria (age of the user, text size, distance of eyesight and the width of the document) can easily be logged manually. In addition, the preference of the user which is one of the alternatives of the overall system was also logged by the agent in a linguistic label format.

Fig. 12 Participants Making Decision on their Preferred Level of Ceiling Lights under Different Criteria

7   Analysis and Results

In this paper, we proposed a higher ordered Fuzzy Logic based approach for Multi-Criteria Group Decision Making (HFL-MCGDM) system which provides a comprehensive valuation from a group of users/decision makers based on the aggregation of the latter's opinions and preferences in the intelligent environment. The theories of intuitionistic fuzzy sets and type-2 fuzzy logic are well suited when dealing with imprecision and vagueness. Consequently, in this paper we presented the hybrid concepts between both theories to employ in MCGDM models. Intelligent shared spaces, such as homes, classrooms, offices, libraries, etc., need to consider the preferences of users who come from diverse backgrounds. However, there are high levels of uncertainties faced in intelligent shared spaces. Hence, there is a need to employ intelligent decision making systems which can consider the various users preferences and criteria in order to offer convenience to a variety of users while handling the faced uncertainties. We carried out experiments in the intelligent apartment (iSpace) located at the University of Essex so as to evaluate various approaches employing group decision making techniques for illumination selection in an intelligent shared environment. It was found that the Footprint of Uncertainty (FOU) (of interval type-2 fuzzy sets) and hesitation index (of intuitionistic fuzzy sets (IFSs)) are able to provide a measurement of the uncertainties present among the various decision makers.

In order to visualise the agreement, we used the Pearson Correlation values (refer to 37) for the crisp output to investigate the diagnostic agreement and correlation between real output data with the HFL-MCGDM and HFL-TOPSIS. Through the experiments, it was found that overall the HFL-TOPSIS systems give lower correlation values compared to our proposed HFL-MCGDM. HFL-TOPSIS systems which employed interval type-2 fuzzy sets were able to give a correlation value with the users' decisions of 0.3870 which is lower when compared to HFL-MCGDM based on type-1 fuzzy sets which is 0.5380. This clearly shows that our proposed HFL-MCGDM system is able to produce better agreements compared to HFL-TOPSIS. Thus the proposed system (HFL-MCGDM) which is based on the 'fuzzy weighted average' method has a better ranking compared to decision systems based on the HFL-TOPSIS method. In Table 4 one can observe that the type-1 fuzzy sets in FL-MCGDM gives 0.5380 correlations to the linguistic appraisal of the user (i.e. the user's decision) whereas the type-2 fuzzy sets in FL-MCGDM gives 0.5555 of correlation values. Markedly, the proposed novel concept, which is based on interval type-2 with hesitation index, gives the highest correlation value of 0.6321. Thus the proposed Type 2-Hesitation FL-MCGDM system better agrees with the users' decision compared to existing fuzzy MCDM including the Fuzzy Logic based TOPSIS (Technique for Order of Preference by Similarity to Ideal Solution), type-1 FL-MCGDM and interval type-2 in FL-MCGDM.

For all the correlation values, we saw that the proposed method which employs type-2 fuzzy sets with hesitation index gives a higher Pearson correlation value than when using type-1 Fuzzy Sets or Interval Type-2 Fuzzy Sets. This shows that our proposed system which uses the hybrid fuzzy theories provides a better correlation by having a much closer group decision to the human decision makers when compared to the other fuzzy theories (where the higher the correlation value, the closer the user's decision to the output from the proposed system.

Table 4 Correlation Values between the Linguistic Decisions from User with Output Ranking from Different Membership Functions in HFL-TOPSIS and HFL-MCGDM

Membership Function	*HFL-TOPSIS*	*HFL-MCGDM*
Type-1 Fuzzy Sets	0.0747	0.5380
Interval Type-2 Fuzzy Sets	0.3870	0.5555
Interval Type-2 Fuzzy Sets with Hesitation Index (IFS2)	0.3870	0.6321

The idea to define hesitation values from the interval membership degree permits the system to capture more uncertainties in the evaluations and provide the highest correlation value. The results provided clearly show that the more the theory can evaluate uncertainties, vagueness and conflict, the higher correlation value can be determined between the real output data. After all, this shows that interval type-2 fuzzy sets combined with IFSs can play an important role in the production of enhanced HFL-MCGDM systems. Finally, this case study shows that our proposed system is capable of handling real world decision problems. The efficiency of the proposed system can be evaluated by the correlation values between the users' decision and output ranking. The higher the correlation values, the closer the user's decision to the output from the proposed system. In this part, we present two data sets applying a different type of real decision problems. These are the problem of finding the postgraduate study by 10 DMs and the preferred location for preferred lighting level for reading in the intelligent environments by 15 DMs. Both data sets show that by utilising BB-BC algorithm we are able to increase the agreements between the DMs and the system.

Table 5 Pearson correlations for Preferred Locations for Postgraduate Study from 10 DMs

Methods	Pearson Correlation without BBBC optimisation	Pearson Correlation with BBBC optimisation
Type-1 Fuzzy Sets	0.1299	-
Interval Type 2 Fuzzy Sets	0.3928	0.4784
Interval Type 2 Fuzzy Sets with Hesitation Index	0.4787	0.4784
General Type 2 Fuzzy Sets	0.5148	**0.5313**

Table 6 Pearson correlations for Preferred Lighting Level for Reading in the Intelligent Environments from 15 DMs

Methods	Pearson Correlation without BBBC optimisation	Pearson Correlation with BBBC optimisation
Type-1 Fuzzy Sets	0.5380	-
Interval Type 2 Fuzzy Sets	0.5555	0.5555
Interval Type 2 Fuzzy Sets with Hesitation Index	0.6338	0.6338
General Type 2 Fuzzy Sets	0.6456	**0.6520**

According to Table 5, data set from the 'Postgraduate Study by 10 DMs' , type-1 and interval type-2 fuzzy logic based MCGDM gives 0.1299 and 0.3928 correlations to the linguistic appraisal of the DMs whereas interval type-2 fuzzy logic with hesitation index based MCGDM gives a correlation value of 0.4748. Markedly, the HFL-MCGDM system without using BB-BC gives a correlation value of 0.5148. When using BB-BC (refer to Table 5), interval type-2 fuzzy based MCGDM gives a similar correlation value to the interval type-2 fuzzy logic with hesitation index based MCGDM which is 0.4784 (also similar to type-2 with hesitation index without BB-BC). However, the proposed HFL-MCGDM based on BB-BC gives the highest correlation value of 0.5315 and outperformed the HFL-MCGDM without the BB-BC optimisation.

According to Table 6, data set from the 'Preferred Lighting Level for Reading in the Intelligent Environments by 15 DMs', it can be observed that type-1 and interval type-2 fuzzy logic based MCGDM gives 0.5380 and 0.5555 correlations to the linguistic appraisal of the DMs whereas interval type-2 fuzzy logic with hesitation index based MCGDM gives a correlation value of 0.6338. Markedly, the HFL-MCGDM system without using BB-BC gives a correlation value of 0.6456. While using BB-BC (refer to Table 6), interval type-2 fuzzy based MCGDM gives a similar correlation value of 0.5555. At the same time, the interval type-2 fuzzy logic with hesitation index based MCGDM also gives the same correlation values. However, the proposed HFL-MCGDM based on BB-BC gives the highest correlation value of 0.6520.

In some cases the BB-BC may find better parameters and in other cases the BB-BC cannot find better parameters than the default parameters that we have supplied for the system (example for interval type-2 at Table 6). However, the overall BB-BC optimisation is always able to find the best parameters based on HFL-MCGDM. Hence, the proposed system, HFL-MCGDM based on BB-BC is able to model the variation in the group decision making process exhibited by the various decision makers' opinion. In addition, the BB-BC optimised HFL-MCGDM system shows the highest agreement with the real decision outputs from the DMs. The proposed system GFL-MCGDM outperforms the MCGDM systems based on type-1 fuzzy sets, interval type-2 fuzzy sets and interval type-2 with hesitation index with or without using the optimised membership function by BB-BC algorithm. The BB-BC optimised GFL-MCGDM system also outperforms the GFL-MCGDM system without the BB-BC optimisation method.

8   Conclusions

Our 1st component, we presented a method that focuses on interval type-2 fuzzy logic MCGDM system with intuititonistic evaluation (from IFSs). The proposed method is expected to handle the linguistic uncertainties and conflicting decision among DMs in MCGDM and give comprehensive evaluation for the membership values. Using type-2 fuzzy logic and intuitionistic fuzzy sets separately in multi-criteria decision making model in representing the inputs from experts' judgment does not seem sufficient in showing the uncertainties and conflicting human-being evaluation. The intuitionistic evaluation in an interval type-2 membership function has been derived in the proposed method which includes seven steps for the aggregation and ranking of the preferred alternatives. In addition, we proposed an approach to extend the interval type-2 fuzzy values into intuitionistic values which will help to evaluate the hesitation values which are not present in type-2 and other fuzzy set theories. This combination clearly handles the linguistic uncertainties by the interval type-2 membership function and simultaneously computes the hesitation degree from the intuitionistic evaluation.

According to Garibaldi (2004), it is possible that the type-2 inferencing used may not hold in all cases especially the umbilical code acid-base balance study [76]. However, the fact that the type-2 system produced the same result as the type-1 FES for this study case (umbilical code acid-base balance) when the intervals were reduced to zero, is a hopeful sign that the inference may be valid in general. In order to extend the study as as to find agreement values between the system and DMs/experts, we created a new decision problem in the intelligent environment which involved more uncertainties involving the variety and inconsistency of human behaviour. We also increased the number of users to 15 and showed that the different background, age, culture and experience of all 15 users are counted in the experiment in order to show the high level of uncertainties compared to umbilical cord acid-base analysis where only 5 clinical experts have been included.

Through the experiments, it was found that overall, FL-TOPSIS systems give lower correlation values (agreements between the system and DMs based on Pearson Correlations) compared to our proposed HFL-MCGDM. Thus the proposed interval Type 2-Hesitation FL-MCGDM system better agrees with the users' decision compared to existing fuzzy MCDM including the Fuzzy Logic based TOPSIS (Technique for Order of Preference by Similarity to Ideal Solution), Type-1 FL-MCGDM and interval Type-2 in FL-MCGDM.

In order to extend the efficiency of the HFL-MCGDM system, for the 2nd component, we presented a General Type-2 Fuzzy Logic based approach for MCGDM (GFL-MCGDM) in part VI. The proposed system aimed to handle the high levels of uncertainties which exist due to the hesitancy, conflicts, ambiguity and vagueness among DMs' opinions. In order to demonstrate the effectiveness of the proposed system, we evaluated the system with a survey conducted randomly among 10 postgraduate students at the University of Essex. The analysis investigated the preferred location of postgraduate study which is: local, abroad or internship. Therefore, the proposed system GFL-MCGDM was able to give a better agreement with the human decision compared to type-1 and interval type-2, interval type-2 with hesitation fuzzy systems. In addition, GFL-MCGDM system was also implemented to the reading data sets for the preferred lighting level in the intelligent environment from 15 users. The increased correlation value for both study cases showed that the proposed method is considered to be effective in handling the high level of uncertainties among the big group of DMs (10 DMs to 15 DMs). Hence, this demonstrated that the proposed method can play an important role in the production of better MCGDM which is able to better settle conflicts among the different individual preferences with different alternatives and criteria followed by synthesising the different individual preferences into a unanimous approval.

Finally the 3rd component, we expanded the system to optimise the parameters involving the type-2 membership functions, including interval type-2, interval type-2 with hesitation and general type-2, in order to maximize the agreements between the system and DMs. With the aim of finding the optimal parameters of the type-2 fuzzy sets, we employed the Big Bang-Big Crunch (BB-BC) optimisation to find the optimal FOU parameters for type-2 membership function including interval and general type-2 fuzzy sets. The proposed system showed agreement between the proposed method and the real decision outputs from DMs - as quantified by the Pearson Correlation - which outperformed the MCGDM systems based on type-1 fuzzy sets, interval type-2 fuzzy sets and interval type-2 with hesitation index and also general type-2 without an optimal BB-BC membership function.

Hence, this showed that the proposed method can play an important role in the production of better fuzzy MCGDM which is able to settle conflicts among the different individual preferences with different alternatives and criteria followed by synthesising the different individual preferences into a unanimous approval. The results provided clearly showed that the more the theory can evaluate uncertainties, vagueness and conflict, the higher correlation value can be determined between the real output data. Hence, the increased correlation value revealed that the proposed method is considered to be effective in handling the high level of uncertainties among the DMs and the aggregation phase of the system.

In future work, we intend to modify the HFL-MCGDM system with different kinds of aggregation operators. There are many kinds of operators which have been implemented in decision making. Few have been found unstable to be implemented in decision making processes. The results tend to lose the information of users opinion such as fuzzy majority which has been proved by [Xu 2007]. Nevertheless, there are few which have been found in the literature such as score function, Euclidean distance and choquet integral. Next, we plan to further the research by testing the system with a larger number of DMs. This is a very crucial factor in real world decision problems which have to handle a variety of humans' behaviour. It is very challenging for a system to be able to handle a large volume of uncertainties. Such an implementation will be challenging and in future work we will make an effort to increase the agreement between the system and the DMs.

References

1. Zeleny, M.: Multiple criteria decision making, pp. 199–249. McGraw-Hill (1982)
2. Kahneman, D., Frederick, S.: A model of heuristic judgment. In: Holyoak, K.J., Morrison, R.G. (eds.) The Cambridge Handbook of thinking and reasoning, pp. 267–293. Cambridge University Press, Cambridge (2005)

3. Hatamlou, A., Abdullah, S., Hatamlou, M.: Data clustering using big bang–big crunch algorithm. In: Pichappan, P., Ahmadi, H., Ariwa, E. (eds.) INCT 2011. CCIS, vol. 241, pp. 383–388. Springer, Heidelberg (2011)
4. MakeItRasional, Multi-criteria evaluation – the foundation of rational decision making process, (Copyright © 2009-2013) (2013),
   http://makeitrational.com/multi-criteria-evaluation
5. Triantaphyllou, E., Shu, B., Nieto Sanchez, S., Ray, T.: Multi-criteria decision making: an operations research approach. In: Webster, J.G. (ed.) Encyclopedia of Electrical and Electronics Engineering, vol. 15, pp. 175–186. John Wiley & Sons, New York (1998)
6. Tindale, R.S., Hogg, M.A., Cooper, J. (eds.): Sage handbook of social psychology, pp. 1–66. Sage Publications, Inc., London (2013)
7. Thibaut, J., Walker, L.: Procedural justice: A psychological analysis. Wiley, New York (1975)
8. Folger, R.: Distributive and procedural justice: Combined impact of "voice" and improvement on experienced inequity. Journal of Personality and Social Psychology 35, 108–119 (1977)
9. Tyler, T.R., Blader, S.L.: The group engagement model procedural justice, social identity, and cooperative behavior. Personality and Social Psychology Review 7(4), 349–361 (2003)
10. Zadeh, L.A.: The concepts of a linguistic variable and its application to approximate reasoning, part I. Information Sciences 8, 199–249 (1975)
11. Atanassov, K.: Intuitionistic fuzzy sets. Fuzzy Sets and Systems 110, 87–96 (1986)
12. Tang, H., Zhou, J., Xue, S., Xie, L.: Big Bang-Big Crunch optimization for parameter estimation in structural systems. Original Research Article Mechanical Systems and Signal Processing 24(8), 2888–2897 (2010)
13. Muralidharan, C., Anantharaman, N., Deshmukh, S.G.: A multi-criteria group decision making model for supplier rating. Journal of Supply Chain Management 38(4), 22–33 (2006)
14. Wang, W., Liu, X., Qin, Y.: Multi-attribute group decision making models under interval type-2 fuzzy environment. Knowledge Based Systems 30, 121–128 (2012)
15. Herrera, F., Herrera-Viedma, E.: Linguistic decision analysis: steps for solving decision problems under linguistic information. Original Research Article Fuzzy Sets and Systems 115(1), 67–82 (2000)
16. Hagras, H.: A hierarchical type-2 fuzzy logic control architecture for autonomous mobile robots. IEEE Transactions on Fuzzy Systems 12(4), 524–539 (2004)
17. Indriyati, Surarso, Bayu, Sarwoko, Adi, E.: Sensitivity analysis of the AHP and TOPSIS methods for the selection of the best lecturer base on the academic achievement. In: Proceeding ISNPINSA Seminar International Diponegoro University, pp. 38–50 (2013)
18. Carlsson, C., Fuller, R.: Fuzzy multiple criteria decision making: Recent developments. Fuzzy Sets and Systems 78, 139–153 (1996)
19. Bellman, R.E., Zadeh, L.A.: Decision-making in a fuzzy environment. Management Science 17(4), 141–164 (1970)
20. Hsu, Y., Chen, C.: Aggregation of fuzzy opinions under group decision making. Fuzzy Sets and Systems 79, 279–285 (1996)
21. Karsak, E.E., Tolga, E.: Fuzzy multi-criteria decision-making procedure for evaluating advanced manufacturing system investments. International Journal Production Economics 69, 49–64 (2001)

22. Bozdag, C.E., Kahramana, C., Ruan, D.: Fuzzy group decision making for selection among computer integrated manufacturing systems. Computers in Industry 51, 13–29 (2003)
23. Kahraman, C., Ruan, D., Dogan, I.: Fuzzy group decision-making for facility location selection. Information Sciences 157, 135–153 (2003)
24. Blin, J.M.: Fuzzy relations in group decision theory. Journal of Cybernetics 4, 17–22 (1974)
25. Yager, R.R.: Fuzzy decision making including unequal objectives. Fuzzy Sets and Systems 1, 87–95 (1978)
26. Castillo, O., Alanis, A., Garcia, M., Arias, H.: An intuitionistic fuzzy system for time series analysis in plant monitoring and diagnosis. Original Research Article Applied Soft Computing 7(4), 1227–1233 (2007)
27. Lin, L., Yuan, X., Xia, Z.: Multi-criteria fuzzy decision-making methods based on intuitionistic fuzzy sets. Journal of Computer and System Science 73, 84–88 (2007)
28. Wang, P.: QoS-aware web services selection with intuitionistic fuzzy set under consumer's vague perception. Expert Systems with Applications 36, 4460–4466 (2009)
29. Gong, Z., Li, L., Forrest, J., Zhao, Y.: The optimal priority models of the intuitionistic fuzzy preference relation and their application in selecting industries with higher meteorological sensitivity. Original Research Article Expert Systems with Applcations 38(4), 4394–4402 (2011)
30. Chen, S.M., Yang, M.W., Lee, L.W., Yang, S.W.: Fuzzy multiple attributes group decision-making based on ranking interval type-2 fuzzy sets. Expert Systems with Applications 39, 5295–5308 (2012)
31. Ye, J.: Multi-criteria fuzzy decision-making method based on a novel accuracy function under interval-valued intuitionistic fuzzy environment. Expert Systems with Applications 36(3), 6899–6902 (2009)
32. Zhang, H., Yu, L.: MADM method based on cross-entropy and extended TOPSIS with interval-valued intuitionistic fuzzy sets. Knowledge-Based Systems 30, 115–120 (2012)
33. Su, Z.X., Xia, G.P., Chen, M.Y., Wang, L.: Induced generalized intuitionistic fuzzy OWA operator for multi-attribute group decision making. Expert Systems with Applications 39, 1902–1910 (2012)
34. Liang, Q., Karnik, N.N., Mendel, J.M.: Connection admission control in ATM networks using survey-based type-2 fuzzy logic systems. IEEE Transactions on Systems, Man, and Cybernetics, Part C: Applications and Reviews 30(3), 329–339 (2000)
35. Hasuike, T., Ishii, H.: A Type-2 fuzzy portfolio selection problem considering possibility measure and crisp possibilistic mean value, IFSA-EUSFLAT, 1120–1124 (2009)
36. Chen, S.M., Lee, L.W.: Fuzzy multiple attributes group decision-making based on ranking values and the arithmetic operations of interval type-2 fuzzy sets. Expert Systems with Applications 37(1), 824–833 (2010)
37. Garibaldi, J.M., Ozen, T., Musikasuwan, S.: Effect of type-2 fuzzy membership function shape on modelling variation in human decision making. In: Proceedings of IEEE International Conference on Fuzzy Systems, Budapest, Hungary (2004)

38. Ozen, T., Garibaldi, J.: Investigating adaptation in type-2 fuzzy logic systems applied to umbilical acid-base assessment. In: European Symposium on Intelligent Technologies, Hybrid Systems and Their Implementation on Smart Adaptive Systems, Oulu, Finland (2003)

39. Ozen, T., Garibaldi, J.: Effect of type-2 fuzzy membership function shape on modeling variation in human decision making. Proceedings IEEE International Conference 2, 971–976 (2004)

40. Lee, L.W., Chen, S.M.: Fuzzy multiple attributes group decision-making based on the extension of TOPSIS method and interval type-2 fuzzy sets. In: Proceedings of the Seventh International Conference on Machine Learning and Cybernetics, pp. 3260–3265 (2008)

41. Chen, S.M., Lee, L.W.: Fuzzy multiple criteria hierarchical group decision-making based on interval type-2 fuzzy sets. IEEE Transactions on Systems, Man, And Cybernetics—Part A: Systems and Humans 40(5), 1120–1128 (2010)

42. Chen, S.M., Lee, L.W.: Fuzzy multiple attributes group decision-making based on interval type-2 TOPSIS method. Expert Systems with Applications 37(4), 790–2798 (2010)

43. Chiao, K.: Multiple criteria group decision making with triangular interval type-2 fuzzy sets. In: Proceeding of IEEE International Conference on Fuzzy Systems, Taipei, Taiwan, June 27-30 (2011)

44. Chiao, K.: Trapezoidal interval type-2 fuzzy set extension of analytic hierarchy process. Tamsui Oxford Journal of Mathematical Sciences 16(2), 311–327 (2012)

45. Wagne, C., Hagras, H.: Novel methods for the design of general type-2 fuzzy sets based on device characteristics and linguistic labels surveys, pp. 537–543 (2009)

46. Wagner, C., Hagras, H.: Toward general type-2 fuzzy logic systems based on zSlices. IEEE Transactions on Fuzzy Systems 4, 637–660 (2010)

47. Shortliffe, E.H.: MYCIN: a rule-based computer program for advising physicians regarding antimicrobial therapy selection. In: Proceedings of the ACM National Congress (SIGBIO Session), vol. 739 (1974)

48. Szolovits, P., Patil, R.S., Schwartz, W.B.: Artificial intelligence in medical diagnosis. Annals of Internal Medicine 108(1), 80–87 (1988)

49. Büyüközkan, G., Feyzo, O.: A fuzzy-logic-based decision-making approach for new product development. Original Research Article International Journal of Production Economics 90(1), 27–45 (2004)

50. Dalalah, D., Bataineh, O.: A fuzzy logic approach to the selection of the best silicon crystal slicing technology. Expert Systems with Applications 36, 3712–3719 (2009)

51. Mergias, I., Moustakas, K., Papadopoulos, A., Loizidou, M.: Multi-criteria decision aid approach for the selection of the best compromise management scheme for ELVs: the case of Cyprus. Journal of Hazardous Materials 147, 706–717 (2007)

52. Uzokaa, F.E., Obotb, O., Barkerc, K., Osujid, J.: An experimental comparison of fuzzy logic and analytic hierarchy process for medical decision support systems. Computer Methods and Programs in Biomedicine 103, 10–27 (2011)

53. Alkhawlani, M.M.: Multi-Criteria vertical handover for heterogonous networks. International Journal of Wireless & Mobile Networks (IJWMN) 3(2), 149–163 (2011)

54. Zadeh, L.A.: Fuzzy sets. Information and Control 8, 338–353 (1965)

55. Xu, Z.: Intuitionistic preference relations and their application in group decision making. Information Sciences 177, 2363–2379 (2007)

56. Wang, J., Zhang, J., Liu, S.: A new score function for fuzzy MCDM based on vague set theory. International Journal of Computational Cognition 4(1), 44–48 (2006)

57. Mendel, J.: Fuzzy logic system for engineering: a tutorial. Proceedings of the IEEE 83(3), 345–374 (1995)
58. Kassem, S.: A type-2 fuzzy logic system for workforce management in the telecommunications domain, Ph.D. dissertation, School of Computer Science and Electronic Engineering, University of Essex (2012)
59. Deschrijver, G., Kerre, E.: On the position of intuitionistic fuzzy set theory in the framework of theories modeling imprecision. Information Sciences 177, 1860–1866 (2007)
60. Przemyslaw, G., Edyta, M.: Some notes on (Atanassov's) intuitionistic fuzzy sets. Fuzzy Sets and System 156, 492–495 (2005)
61. Atanassov, K.: Intuitionistics fuzzy sets. Physica-Verlag, Heidelberg (1999)
62. Tamalika, C., Raya, A.K.: A new measure using intuitionistic fuzzy set theory and its application to edge detection. Applied Soft Computing 8, 919–927 (2008)
63. Wagner, C.: Towards better uncertainty handling based on zslices and general type-2 fuzzy logic systems, Ph.D. dissertation, School of Computer Science and Electronic Engineering, University of Essex (2009)
64. Coupland, S., Gongora, M., John, R., Wills, K.: A comparative study of fuzzy logic controllers for autonomous robots. In: Proceedings of IPMU, Paris, France, pp. 1332–1339 (2006)
65. Figueroa, J., Posada, J., Soriano, J., Melgarejo, M., Roj, S.: A type-2 fuzzy logic controller for tracking mobile objects in the context of robotic soccer games. In: Proceeding of the 2005 IEEE International Conference on Fuzzy Systems, Reno, USA, pp. 359–364 (2005)
66. Mendel, J.M., John, R.I.B.: Type-2 fuzzy sets made simple. IEEE Transactions on Fuzzy System 10(2), 117–127 (2002)
67. Liu, H., Wang, G.: Multi-criteria decision-making methods based on intuitionistic fuzzy sets. European Journal of Operational Research 179, 220–233 (2007)
68. Naim, S., Hagras, H.: A Type 2-Hesitation Fuzzy Logic based Multi-Criteria Group Decision Making System for Intelligent Shared Environments. Journal of Soft Computing (2013), doi:10.1007/s00500-013-1145-0
69. Mendel, J.M.: Uncertain rule-based fuzzy logic systems: introduction and new directions. Prentice-Hall, Upper Saddle River (2001)
70. Naim, S., Hagras, H.: A Big-Bang Big-Crunch Optimized General Type-2 Fuzzy Logic Approach for Multi-Criteria Group Decision Making. Journal of Artificial Intelligence and Soft Computing Research 3, 2 (2013)
71. Zhang, H.X., Wang, F., Zhang, B.: Genetic optimization of fuzzy membership functions. In: Proceedings of the 2009 International Conference on Wavelet Analysis and Pattern Recognition, Baoding, pp. 465–470 (2009)
72. Kumbasar, T., Eksin, I., Guzelkaya, M., Yesil, E.: Adaptive fuzzy model based inverse controller design using BB-BC optimization algorithm. Expert Systems with Applications 38, 12356–12364 (2011)
73. Turanoglu, E., Ozceylan, E., Kiran, M.S.: Particle swarm optimization and artificial bee colony approaches to optimize of single input-output fuzzy membership functions. In: Proceedings of the 41st International Conference on Computers & Industrial Engineering, pp. 542–547 (2011)

74. Bilgin, A., Dooley, J., Whittington, L., Hagras, H., Henson, M., Wagner, C., Malibari, A., Al-Ghamdi, A., Al-haddad, M., Al-Ghazzawi, D.: Dynamic profile-selection for zSlices based Type-2 fuzzy agents controlling multi-user ambient intelligent environments. In: Proceedings of the 2012 IEEE International Conference on Fuzzy Systems, Brisbane (June 2012)
75. Tadik, D., Arsovski, S., Stefanović, M., Aleksić, A.: A fuzzy AHP and TOPSIS for ELV dismantling selection. International Journal for Quality Research 4(2), 139–144 (2010)
76. Garibaldi, J.M., Ozen, T.: Uncertain fuzzy reasoning: a case study in modelling expert decision making. IEEE Transactions on Fuzzy Systems 15(1), 16–30 (2007)

Multi-criteria Influence Diagrams – A Tool for the Sequential Group Risk Assessment

Aleksandar Janjić[*], Miomir Stanković, and Lazar Velimirović

Abstract. This chapter describes the use of influence diagrams in the risk assessment and proposes their extension to the group decision making using fuzzy logic, sequential approach and multi-criteria evaluation. Instead of classical Bayesian networks using conditional probability tables that are often difficult or impossible to obtain, a verbal expression of probabilistic uncertainty, represented by fuzzy sets is used in this approach. Influence diagrams are modeling the multistage decision processes and the interrelations among different chance and value nodes as well, enabling the iterative approach to the risk assessment. After the first, independent assessment of the group of experts, this preliminary risk grade is the input in the second step where the adapted risk grade has been adopted based on known evaluations, and interaction among decision makers. The different risk components and decision maker's attitudes are considered by ordered weighted averages (OWA) operators. This inference engine is illustrated through the assessment of risk caused by improper drug storage in pharmaceutical cold chain by the group of experts in the iterative assessment process.

Keywords: Group decision making, Influence diagrams, OWA, Risk.

Aleksandar Janjić
Department of Power Engineering,
University of Niš, Faculty of Electronic Engineering,
Aleksandra Medvedeva 14, 18000 Niš, Serbia
e-mail: Aleksandar.Janjic@elfak.ni.ac.rs

Miomir Stanković
Department of Matematics,
University of Niš, Faculty of Occupational Safety, Niš, Serbia

Lazar Velimirović
Department of Matematics,
Mathematical Institute of the Serbian Academy of Sciences and Arts, Belgrade, Serbia

[*] Corresponding author.

W. Pedrycz and S.-M. Chen (eds.), *Granular Computing and Decision-Making,*
Studies in Big Data 10, DOI: 10.1007/978-3-319-16829-6_7

1 Introduction

Risk management is a concept of Business Operations Management, first developed and applied in finance, and later, in other industrial branches. Global business has brought with itself increased risks, especially in the food and drug industries [1]. Its intensity within these industries has become especially noticeable in the last decade, which resulted in the passing of various standards and recommendations to cover this field [2,3,4,5,6]. Risk became the crucial decision making criteria in evaluation of development alternatives of several complex systems. In such systems, risk has been usually connected to the reliability calculation, assuming that risk and reliability have identical implications [7]. Gradually, the concept of risk has been introduced in different areas, including asset management and maintenance scheduling as most critical processes in terms of the level of investment [8,9,10,11,12].

1.1 Risk Assessment under Uncertainty

Risk Assessment is integral part of Risk management process, and represents the necessary step before proceeding to the treatment of risk by answering the following questions:

• What can happen and why?

• What are the consequences?

• What is the probability of their future occurrence?

• Are there any factors that mitigate the consequence of the risk or that reduce the probability of the risk?

There is a great number of risk assessment techniques, including fault tree, event tree, decision tree analysis, FMECA (failure mode criticality and effect analysis), Markov processes and Bayes nets. The choice of the right methodology depends on various factors, including the complexity of the problem, the nature and degree of uncertainty, the extent of resources (required in terms of time and level of expertise, data needs or cost), and finally, whether the method can provide a quantitative or qualitative output [13].

What is common in all approaches mentioned above is the definition of risk associated with an event E as the product of probability of event $p(E)$ (failure of component or group of components) and consequences of this event $C(E)$:

$$Risk(E) = p(E) \times C(E) \tag{1}$$

However, this simple relationship takes no account of the factors such as non-linear dependence of utility on the value of consequences, which requires other form of relations:

$$Risk(E) = p(E)^x \times C(E)^y \qquad (2)$$

This approach presumes that both probability and consequences can be calculated accurately, which is not always possible, and depends on quality and quantity of available data. Generally, the assessment of risk is faced with lack of data from the past (equipment history), present (on line condition monitoring) and future (uncertainty about operating conditions). Furthermore, the decision making in complex systems is always related to uncertainties, either because of different criteria weightings of several decision makers, either because of stochastic nature of future events.

It is well accepted that Fuzzy Set Theory provides a useful way to deal with ill-defined and complex problems in decision making by quantifying imprecise information and incorporating vagueness. Dikmen et al. [14] proposed a fuzzy risk rating method in international construction projects. The system identifies risks, models the risks using influence diagrams, selects the membership function of each variable, captures the experts' opinions using aggregation rules, aggregates fuzzy rules into a fuzzy cost overrun risk rating and determines the final level of project risk. In [15] fuzzy reasoning has been extended to the AHP approach to handle subjective assessments and prioritize diverse risk factors. The model quantifies the risk magnitude and combines it with the risk likelihood and risk severity into a fuzzy inference system.

Another form of uncertainty in the risk assessment is the group risk assessment. When people are in groups, they make decision about risk differently from when they are alone. In the group, they are likely to make riskier decisions, as the shared risk makes the individual risk less, which requires the new form of problem modeling.

1.2   Group Risk Assessment

In all decision making processes, the procedures for combining opinions about the alternatives with different points of view are established. The assumption is that there is a set of alternatives $X = \{x_1, x_2 \dots, x_n,\}$, chosen by a group of n agents to cooperate in the selection of actions. It is assumed that each agent represents its own preference in the form of the fuzzy subset of X presented by the membership function. Let A_j indicates a preference for agent j, and let each agent to be unaware of the function of the preference of other participants. Approach for obtaining the group decision is to gather individual preference functions A_j to obtain the group preference functions A, and then choose the one alternative that optimizes the group preference function. One of the most important decisions is the choice of operators for aggregating individual preference function.

Wang and Elang [16] proposed a fuzzy multi-criteria group decision making approach, which allows decision makers to rapidly and effectively evaluate multiple fuzzy risk factors using linguistic terms by aggregating the assessments of multiple risk factors. A methodology which produces the appraisal vector of the risky conditions of a construction project by aggregating the weight coefficient of any risk groups and fuzzy risk factors obtained from experts using the AHP technique, a hierarchical structure of risks, and the fuzzy evaluation matrixes of risk factors has been proposed in [17]. Finally, Nieto et al. [18] proposed an algorithm to handle the inconsistency in the fuzzy preference relation when pairwise comparison judgments are necessary, while [19] proposed an extended version of the Technique for Order Preference by Similarity to an Ideal Solution (TOPSIS), which resolves the multi-criteria risk assessment model under a fuzzy environment.

Although the mentioned models take account of uncertainties by the introduction of fuzzy sets, multiple criteria and multiple experts, the danger that some solutions will not be well accepted by some experts in the group is still present [20,21]. To overcome this problem, it is advisable that experts carry out a consensus process, where the experts discuss and negotiate in order to achieve a sufficient agreement before selecting the best alternative [22,23]. A comprehensive presentation of the state of the art of different consensus approaches is given in [24], with the focus on the soft consensus approach.

However, unlike standard group decision making procedure where the group consensus relies on the principle of majority, the risk assessment process outlines some behavioral characteristics that are opposite to this principle. Many studies have led to the conclusion that tendency for group decisions to be riskier than the average decision made by individuals exists and is referred to as "risky shift".

The predicted results were first noted in 1961 and similar results have been obtained since [25, 26]. This shift toward a riskier group decision is an example of a broader result of group decision making called group polarization. In [27], findings that groups communicating via computer produce more polarized decisions than face-to-face groups are elaborated.

To solve this particular problem of multi expert risk assessment, sequential fuzzy influence diagrams are proposed in this approach. Fuzzy logic is introduced in a twofold manner: via fuzzy probability values expressed linguistically, and via fuzzy random variables. Instead of classical Bayesian networks using conditional probability tables that are often difficult or impossible to obtain, a verbal expression of probabilistic uncertainty, represented by fuzzy sets is used in this approach.

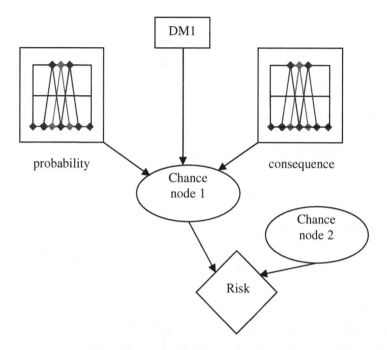

Fig. 1 Risk assessment of individual decision maker (DM 1)

Influence diagrams are modeling the multistage decision processes, and the interrelations among different chance and value nodes as well, enabling the iterative approach to the risk assessment. The first stage of the process – individual risk assessment using fuzzy influence diagram is presented in Figure 1.

After the first, independent assessment of the group of experts, this preliminary risk grade is the input in the second step where based on known evaluations, and interaction among decision makers, the adapted risk grade has been adopted.

This inference engine is illustrated through the assessment of risk caused by improper drug storage in pharmaceutical cold chain by the group of experts in the two step assessment process. The different risk components and decision maker's attitudes are considered by ordered weighted averages operators. This model can be implemented in many complex systems, like the health care (ex. Health Technology Assessment) where is necessary to make decisions based on clinical, social and economic criteria. In power systems criteria are economy, social and environmental, etc.

The objective of this work is twofold: to propose an integrated method for the sequential decision making and the risk assessment in uncertain environment, including decision nodes in the analysis. Second objective is to make this methodology practical, by using robust graphic tools that are not sensitive to the missing or incomplete input values, modeling the interaction among the decision makers and the multi-criteria evaluation.

The rest of the chapter is organized as follows: Section 2 describes the one stage single criteria fuzzy influence diagram. In Section 3, this model is extended with the multi-criteria and sequential group decision making approach. Some simulation results are described in Section 4 and finally the concluding remarks are presented in Section 5.

2 Research Method

2.1 Influence Diagrams in Complex System

Bayesian networks (BN) and Influence diagrams (ID) are graphical tools that aid reasoning and decision-making under uncertainty. The networks represent a system over which a probability distribution is defined, modeling uncertainty both quantitatively and qualitatively. They allow a user to make inferences when only limited information is available. In the artificial intelligence literature, BN and ID are among the most popular types of graphical modeling [28,29,30]. They are used in expert systems involving problem domains in medical diagnosis, map learning, heuristic search, and, very recently, in power systems. Two types of inference support are considered:

- predictive support for node X_i, based on evidence nodes connected to X_i through its parent nodes (also called top-down reasoning), and

- diagnostic support for node X_i, based on evidence nodes connected to X_i through its children nodes (also called bottom-up reasoning).

Generally, in complex diagnostic problems there are usually several symptoms, and each of these symptoms can result from several failures. Probabilistic methods can be used to link symptoms to failures if the necessary failure probabilities can be obtained. BN have been effective in modeling probabilistic relationships in complex diagnostic situations and in providing a framework to identify critical probabilistic mappings. These probabilities can be obtained from operating data, (if available for all prior probabilities and fault mappings over a sufficient period of time), or through the solicitation of subjective probabilities from experts.

A generalization of a BN, in which not only probabilistic inference problems but also decision making problems (following maximum expected utility criterion) can be modeled and solved is the Influence diagram. ID (also called a relevance diagram, decision diagram or a decision network) is a compact graphical and mathematical representation of a decision situation. Initially, they were proposed in [31], as a tool to simplify modeling and analysis of decision trees. Whereas a decision tree shows more details of possible paths, an ID shows dependencies among variables more clearly.

Both crisp BN and ID represent a pair $N = \{(V, E), P\}$ where V and E are the nodes and the edges of a directed acyclic graph, respectively, and P is a probability distribution over V. Discrete random variables $V = \{X_1, X_2, ..., X_n\}$ are assigned to the nodes while the edges E represent the causal probabilistic relationship among the nodes. Each node in the network is annotated with a Conditional Probability Table (CPT) that represents the conditional probability of the variable given the values of its parents in the graph. The CPT contains, for each possible value of the variable associated to a node, all the conditional probabilities with respect to all the combinations of values of the variables associated with the parent nodes. For nodes that have no parents, the corresponding table will simply contain the prior probabilities for that variable.

Building of an ID model is performed with the usage of several graphical elements. A circle depicts an external influence (an exogenous variable) – these are variables whose values are not affected by the decision being made. Rectangle depicts a decision – these are decisions made by the decision maker. Intermediate variable depicts an endogenous variable – these are variables whose values are computed as functions of decision, exogenous and other endogenous variables. Value node, presented as a diamond (objective variable) is a quantitative criterion that is the subject of optimization. Chance node (oval) represents a random variable whose value is dictated by some probability distribution. Arrow shows the influence between variables, and dotted arrow shows information being communicated between elements.

The methods for evaluating and solving ID are based on probabilities and efficient algorithms have been developed to analyze them [31,32,33,34,35]. Like in BN, the input and output values of a node in an ID are based on the Bayesian theorem. The use of probability tables with many elements is, however, very difficult, because of the combinatorial explosion arising from the requirement that the solution must be extracted by the cross product of all probability tables.

Instead of probabilities, solving of an ID can be effectuated using fuzzy reasoning [36,37,38,39]. Each node in the diagram can be represented by appropriate fuzzy set describing the uncertain nature of a given value. A predecessors set of nodes is the set of all nodes having an influence arrow connected directly to the given node. The combination of predecessor nodes fuzzy sets gives the value of resulting node. A commonly used technique for combining fuzzy sets is Fuzzy inference system. Mamdany's type fuzzy inference, used very often, expects the output membership functions to be fuzzy sets. After the aggregation process, a defuzzification is performed on each output variable.

However, both fuzzy logic based approximate reasoning and BN have limitations in the risk assessment. The main limitation of fuzzy reasoning approaches is the lack of ability to conduct inference inversely. Feed-forward-like approximate reasoning approaches are strictly one-way, that is, when a model is given a set of inputs it can predict the output, but not vice versa.

On the other hand, utilization of a probability measure to assess uncertainty in BN is another limitation. It requires too much precise information in the form of prior and conditional probability tables, and such information is often difficult or impossible to obtain. In particular, in dealing with indirect relationships, it is usually difficult to make precise judgments with crisp numbers. In certain circumstances, a verbal expression or interval value of probabilistic uncertainty may be more appropriate than numerical values. In this methodology, linguistic probabilities are used, which will be explained in the sequel.

2.2   Fuzzy Probabilities

The introduction of fuzzy logic in BN has been performed by the fuzzification of random variables [40,41], or by introducing fuzzy probabilities [42,43]. In this paper, both random variables, prior and conditional probabilities are presented by fuzzy sets. Prior to define the inference engine principle, the definition of fuzzy probability that will be used in this paper will be given.

The basic characteristic of a fuzzy set is its membership function. Let U be the universal set, and $\mu_A(x)$ the membership function of the classical subset A of U such that it takes values from the set $\{0,1\}$, where $\mu_A(x) = 1$ if $x \in A$ and $\mu_A(x) = 0$ for $x \notin A$. It may be noted that the boundaries of a subset A sharp and clear and represent two-class classification of elements of the set U. The fuzzy set, on the other hand, introduces ambiguity by canceling sharp boundaries between group members and non-members. In other words, the transition from the members to the 'non-members' is not too sharp. The transition is gradual. It is allowed by the membership function of the A fuzzy subset to the set U, which is defined as follows $\mu_A : U \to [0,1]$. The main difference from the classical subset is now an element $x \in U$ belongs to the A fuzzy subset with a degree between 0 and 1. For easier understanding the function belonging to the A fuzzy subset is denoted by $A(x)$. Precisely $A(x)$ (the membership function) will be the function of the preference in the process of group decision-making, while the alternatives will be shown by variable x.

Based on previous works on linguistic probability [41,42], it is possible to define similar probability measure for fuzzy probabilities.

Definition 1. Given an event algebra ε defined over a set of outcomes Ω, a function *FP*: $\varepsilon \to E$ is termed a fuzzy probability measure if and only if for all $A \in \varepsilon$

$$0_\chi \leq FP(A) \leq 1_\chi$$

$$FP(\Omega) = 1_\chi \ and \ FP(\emptyset) = 0_\chi$$

If $A_1, A_2, \ldots \in \varepsilon$ are disjoint, then   (3)

$$FP(\cup_{i=1}^{\infty} A_i) \subseteq \sum_{i=1}^{\infty} FP(A_i)$$

$$FP(A) = 1_\chi - FP(A^\varepsilon)$$

FP is fuzzy probability measure on (Ω, ε), the tuple $(\Omega, \varepsilon, FP)$ is termed fuzzy probability space. Embedded real numbers are denoted by χ subscript. Expression (3) specify the quantity space in which probabilities will be assessed (linguistic probabilities have zero membership outside the unit interval). Also, expression (3) is intended to capture the intuition one might know the probability of the union of two disjoint events more precisely than the probabilities of either individually.

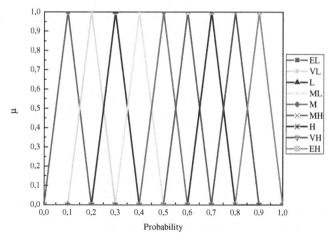

Fig. 2 Fuzzy probabilities

Also, it expresses that knowing something about the probability of an event translates into equally precise knowledge about the probability of its complement.

Based on previous definition, fuzzy probabilities, grouped in several fuzzy sets, are introduced and denoted with linguistic terms (extremely low, very low, low, medium low, medium, medium high, high, very high and extremely high). Appropriate fuzzy sets are presented on Figure 2.

The extension principle may also be used to define fuzzy counterparts to the standard arithmetic operators. The extension of a real arithmetic operator will be denoted by circling its usual symbol. It is also possible to derive these operators by examining the effects of interval based calculations at each α – cut. The extended operators are defined by:

Definition 2. For all $a,b \in R^F$ the extended operators are defined by:

$$\mu_{a\oplus b}(z) = \sup_{x+y=z} \min\left(\mu_a(x), \mu_b(y)\right)$$

$$\mu_{a\otimes b}(z) = \sup_{xy=z} \min\left(\mu_a(x), \mu_b(y)\right)$$

$$\mu_{a-b}(z) = \sup_{x-y=z} \min\left(\mu_a(x), \mu_b(y)\right) \quad (4)$$

$$\mu_{a\oslash b}(z) = \sup_{x/y=z} \min\left(\mu_a(x), \mu_b(y)\right)$$

From previous definition, two fuzzy Bayes rules analogue to classical crisp number relations are formulated. Operator "≅" stands for "=" operator.

Fuzzy joint probability

$$FP(Y = y_j, X = x_i) \cong FP(X = x_i) \otimes FP(Y = y_j \setminus X = x_i) \tag{5}$$

Fuzzy Bayes rule

$$FP(X = x_i \setminus Y = y_j) \cong \frac{FP(X = x_i) \otimes FP(Y = y_j \setminus X = x_i)}{FP(Y = y_j)} \tag{6}$$

Based on the law of total probability another rule for the fuzzy marginalization can be added, represented by the expression (7).

Fuzzy marginalization rule

$$FP(Y = y_j) \cong \sum_i FP(X = x_i) \otimes FP(Y = y_j \setminus X = x_i) \tag{7}$$

Using the above equations, fuzzy BN inference can be conducted. Operations on fuzzy numbers are defined as operations in terms of arithmetic operations on their α – cuts (arithmetic operations on closed intervals).

The overall risk is calculated with exhaustive enumeration of all possible states of nature, and their expected value of risk. Let suppose a system in which risk value node has X_n parent nodes, with different number of discrete states. According to notation explained in the section 2.1, fuzzy probability of the chance node X_i being in the state j is expressed as $FP(X_i = x_{ij})$. Fuzzy value of possible consequences in the state x_{ij} is represented by $FD(X_i = x_{ij})$. The expected value of risk is then calculated:

$$R \cong \sum_j \sum_i FP(X = x_i) \otimes FD(X = x_{ij}) \tag{8}$$

Example 1. The illustration of ID for the power transformer risk assessment is given in this example (see Figure 3).

Power transformer in one transformer substation is planned for the replacement, because of its age and unsatisfying diagnostic test results. Node A is decision node, bound to two outcomes: whether to replace, or keep the existing power transformer in use. Node B is describing the condition of transformer with discrete deterioration states. This node represents the chance node, because of stochastic nature of transformer condition which cannot be fully determined by transformer diagnostic. Node B has the parent node A, because the state of transformer health is directly influenced by the decision of replacement.

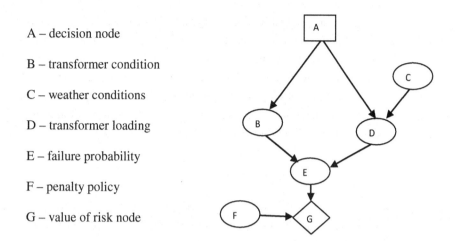

A – decision node

B – transformer condition

C – weather conditions

D – transformer loading

E – failure probability

F – penalty policy

G – value of risk node

Fig. 3 Influence diagram for the risk assessment of decreased reliability

Increased number of transformer outages is expected, but that number, together with consequences that these outages will produce can vary depending on uncertain parameters in the future, including weather conditions, loading of the transformer, and level of penalties imposed by the regulator. Therefore, one has to investigate the possibility of keeping it in service one more year, and to check whether this decision greatly increases the risk of surpassing required values for system reliability, imposed by the regulator.

Probabilities for nodes E and F are given in Table 1, while the complete tables of prior and conditional probabilities can be found in [44].

Table 1 Fuzzy probabilities for chance nodes E and F

Chance node E				
	Alternative 1(replace)		Alternative 2 (not replace)	
FP(E_1) – low level of power outages	H		VL	
FP(E_2) – medium level	VL		VL	
FP(E_3) – high level of power outages	EL		M	
Chance node F				
FP(F_1) severe penalties	H	PEN$_1$	[0.8 0.9 1]	
FP(F_2) mild penalties	L	PEN$_2$	[0.2 0.3 0.4]	

The value node G is the risk node, and as was described earlier, standard definition of risk as the product of probability and consequence (financial penalty) has been used. *FP(E)* is fuzzy probability calculated for the node *E*, and final value of risk is the expected value of risk for all combinations of event *E* over *n* possible outcomes of event *F*. *PEN_j* denotes penalties in the case of the *j*-th outcome of event *F*. Penalties are also represented as fuzzy numbers, and they can be represented with per unit values, relative to the maximal possible penalty. Using expressions (5) – (8), we are calculating the value of node G:

$$Risk \cong \sum_{i,j} FP\left(E_i\right) \otimes FP\left(F_j\right) \otimes PEN_j \qquad (9)$$

Results are presented on Figure 4. Different methods of fuzzy number ordering can be used, but it's obvious that by replacing the transformer the overall risk could be reduced in a great extent.

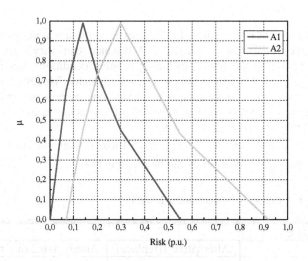

Fig. 4 Fuzzy values of risks for value node G for alternatives A1- replace and A2 - not replace

Single criteria ID can be extended to the case of multi-objective analysis: instead of single value node, the aggregated value of several nodes [45], or Pareto analysis of optimal solution space can be used [46]. In the next section, we will present the use of multi-criteria ID to the case of multi expert risk assessment where the final solution has been obtained by the iterative process.

3   Research Design

3.1   Multi Expert Sequential Influence Diagram

In this section, we are elaborating the extension of IDs to the case of group decision making. The extension is performed on two levels: "horizontal" level of the overall risk assessment by introducing appropriate aggregation function and on the "vertical" level of sequential approach by introducing several stages and reevaluation of alternatives in decision process.

Group decision making situation refers to the set of possible alternatives, $X = \{x_1, x_2, ..., x_n\}, n \geq 2$ and a group of experts $E = \{e_1, e_2, ..., e_m\}, m \geq 2$, who express their opinions about X to achieve a common solution. In a fuzzy context, the objective is to classify the alternatives from best to worst, associating with them some degrees of preference expressed in [0, 1] interval. After the aggregation of preferences, the final solution has been selected. To overcome the problems of single stage decision process, it is advisable that experts carry out a consensus process, where the experts can discuss and share their knowledge about the problem and alternatives in order to facilitate the process of alternative selection [24]. The risk assessment process however, imposes some restriction to the application of majority principle out of two main reasons.

The first one is the so called "risky shift" explaining the effect that people in groups make decision about risk differently from when they are alone. In the group, they are likely to make riskier decisions, as the shared risk makes the individual risk less [25,26]. The second reason - the group polarization is the solidification and further strengthening of a position as a consequence of group discussion. That is, the opinions or positions of individuals, once they meet as a group, tend to become more "polarized" and move more toward the extremes. Out of these reasons, we used the iterative sequential approach of decision making in our methodology, allowing the aversion to opinion changing, considering only aggregated risk indicators.

3.1.1   Individual Risk Aggregation

It is widely accepted that any suitable aggregation of fuzzy sets may be used in decision making, modeling different types of decision behavior and choosing a decision function that best reflects the goals of the decision. Many methods for aggregation can be found in [47,48,49,50,51,52]. Regardless whether the weights and scores are linguistic or numeric, the general form of the aggregation function is:

$$f = \sum_{k=1}^{g} u_k \otimes x_{ik} \tag{10}$$

where f_i is the final score for alternative i. The weights of experts u_k could have quantitative or qualitative values.

The following three types of aggregation are used most commonly in decision making: conjunctive, disjunctive and compensatory aggregation of criteria. Conjunctive aggregation of criteria implies simultaneous satisfaction of all decision criteria, while the disjunctive aggregation implies full compensation amongst the criteria. The compensatory aggregation is more suitable for conflicting criteria or the human aggregation behavior, because human beings tend to partially compensate between criteria, instead of trying to satisfy them simultaneously.

The process of group decision making is divided into homogeneous and heterogeneous processes. The process is homogeneous when no importance score is related to the importance of the agent, while it is the opposite case with the heterogeneous process. However, when there is not any importance score provided, many problems of the group decision making should be classified as heterogeneous processes. Indeed, in many cases, it is not necessary to determine the presence of importance scores associated to the agent, to be absolutely sure that no agent is treated equally. In the case where agents provide information to resolve things, this information can be used as a tool for discrimination of agents so the agents will not have the same importance. In this case, agents with consistent information are given the highest grades. One way of implementing the importance scores in the decision process is to encourage the order of value preferences before their aggregation. For this to be implemented, the OWA operator is used.

A very efficient for information combination method OWA was suggested by R. Yager [53]. Since then OWA operators are studied from different aspects, and applied in engineering and different fields of artificial intellect [54,55,56]. An OWA operator of dimension n with an associated vector $W = (w_1, w_2, ..., w_n)$ is a mapping $F:R^n \to R$ defined as:

$$F(x_1, x_2, ..., x_n) = \sum_{i=1}^{n} w_j x_{\sigma(j)} \qquad (11)$$

where σ is a permutation that orders the elements : $x_{\sigma(1)} \leq x_{\sigma(2)} \leq \cdots \leq x_{\sigma(n)}$. The weights are all non negative ($w_i \geq 0$) and $\sum_{i=1}^{n} w_i = 1$.

The OWA operators provide a parameterized family of aggregation operators, which include many of the well-known operators such as the maximum, the minimum, the k-order statistics, the median and the arithmetic mean. In order to obtain these particular operators we should simply choose particular weights. Since this operator generalizes the minimum and the maximum, it can be seen as a parameterized way to go from the min to the max. In this context, Yager introduces the operator *maxness* (initially called *orness*) which is defined with:

$$maxness(w_1, w_2, ... w_n) = \sum_{j=1}^{n} w_{n-1+j} \frac{n-j}{n-1} = \sum_{j=1}^{n} w_j \frac{j-1}{n-1} \qquad (12)$$

The minimal value is obtained for *maxness*$(1,0,...,0) = 0$, and the maximal value is obtained for *maxness*$(0,0,...,1) = 1$.

The Fuzzy OWA Operator (FOWA) is an extension of the OWA operator that uses uncertain information represented in the form of fuzzy numbers. FOWA provides a parametrized family of aggregation operators that include the fuzzy maximum, the fuzzy minimum and the fuzzy average criteria, among others. FOWA operator of dimension n is a mapping that has an associated weighting vector W, such that:

$$FOWA(\bar{a}_1, \bar{a}_2, \dots, \bar{a}_n) = \sum_{i=1}^{n} w_j \, b_j \qquad (13)$$

Where b_j is the j-th largest of the $\bar{a}_i$, and $\bar{a}_i$ are fuzzy numbers.

In [52] Chiclana et al. considered group decision making problems where the information about the alternatives was represented using fuzzy preference relations and a fuzzy majority guided choice scheme based on OWA operator was designed. In [53], set of IOWA operators is introduced, guided by fuzzy linguistic quantifiers, allowing the introduction of some semantics or meaning in the aggregation.

In the proposed methodology, both OWA and FOWA aggregation is enabled on the level of ID value nodes describing individual experts risk assessments. Each step of the algorithm will be described in the next section.

3.1.2 Algorithm of Multi Stage Risk Assessment

According to the definition of risk as the product of state probability and possible damage in that state of nature, the group has to evaluate both values: probabilities and damages. After the expert's individual risk assessments, experts are reevaluating their starting evaluations in the second stage of the decision process. Figure 5 shows the ID for a two-stage version of the risk assessment problem. The variables X_1 and X_2 represent possible states of nature with probabilities and possible damages assessed by each expert. These values represent the input to the second stage of decision making process, where the experts reevaluate their starting assessments. Since the same variables occur at each stage, we can make the ID arbitrarily large by increasing the number of stages.

The methodology for the sequential risk assessment consists of several steps:

Step 1. In this step, experts are evaluating prior and conditional probabilities of each chance node, expressing these values in linguistic terms, according to notation presented on Figure 2. Experts are expressing possible consequences and damages of each state of nature by appropriate fuzzy sets.

Step 2. Overall value of risk for each expert individually is calculated by the expression (8) using the rules of fuzzy arithmetic.

Step 3. Using the OWA or FOWA operator, the aggregated value of risk is calculated.

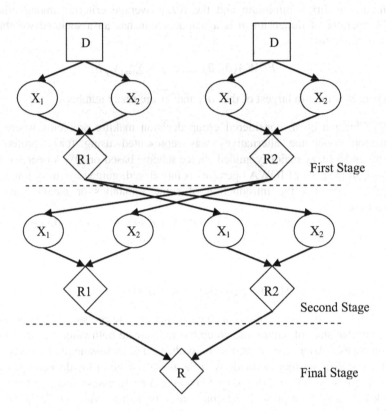

Fig. 5 Two stage risk assessment approach

Step 4. Experts are discussing these values, share their knowledge and revaluate their initial attitudes. In the case of presence of moderator, the advice is given to the experts and the consensus phase is finished.

Step 5. In this step, experts are reevaluating prior and conditional probabilities of each chance node, expressing these values in linguistic terms, consequences and damages of each state of nature by appropriate fuzzy sets.

Step 6. If the difference between new and previous aggregated value is below some threshold value δ, the iterative process ends and the selection of alternative with the lowest risk is made. If not, the process returns to the Step 4.

The graphical representation of the algorithm is given on Figure 6. The crucial step in the algorithm described above is the preference aggregation – step 3. In this stage, all experts' opinions are combined to get a final rating for each alternative. The selection of aggregation function plays an important role in the accuracy of the final solution. Although the use of linguistic variables makes decision makers' evaluations more flexible and reliable, the aggregation of linguistic labels is rather complicate, especially when applies the weighting associated with the evaluations.

Furthermore, it is not clear how to reorder the arguments, and it is necessary to establish a criterion for comparing them. The most simple and practical way is to select the fuzzy number with the highest value in its highest membership level ($\alpha=1$). If the membership level is interval, the average of interval will be calculated. Out of these reasons, both crisp and fuzzy OWA values are calculated in this methodology.

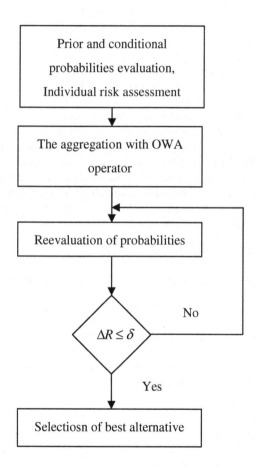

Fig. 6 Two stage risk assessment approach

The weighting vector W (the same in both OWA and FOWA approach) has been selected according to the aggregator risk preferences, following the expression (12). Another important issue is the selection of the threshold value δ, that has to be carefully selected based on the past experience. This method of risk assessment is illustrated on an example taken from the supply chain management of pharmaceutical cold chain.

4   Case Study

The pharmaceutical cold chain concerns the pharmaceuticals that must be distributed at temperature between 2 and 8 °C [51]. Stability data from both accelerated and long-term studies are used to establish recommended storage conditions and expiration dating for drug products. As long as the product remains in its approved container within the specified temperature range, the quality of the product is assured until the date of expiration. However, it is likely that a product will be exposed to temperatures outside of its specified storage range as it passes through the distribution chain from the manufacturer to the final customer.

Although this calculated temperature should not be over 8 °C, some transient spikes up to 25 °C are allowed if the manufacturer so instructs. Therefore, some uncertainty about the drug efficiency still exists, and that is the reason why influence diagrams, with probability chance nodes are used for modelling risk. In this case study, a decision should be taken about the shipment of drugs with mean kinetic temperature slightly above 8°C. In that case, three possible outcomes are present directly affecting patient safety:

- drug will not lose its efficiency at all,

- drug will lose its efficiency but not provoking any harmful effect,

- drug will lose its efficiency with   provoking some harmful effect.

Taking decisions about recall of whole shipment is obviously very hard, and for that reason, besides patient safety risk, one more criteria is introduced: risk of possible competitor market takeover. This risk is represented with chance node with three possible discrete states:

- no market takeover,

- 50% market takeover,

- 100% market takeover.

The possible market takeover depends on patient safety chance node, because possible harmful effects or even drug inefficiency could provoke the customer switch to other supplier. In the case of second alternative – shipment withdrawal, the competitor access does not depend on any other factor then the market itself. Complete ID based on these assumptions is presented on Figure 7, with following notation: D – decision node with 2 alternatives: to not withdraw or withdraw the shipment; S – chance node of patient safety with three possible states; C – competitor access chance node with three possible states; R – risk value nodes for three different experts.

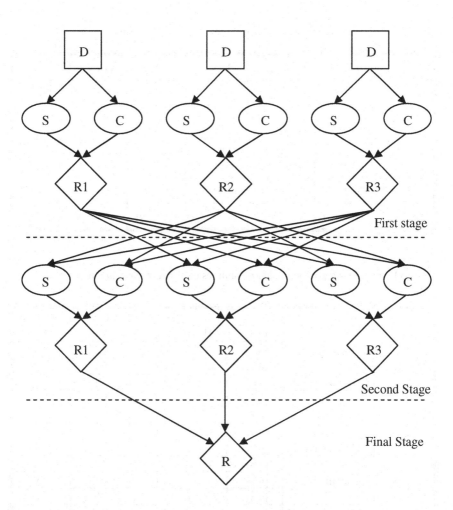

Fig. 7 Multistage ID for the supply chain risk assessment

After the first stage, when all experts perform the risk assessment separately, their assessment of possible damages, probabilities and conditional probabilities is influenced by the results of other experts in the second stage of the risk assessment process. If the difference between new and previous aggregated value is below some threshold value δ, the iterative process ends and the selection of alternative with the lowest risk is made. If not, the process returns to the previous step. Triangular fuzzy values of possible damages in different states of nature are denoted as $D(S_i)$ and $D(C_i)$ and presented in Table 2. The financial loss provoked by the shipment withdrawal is not taken into account because of its negligible values towards the possible market takeover risk.

Table 2 Fuzzy values of possible damages in different states of nature

	D (S_1)	**D(S_2)**	**D(S_3)**	**D(C_1)**	**D(C_2)**	**D(C_3)**
In mil. €	No loss of effectivity	Loss of effectivity	Harmfull effect	No market takeover	50% market takeover	100% market takeover
Expert 1	(0, 0, 0)	(0, 0, 0)	(0, 1, 2)	(0, 0, 0)	(0, 1, 2)	(1, 2, 3)
Expert 2	(0, 0, 0)	(0, 0.5, 1)	(1, 2, 3)	(0, 0, 0)	(0, 1, 2)	(1, 2, 3)
Expert 3	(0, 0, 0)	(0, 0.5, 1)	(1, 2, 3)	(0, 0, 0)	(1,1.5,3)	(2,3,4)

Table 3 Probabilities and conditional probabilities of different states in the first evaluation stage

			p(C_1\\S_i)	**p(C_2\\S_i)**	**p(C_3\\S_i)**
Expert 1	**Alternative 1**	**p(S_1)**	H	H	VL	EL
		p(S_2)	L	L	M	VL
		p(S_3)	EL	EL	VL	H
	Alternative 2	**p(C_i)**		EL	ML	M
Expert 2	**Alternative 1**	**p(S_1)**	L	H	VL	EL
		p(S_2)	ML	H	VL	EL
		p(S_3)	L	VH	EL	EL
	Alternative 2	**p(C_i)**		EL	VL	H
Expert 3	**Alternative 1**	**p(S_1)**	L	L	M	VL
		p(S_2)	M	EL	VL	H
		p(S_3)	VL	EL	EL	VH
	Alternative 2	**p(C_i)**		EL	VL	H

Probabilities and conditional probabilities are expressed by linguistic terms, with following fuzzy sets: extremely low (EL), very low (VL), medium to low (ML), medium (M), medium high (MH), high (H), very high (VH) and extremely high (EH) presented in Figure 2. Expert's opinions in the first stage of risk assessment process are presented in Table 3.

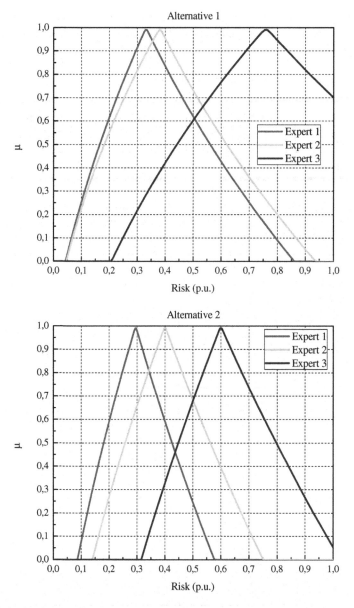

Fig. 8 Calculated values for the overall risk for alternative 1 and alternative 2

Expected value of risk for alternative 1 and 2 are calculated in MATLAB Fuzzy toolbox package. Calculated individual values of risk for both alternatives are presented in Figure 8.

Values that are defuzzified by the centroid method are presented in Table 4. These values represent the values of risk expressed in absolute values (10^6 €), while the aggregated value represents the relative value of risk in the interval [0, 1].The adopted threshold value for the difference of calculated risk between two successive iteration is $\delta = 3\%$.

Table 4 Calculated risk values in the first evaluation stage

	Expert 1	Expert 2	Expert 3
Alternative 1	1.11	1.51	2.97
Alternative 2	1.48	1.68	2.55

As stated in previous section, the *maxness* characterizes the degree to which the aggregation is like an *or* (or *and*). Therefore, when *maxness* = 0, the OWA becomes a "minimum" operator and, conversely, when *maxness* = 1, the operator becomes a "maximum" operator. In this study following OWA values are used: [0.5 0.3 0.2], with the *maxness* = 0.35.

Using the ordered weighted average operator (OWA), final risk values for alternatives 1 and 2 are given below:

$$OWA_1 = 0.5 \times 0.37 + 0.3 \times 0.51 + 0.2 \times 1 = 0.535$$

$$OWA_2 = 0.5 \times 0.58 + 0.3 \times 0.65 + 0.2 \times 1 = 0.685$$

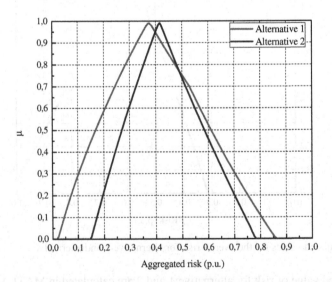

Fig. 9 Calculated values for the overall risk for alternative 1 and alternative 2

For the sake of comparison, the aggregated risks for both alternatives using FOWA are presented on figure 9. After the first stage of risk assessment, experts are reevaluating both estimated damages (Table 5) and associated probabilities (Table 6).

Table 5 Possible damages in different states of nature in the second stage

	D (S_1)	D(S_2)	D(S_3)	D(C_1)	D(C_2)	D(C_3)
In mil. €	No loss of effectivity	Loss of effectivity	Harmfull effect	No market takeover	50% market takeover	100% market takeover
Expert 1	(0, 0, 0)	(0, 0.5, 1)	(0, 1, 2)	(0, 0, 0)	(0, 1, 2)	(1, 2, 3)
Expert 2	(0, 0, 0)	(0, 0.5, 1)	(1, 2, 3)	(0, 0, 0)	(0, 1, 2)	(1, 2, 3)
Expert 3	(0, 0, 0)	(0, 0.5, 1)	(1, 2, 3)	(0, 0, 0)	(1,1.5,3)	(2,2.5,3)

Table 6 Probabilities and conditional probabilities of different states in the second evaluation stage

				p(C_1\S_i)	p(C_2\S_i)	p(C_3\S_i)
Expert 1	Alternative 1	p(S_1)	M	H	VL	EL
		p(S_2)	L	L	M	VL
		p(S_3)	VL	EL	VL	H
	Alternative 2	p(C_i)		EL	ML	M
Expert 2	Alternative 1	p(S_1)	L	M	VL	EL
		p(S_2)	ML	L	M	VL
		p(S_3)	L	EL	EL	VH
	Alternative 2	p(C_i)		EL	VL	H
Expert 3	Alternative 1	p(S_1)	L	M	L	VL
		p(S_2)	ML	L	M	VL
		p(S_3)	L	EL	VL	H
	Alternative 2	p(C_i)		EL	VL	H

Calculated risk values for both alternatives are given in Table 7.

Table 7 Calculated risk values in the second evaluation stage

	Expert 1	**Expert 2**	**Expert 3**
Alternative 1	1.68	2.13	2.68
Alternative 2	1.48	1.68	2.55

$OWA_1 = 0.5 \times 0.62 + 0.3 \times 0.79 + 0.2 \times 1 = 0.747$

$OWA_2 = 0.5 \times 0.58 + 0.3 \times 0.65 + 0.2 \times 1 = 0.685$

Table 8 Fuzzy values of possible damages in different states of nature in the third evaluation stage

	D (S$_1$)	**D(S$_2$)**	**D(S$_3$)**	**D(C$_1$)**	**D(C$_2$)**	**D(C$_3$)**
In mil. €	No loss of effectivity	Loss of effectivity	Harmfull effect	No market takeover	50% market takeover	100% market takeover
Expert 1	(0, 0, 0)	(0, 0.5, 1)	(0.5,1,2.5)	(0, 0, 0)	(0, 1, 2)	(1, 2, 3)
Expert 2	(0, 0, 0)	(0, 0.5, 1)	(1, 2, 3)	(0, 0, 0)	(0, 1, 2)	(1, 2, 3)
Expert 3	(0, 0, 0)	(0, 0.5, 1)	(1, 2, 3)	(0, 0, 0)	(1,1.5,3)	(2,2.5,3)

Table 9 Probabilities and conditional probabilities of different states in the third evaluation stage

				p(C$_1$\S$_i$)	**p(C$_2$\S$_i$)**	**p(C$_3$\S$_i$)**
Expert 1	**Alternative 1**	**p(S$_1$)**	M	H	VL	EL
		p(S$_2$)	L	L	M	VL
		p(S$_3$)	VL	EL	VL	H
	Alternative 2	**p(C$_i$)**		EL	L	MH
Expert 2	**Alternative 1**	**p(S$_1$)**	L	M	VL	EL
		p(S$_2$)	ML	L	M	VL
		p(S$_3$)	L	EL	EL	VH
	Alternative 2	**p(C$_i$)**		EL	VL	H
Expert 3	**Alternative 1**	**p(S$_1$)**	L	M	L	VL
		p(S$_2$)	ML	L	M	VL
		p(S$_3$)	L	EL	VL	H
	Alternative 2	**p(C$_i$)**		EL	VL	H

Table 10 Calculated risk values in the second evaluation stage

	Expert 1	Expert 2	Expert 3
Alternative 1	1.61	2.13	2.68
Alternative 2	1.58	1.68	2.55

$$OWA_1 = 0.5 \times 0.60 + 0.3 \times 0.79 + 0.2 \times 1 = 0.737$$

$$OWA_2 = 0.5 \times 0.58 + 0.3 \times 0.65 + 0.2 \times 1 = 0.705$$

The value of δ being too great (21.8%), the next iteration is needed, and final results are presented in Tables 8, 9, and 10. The difference between results in actual and previous iteration of 2,4 % is below the threshold value of 3% and the calculation stops. According to calculated values, the alternative 2 is chosen because of lower estimated risk.

5   Conclusion

This chapter described the methodology for the group risk assessment of complex technical systems in uncertain environment. The methodology is based on fuzzification of influence diagrams and their extension to the multi-criteria evaluation, demonstrating their use in risk management and group decision making. The main advantage of this methodology is its practicality, enabled by robust graphic tools that are not sensitive to the missing or incomplete input values, modeling the interaction among the decision makers and the multi-criteria evaluation.

Furthermore, the methodology demonstrated the possibility of influence diagrams for modeling the multistage decision processes and the interrelations among different chance and value nodes as well, enabling the iterative approach to the risk assessment. After the first, independent assessment of the group of experts, this preliminary risk grade is the input in the second step where based on known evaluations, and interaction among decision makers, the adapted risk grade has been adopted. To avoid some behavioral characteristics that are opposite to the principle of majority (group polarization and risky shift), the iterative sequential approach of decision making allowing the aversion to opinion changing, considering only aggregated risk indicators has been used. The fuzzy logic is introduced in a twofold manner: via fuzzy probability values expressed linguistically, and via fuzzy random variables. Calculation of these probabilities is performed with interval based fuzzy arithmetic.

Results presented in case studies proved that this new form of description - influence diagram with linguistic probabilities, that is both a formal description of the problem that can be treated by computers and a simple, easily understood representation of the problem can be successfully implemented for various class of risk analysis problems in complex technical systems.

In future research, methodology will be further improved using generalized, induced and unified aggregation operators. Other applications will be considered as well, analyzing in a more precise and quantitative way the relation between a group's attitude and the different parameters of aggregation operators.

Acknowledgements. This work was supported by the Ministry of Education, Science and Technological Development of the Republic of Serbia under Grant III 42006 and Grant III 44006.

References

[1] Staford, J.: Calculating the Risk of Batch Failure in the Manufacturing of Drug Products. Drug Development and Industrial Pharmacy 25(10), 1083–1091 (2006)

[2] BS 31100:2007, Code of Practice for Risk Management (2007)

[3] AS/NZS 4360, Risk management, Canberra (2004)

[4] FERMA, Risk Management Standard, Bruxelles (2002)

[5] ISO 14121, Safety of Machinery - Application of Risk Management to Medical Devices, Geneva (2007)

[6] ISO Guide 73, Risk Management – Vocabulary Guidelines for Use in Standards, Geneva (2002)

[7] Anders, G.J., Edrenyi, J., Yung, C.: Risk based planner for asset management. IEEE Comput. Applicat. Power 14(4), 718–723 (2001)

[8] Natti, S., Kezunovic, M.: A risk based decision approach for maintenance scheduling strategies for transmission system equipment. In: Proceedings of the 10th International Conference on Probabilistic Methods Applied to Power Systems, Rincon, pp. 1–6 (2008)

[9] Jiang, Y., McCalley, J., Voorhis, T.V.: Risk based resource optimization for transmission system maintenance. IEEE Trans. Power Systems 21(3), 1191–1200 (2006)

[10] Yeddanapudi, S.R.K., Li, Y., McCalley, J., et al.: Risk based allocation of distribution system maintenance resources. IEEE Trans. Power Systems 23(2), 287–295 (2008)

[11] Janjic, A., Popovic, D.: Selective maintenance schedule of distribution networks based on risk management approach. IEEE Trans. Power Systems 22(2), 597–604 (2007)

[12] Li, W.: Risk assessment of power systems: models, methods, and applications. Wiley-IEEE Press (2005)

[13] IEC 31010, Risk Management – Risk assessment techniques (2009)

[14] Dikmen, I., Birgonul, M., Han, S.: Using fuzzy risk assessment to rate cost overrun risk in international construction projects. International Journal of Project Management 25(5), 494–505 (2007)

[15] Zeng, J., An, M., Smith, N.J.: Application of a fuzzy based decision making methodology to construction project risk assessment. International Journal of Project Management 25(6), 589–600 (2007)

[16] Wang, Y.M., Elhag, T.M.S.: A fuzzy group decision making approach for bridge risk assessment. Computers and Industrial Engineering 53(1), 137–148 (2007)

[17] Zhang, G., Zou, P.X.W.: Fuzzy analytical hierarchy process risk assessment approach for joint venture construction projects in China. Journal of Construction Engineering and Management 133(10), 771–779 (2007)

[18] Nieto-Morote, A., Ruz-Vila, F.: A fuzzy approach to construction project risk assessment. International Journal of Project Management 29(2), 220–231 (2011)

[19] Karimiazari, et al.: Risk assessment model selection in construction industry. Expert Systems with Applications 38(1), 9105–9111 (2011)

[20] Butler, C.T., Rothstein, A.: On Conflict and Consensus: A Handbook on Formal Consensus Decision Making, Takoma Park (2006)

[21] Saint, S., Lawson, J.R.: Rules for Reaching Consensus: A Modern Approach to Decision Making, Jossey-Bass (1994)

[22] Kacprzyk, J., Fedrizzi, M., Nurmi, H.: Group decision making and consensus under fuzzy preferences and fuzzy majority. Fuzzy Sets and Systems 49(1), 21–31 (1992)

[23] Carlsson, C., Ehrenberg, D., Eklund, P., et al.: Consensus in distributed soft environments. European Journal of Operational Research 61(1-2), 165–185 (1992)

[24] Herrera-Viedma, E., Cabreizo, F.J., Kacprzyk, J., Pedrycz, W.: A review of soft consensus models in a fuzzy environment. Information Fusion 17(1), 4–13 (2014)

[25] Stoner, J.A.F.: A comparison of individual and group decisions involving risk. MIT School of Industrial Management Unpublished, Master's Thesis (1961)

[26] Stoner, J.A.F.: Risky and cautious shifts in group decisions: The influence of widely held values. Journal of Experimental Social Psychology 4(4), 442–459 (1968)

[27] Lea, M., Spears, R.: Computer-mediated communication, de-individuation and group decision-making. International Journal of Man-Machine Studies 34(2), 283–301 (1991)

[28] Pearl, J.: Probabilistic reasoning in intelligent systems. Networks of Plausible Inference, Morgan Kaufmann, Palo Alto, USA (1988)

[29] Jenssen, F., Nielssen, T.: Bayesian networks and decision graphs. Springer Science (2007)

[30] Howard, R., Matheson, J.: Influence Diagrams. Decision Analysis 2(3), 127–143 (2005)

[31] Shachter, R.D.: Evaluating influence diagrams. Operations Research 34(1), 871–882 (1986)

[32] Kaufmann, M., Shachter, R.D., Peot, M.A.: Decision making using probabilistic inference methods. In: Proc. 8th Conf. Uncertainty in Artificial Intelligence, pp. 276–283. Morgan Kaufmann, San Francisco (1992)

[33] Shenoy, P.P.: Valuation-based systems for Bayesian decision analysis. Operations Research 40(1), 463–484 (1992)

[34] Jensen, F.V., Dittmer, S.L.: From influence diagrams to junction trees. In: Proceedings of 10th Conf. Uncertainty in Artificial Intelligence, pp. 367–373. Morgan Kaufmann, San Francisco (1994)

[35] Zhang, N.L.: Probabilistic inference in influence diagrams. In: Proceedings of 14th Conf. Uncertainty in Artificial Intelligence, pp. 514–522. Morgan Kaufmann, San Francisco (1998)

[36] Mateou, N.H., Hadjiprokopis, A.P., Andreou, A.S.: Fuzzy Influence Diagrams: an alternative approach to decision making under uncertainty. In: Proceedings of the International Conference on Computational Intelligence for Modelling, Control and Automation, Vienna, Austria, pp. 58–64 (2005)

[37] An, N., Liu, J., Bai, Z.: Fuzzy Influence Diagrams: an approach to customer satisfaction measurement. In: Proceedings of the Fourth International Conference on Fuzzy Systems and Knowledge Discovery, Haikou, pp. 493–497 (2007)

[38] Hui, L., Ling, X.Y.: The traffic flow study based on Fuzzy Influence Diagram theory. In: Proceedings of the Second International Conference on Intelligent Computation Technology and Automation, Changsha, Hunan, pp. 845–848 (2009)

[39] Yang, C.C., Cheung, K.: Fuzzy Bayesian analysis with continuous valued evidence. In: IEEE International Conference on Systems, Man and Cybernetics, Intelligent Systems for the 21st Century, Vancouver, Canada, pp. 441–446 (1995)

[40] Yang, C.C.: Fuzzy Bayesian inference. In: IEEE International Conference on Systems, Man, and Cybernetics, Computational Cybernetics and Simulation, Orlando, USA, pp. 2707–2712 (1997)

[41] Halliwell, J., Keppens, J., Shen, Q.: Linguistic Bayesian networks for reasoning with subjective probabilities in forensic statistics. In: Proceedings of the 5th International Conference of AI and Law, Edinburgh, Scotland, pp. 42–50 (2003)

[42] Halliwell, J., Shen, Q.: Towards a linguistic probability theory. In: Proceedings of the 11th International Conference on Fuzzy Sets and Systems, Honolulu, USA, pp. 596–601 (2002)

[43] Jenkinson, R., Wang, J., Xu, D.L., Yang, J.B.: An ofshore risk analysis method using fuzzy Bayesian network. J. Offshore Mech. Arct. Eng. 131(4), 1–12 (2009)

[44] Janjic, A., Stajic, Z., Radovic, I.: A practical inference engine for risk assessment of power systems based on hybrid fuzzy influence diagrams. In: Proceedings of the 13th WSEAS International conference on Fuzzy Systems, Iasi, Romania, pp. 29–34 (2012)

[45] Watthayu, W.: Representing and Solving Influence Diagram in Multi-Criteria Decision Making: A Loopy Belief Propagation Method. In: Proceedings of the International Symposium on Computer Science and its Applications, Hobart, Australia, pp. 118–125 (2008)

[46] Diehl, M., Haimes, Y.Y.: Influence diagrams with multiple objectives and tradeoff analysis. IEEE Transactions on Systems, Man and Cybernetics, Part A: Systems and Humans 34(3), 293–304 (2004)

[47] Baas, S.M., Kwakernaak, H.: Rating and ranking of multiple-aspect alternatives using fuzzy sets. Automatica 13(1), 47–58 (1977)

[48] Chen, S., Hwang, C.L.: Fuzzy multiple attribute decision-making. Springer (1992)

[49] Cheng, C.-H.: A simple fuzzy group decision making method. In: IEEE International Conference on Fuzzy Systems, Seoul, South Korea, pp. 910–915 (1999)

[50] Delgado, M., Herrera, F., Herrera-Viedma, E., Martinez, L.: Combining Numerical and Linguistic Information in Group Decision Making. Journal of Information Sciences 107(1), 177–194 (1998)

[51] Herrera, F., Herrera-Viedma, E.: Aggregation operators for linguistic weighted information. IEEE Transactions on Systems, Man, and Cybernetics Part A: Systems and Humans 27(5), 646–656 (1997)

[52] Chiclana, F., Herrera, F., Herrera-Viedma, E.: Integrating three representation models in fuzzy multipurpose decision making based on fuzzy preference relations. Fuzzy Sets and Systems 97(1), 33–48 (1998)

[53] Chiclana, F., Herrera-Viedma, E., Herrera, F., et al.: Some induced ordered weighted averaging operators and their use for solving group decision-making problems based on fuzzy preference relations. European Journal of Operational Research 182(1), 383–399 (2007)

[54] Yager, R.R.: On ordered weighted averaging aggregation operators in multi-criteria decision making. IEEE Transactions on Systems, Man and Cybernetics 18(1), 183–190 (1988)

[55] Yager, R.R.: Fuzzy Screaning Systems. Kluwer, Dordrecht (1993)

[56] Yager, R.R.: Families of OWA operators. Fuzzy Sets and Systems 59(2), 125–148 (1993)

Consensus Modeling under Fuzziness – A Dynamic Approach with Random Iterative Steps

Pasi Luukka, Mikael Collan[*], and Mario Fedrizzi[*]

Abstract. This chapter presents a new dynamic model for consensus reaching under fuzziness that uses randomness in the modeling of the individual process iterations. Repeating the process multiple times leads to many singular (different) consensus process paths. The imprecision of the overall result, the resulting different consensus outcomes, is captured by introducing a simple process to form an overall distribution of the outcomes. The model uses a random term that is drawn from a uniform distribution to introduce randomness in the consensus reaching process, and allows for the modeling of real-world behavioral aspects of negotiations, such as negotiator "power" issues by tuning the "amount" of randomness used for each negotiation participant. The new model is numerically illustrated.

Keywords: Fuzzy preferences, Consensus modeling, Random steps iteration, Negotiation dynamics.

1   Introduction

The notion of consensus plays an important role in the theory of group decisions, particularly when the collective preference structure is generated by a dynamical process of aggregation of the single individual preference structures. In this

Pasi Luukka · Mikael Collan
Department of Business Economics and Law,
LUT School of Business, P.O. Box 20, Lappeenranta, Finland
e-mail: {pasi.luukka,mikael.collan}@lut.fi

Mario Fedrizzi
Department of Industrial Engineering,
University of Trento, Via Sommarive 9, 38123, Trento, Italy
e-mail: mario.fedrizzi@unitn.it

[*] Corresponding authors.

© Springer International Publishing Switzerland 2015
W. Pedrycz and S.-M. Chen (eds.), *Granular Computing and Decision-Making*,
Studies in Big Data 10, DOI: 10.1007/978-3-319-16829-6_8

process of aggregation each single decision maker gradually transforms his/her preference structure by combining it, through iterative weighted averaging, with the preference structures of the other decision makers. In this way the collective decision emerges dynamically, as a result of the consensual interaction among the various decision makers in the group. These issues and the importance of reaching a consensus is especially relevant, for example, for large industrial investments that tie-up large amounts of capital and may carry important risks, and whose estimation is mostly based on experts' judgments.

From the point of view of applied mathematics, the models of consensual dynamics stand in the context of multi-agent complex systems, with interactive and nonlinear dynamics. The consensual interaction among the various agents (decision makers) acts on their state variables (the preferences) in order to optimize an appropriate measure of consensus, which can be of type 'hard' (unanimous agreement within the group of decision makers) or 'soft' (partial agreement within the group of decision makers).

The problem was addressed for the first time by [1] with a mechanism for reaching a rational consensus, i.e. the question of which sequence of weights to employ in averaging of individual opinions. The consensus modeling framework introduced by Lehrer was further extended in the following 1980's by, among others, [2], [3], [4], [5], [6], [7], and [8], mostly in the probabilistic framework.

An alternative approach, based on ordinal preference aggregation was proposed by [9] and [10], as inspired by the Borda-Kendall rule (see [11]), and by introducing a distance consensus, measuring the sum for all the individuals of the number of pairs of alternatives, on which the relative position is different in the individual's and in the group's ordinal preferences.

The complexity of these approaches, in combination with the presentation of the related mathematical programming formulations, and the introduction of conditions under which the various methods yield the same consensus ranking, have been discussed in [12]. [13] provides an interesting overview of distance minimizing methods, introducing a way of measuring the degree of disagreement prevailing in the profile.

The approaches previously described can be basically classified in two groups, those assigning to each alternative and aggregated value of the votes obtained in different rank positions, and those based on the minimization of a distance measure, aiming at finding the ranking that maximizes consensus. [14] proposed an integration of the two approaches by generating a collective order from a set of individual rankings, by associating to each alternative an aggregated value of the votes received positions. The consensual solution is obtained by solving a goal programming problem.

Adopting a similar approach, some classes of consensus measures, based on metrics on weak orders and indices of contribution to consensus for each decision maker, for prioritizing them in order of their contributions to consensus, have been introduced by [15].

Other approaches to consensus reaching have been developed, assuming that individuals are expected to modify their opinions in order to increase the consensus level, using, e.g., a mediation process such as Delphi, see [16]. During this consensus process a significant amount of time and resources is used in order to move the individuals' opinions towards a shared group opinion, and therefore the problem of minimization of costs becomes relevant. The problem has been addressed by [17], assuming that individuals of unequal importance, and with a linear cost of changing their opinion are involved in the consensus reaching process, and then further extended by [18] for finding the group opinion that minimizes a quadratic cost function. A novel framework for achieving minimum cost consensus, and extending the previous approaches, has been introduced in [19] assuming that collective opinion is obtained from individual opinions, using the weighted averaging and OWA operators.

However, since decision makers typically have different and conflicting opinions, to a lesser or greater extent, the traditional strict meaning of consensus is often unrealistic. The human perception of consensus is typically 'softer', and people are generally willing to accept that consensus has been reached, when most actors agree on the preferences associated to the most relevant alternatives, or as the case is here, on the form and the size of cash-flow estimates.

Combining the fuzzy notion of consensus with the expressive power of linguistic quantifiers [20, 21] the so-called "soft consensus measure" in the context of fuzzy preference relations has been discussed in [22-27], and various interesting implications of the model in the context of decision support have been developed in [28, 29].

The soft consensus paradigm proposed in [22] was then reformulated in [30-36]. The linguistic quantifiers in the original soft consensus measure were substituted by smooth scaling functions with a similar role, and a dynamic model was obtained from the gradient descent optimization of a soft consensus cost function that combines a soft measure of collective dissensus with an individual mechanism of opinion changing aversion, or "resistance". The resulting soft consensus dynamic acts on the network of single preference structures by a combination of a collective process of dissensus, and an individual mechanism of resistance to change.

Introduced as an extension of the crisp model of consensus dynamics described in [33], the fuzzy soft consensus model in [35] substitutes the standard crisp preferences by fuzzy triangular preferences. The fuzzy extension of the soft consensus model is based on the use of a distance measure between triangular fuzzy numbers. Similarly with the standard crisp model, the fuzzy dynamics of preference change towards consensus derives from the gradient descent optimization of the new cost function of the fuzzy soft consensus model.

When consensus reaching processes are modeled with the help of linguistic scales they are most often referred to as soft consensus models, if the linguistic scales are mapped into fuzzy number scales the term used is fuzzy soft consensus, in this chapter we discuss the use of "directly" fuzzy estimates, given as triangular fuzzy numbers, and we thus call such consensus models "fuzzy consensus models". For the interested reader we suggest to refer to the surveys in [37, 38], [39], [40, 41].

It is well known that the iterative steps that negotiators take, when trying to reach a consensus, are not necessary uniform in size, or even in the direction to which the steps are taken. In fact, the size and the direction may vary, depending on the negotiation situation and the negotiators. Different negotiators may act differently in "same" situations, and thus generate different new situations. The dynamics of the iterations (or situations) of a negotiation can be understood as paths to reach consensus – as each iteration may be different, there may be many different paths, and many different consensus outcomes.

It has been the norm that modeling this randomness has been omitted in previous consensus models under fuzziness, here we present a new model that takes randomness into consideration. The model structure is based on a previous approach by Fedrizzi and others from 2008 [37] that uses a two-part formula for each iteration step, and that consists of a dissensus component that "drives" the expert judgments closer to each other, and of a resistance component that limits the speed of the process. The resistance component of the formula can be understood as the aversion of the decision-makers to change their judgments. The inclusion of randomness causes the new model results not to converge to a single solution, this is different from many previous consensus dynamics models that do not include randomness, and with a large number of iterations converge to a specific single consensus result.

In the new model we position a component within the resistance component of the model that randomizes the size of the resistance component: we use a uniform distribution to model the random size of resistance steps (random terms are drawn from uniform distributions). When uniform, or flat, distributions are used, the probability and size of possible backwards (negative) iteration steps can easily be adjusted, by adjusting the probability of large resistance terms. In fact, randomness used in the iterations can be tuned differently for each decision-maker, to allow for "behavioral aspects".

Aspects of the behavioral modeling include issues such as modeling: cultural aspects, power aspects, negotiation strategy aspects, and other aspects relevant to negotiations. For example, a high-power "leader" decision-makers may be modeled to have a larger probability to cause a backwards negotiation step than "followers", or "leaders" may be modeled to have a different "speed" of dissensus, in other words, they move their position slower towards the others negotiators' position than "follower" decision-makers. Also issues relevant to the "negotiation evolution", such as different size of resistance to change of the negotiators in different phases of the negotiation can be modeled, meaning that the further the consensus negotiation advances, the smaller the randomness can be modeled to become (as a function of the number of iterations). In this chapter we also shortly discuss what amount of randomness is practical for this type of modeling.

The rest of this chapter is organized as follows: in the next three sections we present the framework for consensus modeling processes under fuzziness, continue with the mathematical presentation of our new model, and present a technique for handling the imprecision emanating from the new process. Then we illustrate the new model with numerical simulation examples that demonstrate the effect of the introduction of randomness into the model, the effect of using different negotiation profiles for the different decision-makers, and the method used to handle the added imprecision growing from the use of randomness in the process. Finally we conclude the chapter with a short summary and a discussion.

2   The Consensus Modeling Process Under Fuzziness

Consensus models start from a set of diverging (different) estimates (opinions, preferences, or alternatives) from multiple experts (managers, decision-makers), whose estimation depends on the ideas, knowledge, attitudes, experiences, and motivations of the experts doing the estimation, and therefore the estimates are considered as individual normative judgments. The estimates are often made with regards to future events or values and therefore contain imprecision, in such cases they can be modeled with fuzzy numbers. If fuzzy numbers (possibility distributions) are used as the initial estimates and the consensus is reached in a way that the end result is also a fuzzy number, the consensus can be, as was discussed above, called fuzzy consensus.

Let us remark that when dealing with the representation of uncertain experts' judgments in decision making, triangular [42] and trapezoidal [43] shaped fuzzy numbers are the most widely used. The main reason is that their membership functions are piecewise linear functions, allow simpler calculations, and make concrete applications more easily realizable. Then, as the state of the art suggests [43-45], they seem be able to represent uncertainty effectively in the majority of real applications.

Consensual dynamics modeling is commonly based on introducing an iterative mechanism, based on a "consensus creating" algorithm that drives the experts´ (estimates), towards a consensus estimate. The process starts with the representation of each expert's initial estimates with regards to the relevant phenomenon under analysis as a fuzzy preference.

Starting from the results obtained in [33, 35, 36] we introduce the type of dynamic process for finding consensus that is used here, the process is based on using a cost function C as the driver of the consensus process that is defined as a convex linear combination of two components: a measure of collective dissensus D, and a component of opinion-changing resistance, R. Here we consider the case, where the initial expert estimates are given as triangular fuzzy numbers, and the context underlying this paper is the creation of consensual pay-off distributions [46, 47] for large industrial investments.

The process requires the difference between the expert estimates to be calculated as a distance, as the estimates are given as triangular fuzzy numbers the distance between these must be calculated. In the literature, several definitions exist for the calculation of the distance between fuzzy numbers, for example, see [48-51]. Here a distance belonging to a family of distances introduced in [48] is adopted.

Given the two triangular fuzzy numbers $\mathbf{x}=(\delta_L, x, \delta_R)$ and $\mathbf{y}=(\varepsilon_L, y, \varepsilon_R)$, where x and y are the central values and δ_L, ε_L and δ_R, δ_R are the left and right spreadsrespectively. The distance is calculated as

$$d(\mathbf{x}, \mathbf{y})=(d_L+d_R)/2 \qquad (1)$$

where d_L and d_R are computed using integrals

$$d_L = \int_0^1 (x_L(\alpha) - y_L(\alpha))^2 d\alpha, \ d_R = \int_0^1 (x_R(\alpha) - y_R(\alpha))^2 d\alpha \tag{2}$$

where the difference is calculated using left and right side α-cuts of the triangular fuzzy numbers which are given as

$$[x_L(\alpha), x_R(\alpha)] = [x - \delta_L + \delta_L \alpha, x + \delta_R - \delta_R \alpha], \ [y_L(\alpha), y_R(\alpha)] = [y - \varepsilon_L + \varepsilon_L \alpha, x + \varepsilon_R - \varepsilon_R \alpha] \tag{3}$$

for each $\alpha \in [0, 1)$.

Let us remark that d is not exactly a distance because it does not always satisfy the transitivity axiom, nevertheless for the sake of simplicity, the term distance is used when referring to d. By solving the integrals d_L and d_R we get

$$d(x, y) = \Delta_D^2 + \Delta_L^2/6 + \Delta_R^2/6 + \Delta_D(\Delta_R - \Delta_L)/2 \tag{4}$$

where $\Delta_D = x - y, \ \Delta_L = \delta_L - \varepsilon_L, \text{ and } \Delta_R = \delta_R - \varepsilon_R.$

Assuming now, for the sake of simplicity, that only two alternatives are involved, we indicate with $\boldsymbol{p}^{(i)} = (\delta_L^{(i)}, p^{(i)}, \delta_R^{(i)})$ and $\boldsymbol{p}^{(j)} = (\varepsilon_L^{(i)}, p^{(j)}, \varepsilon_R^{(j)})$ the preferences expressed by the decision makers i and j respectively. Following the dynamic consensus process introduced in [33], we determine the global dissensus measure of the group of decision maker.

$$D = \frac{1}{4}\sum_{i=1}^n C_1(i), \tag{5}$$

where $C_1(i)$ is given as

$$C_1(i) = (\sum_{j \, (\neq i)=1}^n C_1(i,j))/(n-1) \tag{6}$$

and

$$C_1(i, j) = f(d(\boldsymbol{p}^{(i)}, \boldsymbol{p}^{(j)})). \tag{7}$$

Here $f(\cdot)$ is a scaling function defined as $f(x) = x - \frac{1}{\beta}\ln(1 + e^{\beta(x-\alpha)})$, where $\alpha \in (0, 1)$ is a threshold parameter and $\beta \in (0, \infty)$ is a free parameter controlling the polarization of the sigmoid function $f'(x) = 1/(1 + e^{\beta(x-\alpha)})$. The cost function for changing the initial preference $\boldsymbol{\pi}^{(i)}$ of decision maker i into the new preference $\boldsymbol{p}^{(i)}$ is given as

$$C_2(i) = f(d(\boldsymbol{p}^{(i)}, \boldsymbol{\pi}^{(j)})) \tag{8}$$

Accordingly, the global opinion changing aversion of the group of decision makers is given by summing up the individual one to C_2 by

$$R = \frac{1}{2}\sum_{i=1}^n C_2(i) \tag{9}$$

The global cost function C can now be given as

$$C = (1-\lambda)D + \lambda R \tag{10}$$

where $\lambda \in [0,1]$ is a parameter representing the relative importance of the resistance component R in relation to the dissensus component D. The selection of a proper value for the parameter lambda, can be interpreted in a way that for "ex-ante difficult negotiations" the weight of the dissensus component D should be lower than for "ex-ante straight-forward negotiations", because component D determines a "constant speed" at which the algorithm drives the process towards consensus. We leave any further discussion about the size of lambda outside the scope of this chapter, however we observe that lambda must be larger than zero.

We can now clearly see the two-component nature of the cost function. The consensual dynamic is based on the minimization of the cost function $C(p^{(i)})=C(\delta_L^{(i)}, p^{(i)}, \delta_R^{(i)})$ through the standard gradient method. The new estimate, for any decision maker, is obtained from the previous one according to the following iterative process (index i is skipped for simplicity)

$$\mathbf{p} \rightarrow \mathbf{p^*} = \mathbf{p} - \gamma \nabla C. \tag{11}$$

where γ is the iteration step size.   More exactly, new estimate is obtained by

$$p \rightarrow p^* = p - \gamma \frac{\partial C}{\partial p}$$

$$\delta_L \rightarrow \delta_L^* = \delta_L - \gamma \frac{\partial C}{\partial \delta_L}$$

$$\delta_R \rightarrow \delta_R^* = \delta_R - \gamma \frac{\partial C}{\partial \delta_R}$$

The components of the gradient ∇C are obtained deriving C, with respect to $\delta_L^{(i)}$, $p^{(i)}$, and $\delta_R^{(i)}$. For closer formulation of the iterative process we refer to [36]. Using the above described process for reaching consensus we can create a triangular consensus estimate.

3   Extending the Consensus Algorithm with a Random Component Drawn from a Uniform Distribution

The above-presented "general" consensus reaching algorithm under fuzziness converges to the mean of the original expert estimates, when a large number of iterations is run, even though the convergence does not take place with a small number of iterations this is still not a desirable property from the model. The reason for the undesirability of the converging result is that it is not realistic, negotiations do not always have the same result, in fact if they did negotiations would not be needed.

Also, the iteration step size in the above presented algorithm is constant, this may be unrealistic, as in real life negotiations and group discussion processes the step size (the amount of how much closer to consensus the participants move during a single iteration) may vary from iteration to iteration, and sometimes even the direction of movement may change, it is not unheard of that a discussant may even reverse a position, or "back-up" in a negotiation. Differences between negotiators may stem from multiple sources such as cultural background [52, 53], different negotiation strategies [54, 55], and asymmetric information or power between the negotiators [56], and/or any combination of these and other sources. The cited references are just a drop in the ocean of negotiation related literature on these topics.

To make the algorithm more realistic, we introduce randomness into the resistance component R of the cost function C. First, let us look at the composition of the cost function in more detail. The two components of the cost function C, the dissensus component D and the resistance component R, can be considered separately due to the fact that ∇C is a linear combination of ∇D and ∇R,

$$\nabla C = (1 - \lambda)\nabla D + \lambda \nabla R \tag{12}$$

For the component D we compute

$$\frac{\partial D}{\partial p^i} = d_i \left((p^i - \bar{p}^i) + \frac{1}{4}\left(\delta_R^i - \overline{\delta_R^i} - \delta_L^i + \overline{\delta_L^i}\right) \right) \tag{13}$$

where

$$d_i = \sum_{j(\neq i)=1}^{n} \frac{d_{ij}}{n-1}, \quad d_{ij} = f'(d(p^i, p^j))$$

and

$$\bar{p}^i = \frac{\sum_{j(\neq i)=1}^{n} d_{ij} p^j}{\sum_{j(\neq i)=1}^{n} d_{ij}}$$

$$\overline{\delta_L^i} = \frac{\sum_{j(\neq i)=1}^{n} d_{ij} \delta_L^j}{\sum_{j(\neq i)=1}^{n} d_{ij}}$$

$$\overline{\delta_R^i} = \frac{\sum_{j(\neq i)=1}^{n} d_{ij} \delta_R^j}{\sum_{j(\neq i)=1}^{n} d_{ij}}$$

Analogously we compute

$$\frac{\partial D}{\partial \delta_L^i} = d_i \left(\frac{1}{6}\left(\delta_L^i - \overline{\delta_L^i}\right) - \frac{1}{4}(p^i - \bar{p}^i) \right) \tag{14}$$

$$\frac{\partial D}{\partial \delta_R^i} = d_i \left(\frac{1}{6}\left(\delta_R^i - \overline{\delta_R^i}\right) + \frac{1}{4}(p^i - \bar{p}^i) \right) \tag{15}$$

Let us next consider our "new" resistance component R, where we introduce the randomness into the process. First we have

$$\frac{\partial R}{\partial p^i} = r_i \left((p^i - \pi^i) + \frac{1}{4}\left(\delta_R^i - \theta_R^i - \delta_L^i + \theta_L^i\right) \right) \tag{16}$$

where

$$r_i = f'\left(d(p^i, \pi^i)\right),$$

and for left and right side components we have

$$\frac{\partial R}{\partial \delta_L^i} = r_i\left(\frac{1}{6}(\delta_L^i - \theta_L^i) - \frac{1}{4}(p^i - \pi^i)\right) \tag{17}$$

$$\frac{\partial R}{\partial \delta_R^i} = r_i\left(\frac{1}{6}(\delta_R^i - \theta_R^i) + \frac{1}{4}(p^i - \pi^i)\right) \tag{18}$$

Now, when we combine the two components we get

$$\frac{\partial C}{\partial p^i} = \left((1 - \lambda)d_i + \lambda r_i\right)\Delta p^i - (1 - \lambda)d_i\Delta \bar{p}^i - \lambda r_i \Delta \pi^i \tag{19}$$

where

$$\Delta p^i = p^i + \frac{1}{4}(\delta_R^i - \delta_L^i)$$

$$\Delta \bar{p}^i = \bar{p}^i + \frac{1}{4}(\bar{\delta}_R^i - \bar{\delta}_L^i)$$

$$\Delta \pi^i = \pi^i + \frac{1}{4}(\theta_R^i - \theta_L^i)$$

and by adding the random component we get this into form

$$\frac{\partial C}{\partial p^i} = \left((1 - \lambda)d_i + \lambda r_i\right)\Delta p^i - (1 - \lambda)d_i\Delta \bar{p}^i - \lambda r_i(\Delta \pi^i + \varphi^i) \tag{20}$$

where φ^i is a uniformly distributed random component, divided by the number of iterations, creating the effect of changing step size and allowing experts to (randomly) "back-up" in the negotiation. It must be noted that the index i refers to each individual expert giving their estimates, thus for each expert the random term can be assigned to be different, or individual. The backing-up effect takes place in the case that φ^i is negative and large enough (larger than the effect of the D component).

For the derivative of C, with regards to the left spread of the triangular fuzzy number, we have

$$\frac{\partial C}{\partial \delta_L^i} = \left((1 - \lambda)d_i + \lambda r_i\right)\Delta \delta_L^i - (1 - \lambda)d_i\Delta \bar{\delta}_L^i - \lambda r_i\left(\Delta \theta_L^i - \varphi^i\right) \tag{21}$$

where

$$\Delta\delta_L^i = \frac{1}{6}\delta_L^i - \frac{1}{4}p^i$$
$$\Delta\overline{\delta_L^i} = \frac{1}{6}\overline{\delta_L^i} - \frac{1}{4}\overline{p^i}$$
$$\Delta\theta_L^i = \frac{1}{6}\theta_L^i - \frac{1}{4}p^i$$

and φ^i is the random term. The derivative of C with regards to the right spread of the triangular fuzzy number is similarly

$$\frac{\partial C}{\partial \delta_R^i} = \left((1 - \lambda)d_i + \lambda r_i\right)\Delta\delta_R^i - (1 - \lambda)d_i\Delta\overline{\delta_R^i} - \lambda r_i\left(\Delta\theta_R^i + \varphi^i\right) \qquad (22)$$

where

$$\Delta\delta_R^i = \frac{1}{6}\delta_R^i - \frac{1}{4}p^i$$
$$\Delta\overline{\delta_R^i} = \frac{1}{6}\overline{\delta_R^i} - \frac{1}{4}\overline{p^i}$$
$$\Delta\theta_R^i = \frac{1}{6}\theta_R^i - \frac{1}{4}p^i$$

and φ^i is the random term, drawn from a uniform distribution.

By having introduced the random term for each expert that affects the resistance component of the cost function we have achieved a situation, where the model no longer converges to one result, and where the iteration step size is no longer constant. The non-convergence of the model is a more realistic representation of reality, however it has its own challenges, e.g., how can one treat and understand a situation, where each "run" of the consensus model may bring a different outcome. There is also the issue of "how much" randomness is suitable for realistic modeling of consensual dynamics. In the following section we shortly discuss these issues and present potential ways to handle the newly arisen challenges.

4 Handling Randomness in Consensus Dynamics and in the Results

By adding the random component into the consensus forming algorithm we create an "effect" that causes the final consensus reached to be partly random, and each reached consensus result is likely to be different from the ones created before (for the same starting situation). This is realistic, as in reality each negotiation, even with the same starting situation, may also be different from each other negotiation, if we consider a multiple iteration negotiation process with different negotiators it is likely that every negotiation "path" or process is unique and each result is most likely different from all the other results. As referred to above, the fact that we get differing outcomes can also be considered a handicap as we are trying to reach decision-supporting information with the model – if the result, each time the iterative process is run, is unique and different from all other times, the process is run the obtained result is not "stable", or in other words, the outcome is imprecise. To capture this imprecision and to make the results usable we propose the following simple three-step procedure.

First, we form (simulate, run) N different triangular consensus estimates and see that the result is a distribution of triangular fuzzy estimates, N is a large number, such as one hundred or more.

Second, we rank all the triangular consensus estimates from the smallest to the largest by applying the ranking method for fuzzy numbers proposed by Kaufman and Gupta [57].

Third, after having ranked the triangular consensus estimates, each defined as $\{a_{tri}, b_{tri}, c_{tri}\}$ we take the minimum and the maximum triangular consensus solutions and use them to construct an overall trapezoidal consensus solution so that:

If we define the trapezoid by $\{a_{tra}, b_{tra}, c_{tra}, d_{tra}\}$, and we take the trapezoid $\{a_{tra}, b_{tra}\}$ from the minimum triangular estimate by assigning the minimum triangular estimate's $\{a_{tri}, b_{tri}\}$ as the $\{a_{tra}, b_{tra}\}$ of the resulting trapezoid and similarly by assigning the maximum triangular estimate's $\{b_{tri}, c_{tri}\}$ as the $\{c_{tra}, d_{tra}\}$ of the resulting trapezoid. This way, our trapezoidal overall consensus estimate is of the general form $(\min(\delta_L), \min(p), \max(p), \max(\delta_R))$, and it includes all the triangular consensus solutions. We consider this trapezoidal consensus solution to be a stable enough consensus estimate solution for any practical purposes, while keeping in mind that the exact shape of the trapezoid may vary with different quantity (N) of simulation runs that result in random triangular consensus estimates.

The spread of the triangular consensus estimate results depends on the amount of randomness that is used in the simulations, generally, the more randomness is used the wider the end-result trapezoid, and when very little randomness is used the trapezoid core is very narrow.

In the new model the randomness term is included in the resistance component of the cost function and for each iteration the random term is drawn from a uniform distribution, the size of the random effect for each iteration is a part of the "speed of change" component ∂C (see equation 11) that is multiplied with gamma (γ) that has been set in our simulations to 0.005. This means that the effect of the randomness term for a single iteration of the algorithm is not necessarily very large, it will however accumulate, when iterations are run.

The uniform distribution, from which the randomness term is drawn, or more contextually precisely, the distribution from which the term that determines the amount and the direction of the speed of resistance to change is randomly drawn from, can be separately selected for each expert. This uniform (interval) distribution is defined by the points $\{a_{uni}, b_{uni}\}$, and it uses the same units that are used in the original estimates.

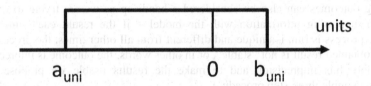

Fig. 1 Uniform distribution $\{a_{uni}, b_{uni}\}$ that uses the same units of observation in which, e.g., estimates are made and where $(a_{uni}, b_{uni}) = (-3, 1)$. Index "i" omitted for simplicity.

Figure 1 shows a uniform distribution that has been tuned for an expert, for whom it is three times as likely to draw a negative value, than a positive value. This is the expert's negotiation profile and it is denoted as $(a_{uni}, b_{uni}) = (-3, 1)$. When a positive value is drawn the expert accelerates the speed at which the process moves towards consensus, and when a negative value is drawn the expert "pushes the breaks" and decelerates the speed at which the process moves towards consensus. The larger the absolute value of the drawn value the larger is the effect. If the deceleration effect caused by the random term to the resistance component makes the effect of the resistance component to be larger than the effect of the dissensus component a single iteration will move the estimate of the expert "backwards".

The above issues are further discussed in the next chapter, where the new method is illustrated with numerical experiments.

5   Illustrating the New Method with Some Numerical Simulations

We illustrate the new model with a number of numerical experiments, where we show the effect of introducing randomness to the end result, present examples of how tuning individual experts' behavior affects the results, and illustrate how using different quantities, or magnitude, of randomness changes the end results. The starting point are the expert given initial estimates of cash-flows relevant for large industrial investments, these have been given as pay-off distributions that are treated as triangular fuzzy numbers. We used 20000 iterations (K) in each simulation, and we ran 200 simulation runs (N) for each experiment.

In the first numerical simulation, we consider two experts with different initial estimates in the beginning stage. In Figure 2, the initial estimates (black triangles) are clearly apart, after running the consensus model without any randomness the resulting consensus triangle (red) can be seen in the middle of the original estimates. When the "base model" without the random term is run, the consensus end-result converges always to the same resulting triangle.

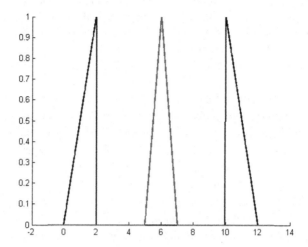

Fig. 2 Two experts´ initial estimates on the sides (black triangles) and the consensus end result (red triangle) in the middle, when no randomness is introduced to the model.

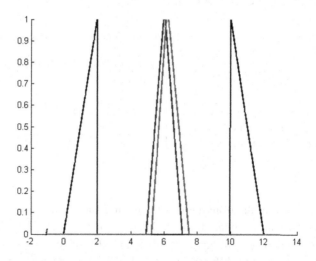

Fig. 3 Consensus trapezoids resulting from using randomness in the model, results from using the same and individual negotiation profiles for the experts. Initial estimates are visible on the sides, and resulting consensus trapezoids in the center

The second and the third numerical simulations also consider a two expert case, they present the effect of including randomness into the model, when both experts have the same negotiation profile, and the effect of considering individual negotiation profiles for the two experts.

The same negotiation profile in the second numerical simulation means specifically that the experts have the profile $(a_{uni1}, b_{uni1}) = (-1,1)$ and $(a_{uni2}, b_{uni2}) = (-1,1)$. Generally, this is $\varphi^i = U(a_{inti}, b_{inti})/K$ where for expert i $a_{inti} = -1$ and $b_{inti} = 1$ and where the K is the number of iterations applied in the computations. In the third simulation the negotiation profiles are individual for the two experts, specifically $(a_{uni1}, b_{uni1}) = (-1,1)$ and $(a_{uni2}, b_{unu2}) = (-1,3)$. Expert "b" is more drawn towards consensus as the likelihood of drawing an "accelerating" value for the random term is 3:1, while expert "a" is equally likely to draw an accelerating and a decelerating value for the random term.

In Figure 3, the two experts' initial triangular estimates can be seen to the left and to the right and the resulting trapezoidal consensus result are visible in the middle. The red trapezoid presents the consensus result, when both experts have the same negotiation profile, the blue trapezoid that is slightly wider and to the right of the red trapezoid, presents the consensus result, when individual negotiation profiles are used for the two experts.

Table 1 Consensus end-result reached with the first three simulations, two expert cases

Method	Reached consensus end result
Without any randomness (base model)	(5,6,6,7)
Uniform distribution with the same negotiation profile for both experts	(4.925,5.963,6.044,7.089)
Uniform distribution with an individual negotiation profile for both experts	(5.241,6.123,6.250,7.495)

The consensus results from the first three simulations, shown in Figures 2 and 3, are presented in a tabular form in Table 1, and clearly show the effect of introducing randomness in the algorithm and the effect of using individual negotiation profiles for different experts. Effect of increasing the amount of randomness is also examined in Table 2 and in Figure 4.

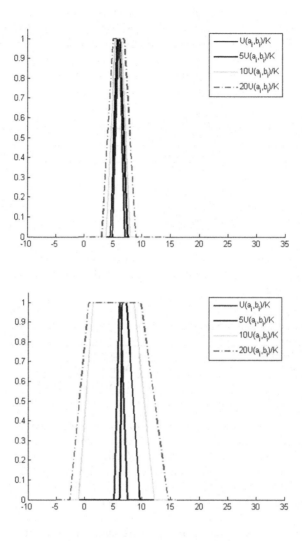

Fig. 4 The effect of increasing the magnitude of randomness in the two expert case. Above, when using the same negotiation profile. Below, when using expert specific profiles.

Table 2 Effect of using behavioral profiles, two expert cases

Magnitude of randomness	Profile 1	Profile 2
$U(a_i, b_i)/K$	(4.93,5.96,6.04,7.09)	(5.24,6.12,6.25,7.49)
$5U(a_i, b_i)/K$	(4.58,5.79,6.27,7.53)	(6.19,6.61,7.34,9.66)
$10U(a_i, b_i)/K$	(3.87,5.44,6.47,7.93)	(-1.01,1.52,8.66,12.16)
$20U(a_i, b_i)/K$	(3.09,5.05,6.98,8.96)	(-2.55,0.73,9.85,14.57)

Our fourth and fifth simulation experiment both used estimates from four experts in a setup with the randomness term. The fourth simulation experiment was run with all the experts set to have the same negotiation profile and the same random term draw, namely $\varphi^i = U(a_{unii} = -1, b_{unii} = 1)/K$. The fifth experiment used individual negotiation profiles for all four experts, these are visible in Table 3, together with the experts' initial estimates.

Table 3 Experts' behavioral profiles and initial estimates

Expert	Initial estimates	Profile 2: $U(a_i, b_i)$
1	(-4,4,16)	$U(-1,1)$
2	(-3.85,7,25)	$U(-1,3)$
3	(-1,6,19)	$U(0,1)$
4	(-8,6,20)	$U(-2,1)$

Figure 5 and Table 4 present the results of simulation experiments four and five. The reached consensus results are rather similar for the two cases, yet they are clearly different.

Table 4 Reached consensus results from simulation experiments 4 and 5, four expert cases

Behavioral profile	Reached consensus
Profile 1	(-3.00,6.38,7.51,23.45)
Profile 2	(-3.41,6.18,7.69,23.81)

The amount of randomness used in simulations four and five in the two negotiation profiles is in the same order of magnitude, we feel this explains why the obtained results are rather similar. When a large number of iterations ($K=20000$) and a relatively large number of simulation runs ($N=200$) is used, the minimum and the maximum triangles that are used for the construction of the resulting consensus trapezoid are likely to be "rather stable" for the lack of a better word, and reflect the difference in the amount of randomness in the different negotiation profiles used, there is slightly more randomness in "profile 2" and thus the resulting consensus trapezoid is slightly wider.

To study the effect of increasing randomness in the iterations we tested the two negotiation profiles with four different magnitudes of randomness (simulation experiments six and seven). The same initial estimates and users profiles were used in these experiments. Table 5 lists the different amounts of randomness used, and the obtained results for both profiles. Figure 6 presents the same results graphically.

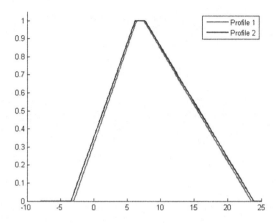

Fig. 5 Simulation experiments 4 and 5 with four different decision makers. Consensus trapezoids from two different negotiation profiles, four-expert cases

Table 5 Two behavioral profile comparison within reached consensus when magnitude of randomness is increased, four expert cases

Magnitude of randomness	Profile 1	Profile 2
$U(a_i, b_i)/K$	(-3.95,5.88,6.10,20.69)	(-3.95,5.88,6.11,20.72)
$5U(a_i, b_i)/K$	(-3.00,6.38,7.51,23.45)	(-3.41,6.18,7.69,23.81)
$10U(a_i, b_i)/K$	(-2.30,6.77,9.46,27.25)	(-1.66,7.08,9.43,27.24)
$20U(a_i, b_i)/K$	(1.23,8.62,13.39,34.93)	(1.50,8.74,13.20,34.51)

As was expected, the increase of magnitude to randomness seems to have a clearly noticeable effect on the results and the effect is larger in the simulation experiment seven, run with negotiation profile 2.

Another issue that has bearing on the end-result of the consensus result is the "location" and distribution of the initial expert estimates. The further apart the estimates are initially the wider the resulting consensus trapezoid is likely to be, ceteris paribus.

To study this issue we ran simulation experiments eight thru ten with five experts and with initial estimates more widely dispersed, or in other words further apart from each other. The used initial estimates are visible in Table 6. Simulation experiment eight was run with the randomness set to zero and simulations nine and ten used the magnitude of randomness set to $U(a_i, b_i)/K$. To further study the effect of individual negotiation profiles we ran experiment nine with the same negotiation profile for all experts, specifically the profile $\varphi^i = U(a_i = -1, b_i = 1)/K$ and experiment ten with individual profiles and in the second case with individual negotiation profiles or each expert, visible in Table 6.

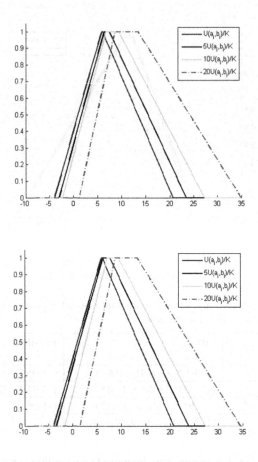

Fig. 6 The effect of increasing the magnitude of randomness to the trapezoidal consensus results with two different negotiation profiles, four expert cases

Table 6 Initial estimates used in experiments 8 - 10, and the individual negotiation profiles used in experiment 10

Expert	Initial estimates	Profile 2: $U(a_i, b_i)$
1	(0,2,2)	$U(-1,1)$
2	(10,10,12)	$U(-1,3)$
3	(-3,1,6)	$U(0,1)$
4	(7,8,9)	$U(-2,1)$
5	(3,6,7)	$U(-2,3)$

Figure 7 shows graphically the consensus results from experiments eight thru ten. When the simulation is run with the randomness term set to zero the end result is a triangular consensus and as expected using a single negotiation profile for all experts results in a narrower consensus trapezoid than using individual negotiation profiles for all experts. The results of experiments nine and ten are also numerically visible in Table 6.

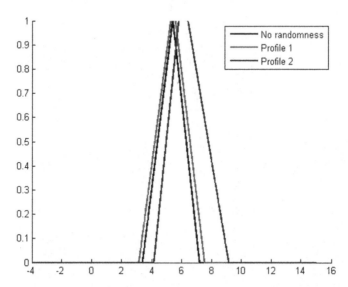

Fig. 7 Results from five experts experiment with two different profiles and five experts.

To further investigate (experiments eleven and twelve) how the larger dispersion of the initial estimates affects the results we ran some more simulations with different magnitudes of randomness for the same five expert case with the same initial estimates, and with the same two negotiation profiles.

Table 7 Two behavioral profile comparison, when the magnitude of randomness is increased, with five experts.

Magnitude of randomness	Profile 1	Profile 2
0	(3.40,5.40,5.40,7.20)	(3.40,5.40,5.40,7.20)
$U(a_i, b_i)/K$	(3.33,5.36,5.43,7.27)	(3.55,5.48,5.60,7.59)
$5U(a_i, b_i)/K$	(3.16,5.28,5.56,7.53)	(4.15,5.80,6.40,9.16)
$10U(a_i, b_i)/K$	(2.68,5.03,5.87,8.14)	(4.98,6.23,7.51,11.34)
$20U(a_i, b_i)/K$	(1.69,4.53,6.11,8.61)	(1.52,4.19,9.42,14.89)

Table 7 presents numerically the resulting consensus trapezoids and shows, as was expected, that as randomness increases the width of the consensus trapezoid increases, ceteris paribus. What is again show is also the fact that using individual negotiation profiles with more imprecision seems to also increase the width of the consensus trapezoids. The results are graphically presented in Figure 8.

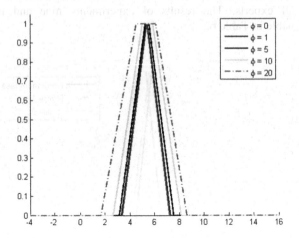

Fig. 8 Effect of increasing the magnitude of randomness, when using a uniform negotiation profile

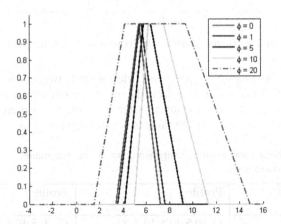

Fig. 9 Effect of increasing the magnitude of randomness, when using individual negotiation profiles for all five experts

It is visible in Figures 8 and 9 that the effect of "strong experts" in the second used negotiation profile is pulling the resulting trapezoidal consensus result to the right and the effect is accentuated when randomness increases. It seems that both increasing the dispersion of the initial estimates and increasing the amount of randomness used causes the end results, the resulting consensus trapezoids to be wider, this is according to our expectations and rather intuitive.

6   Discussion and Conclusion

This chapter has presented a new modeling approach for dynamic consensus modeling under fuzziness that is based on the two-component general construct of previous soft consensus and fuzzy soft consensus models. Models such as the one presented here are relevant, for example, when multiple experts need to find consensus relating to cash-flow estimates for long term industrial investments that are commonly done under imprecise knowledge about the future. This new approach uses a random term drawn from a uniform distribution to model the randomness of human negotiation behavior in a way that has a close resemblance to real-world negotiations. Using a random term in consensus modeling in this way is, to the best of our knowledge, a new contribution.

By adjusting the width and range of the uniform distribution from which the random term is drawn, different types of behavior can be modeled, including issues, such as negotiators sometimes wanting to "go backwards". The randomness term used also changes the iteration step size to random, and changes the step size with the number of iterations, which is also more realistic than always using a pre-defined constant iteration step size. It is realistic to expect that the closer to the final-result the experts are, the less dramatic are their iterations likely to be.

As a result of having included a random term in the process, each simulation run of the new model results in a unique consensus reaching negotiation path, which also means that any results obtained do not converge to a single triangular result – this may be considered a handicap as the resulting randomness in the result can be considered "instability". To remedy this problem we have introduced a simple procedure to create a trapezoidal overall consensus estimate from the distribution of the random triangular consensus estimates to get a "stable" overall consensus result that is representative of all consensus results, and thus usable as input in further analysis, or directly in decision-making. On the other hand this "instability" can be considered to be a more realistic representation of reality – what we see here is the relevance – precision trade-off that commonly haunts all mathematical representations of imprecise reality.

Using increasing amounts of randomness in the consensus process simulations causes the overall trapezoidal consensus estimate to be grow in width, this is in line with logic and reflects the fact that imprecision increases with more randomness. One interpretation of this is that the more volatile the negotiators are the harder it is to ex-ante estimate the consensus with precision. If the model is run for initial estimates that are far apart the resulting consensus results seem to form a wider distribution, which is also intuitive. Also the introduction of individual negotiation profiles causes differences in the end-results.

The new model was illustrated with numerical experiments that show how the final consensus result changes as parameter values are changed. Interesting research avenues for future research include, in addition to enhancing the existing mathematical models, empirical research work for the determination of natural levels for the used parameters.

References

[1] Lehrer, K.: Social Consensus and Rational Agnoiology. Synthese 31, 141–160 (1975)
[2] Kelly, F.P.: How a Group Reaches Agreement: A Stochastic Model. Mathematical Social Sciences 2, 1–8 (1981)
[3] French, S.: Consensus of opinion. European Journal of Operational Research 7, 332–340 (1981)
[4] Lehrer, K., Wagner, K.: Rational Consensus in Science and Society. Reidel, Dordrecht (1981)
[5] Sen, A.: Choice, Welfare, and Measurement. Basil Blackwell, Oxford (1982)
[6] Wagner, C.G.: Allocation, Lehrer models, and the consensus of probabilities. Theory and Decision 14, 207–220 (1982)
[7] Loewer, B.: Special Issue on Consensus. Synthese 62 (1985)
[8] Wagner, C.G.: Consensus for Belief Functions and Related Uncertainty Measures. Theory and Decision 26, 296–304 (1989)
[9] Kemeny, J.G.: Mathematics Without Numbers. Daedalus 88, 577–591 (1959)
[10] Kemeny, J.G., Snell, L.J.: Mathematical Models in the Social Sciences. MIT Press, Cambridge (1973)
[11] Kendall, M.: Rank Correlation Methods. Hafner, New York (1962)
[12] Cook, W.D.: Distance-based and ad hoc Consensus Models in Ordinal Preference Ranking. European Journal of Operational Research 172, 369–385 (2006)
[13] Nurmi, H.: Distance from Consensus: A Theme and Variations. In: Simeone, B., Pukelsheim, F. (eds.) Mathematics and Democracy, Springer, Heidelberg (2006)
[14] Contreras, I.: A Distance-based Consensus Model with Flexible Choice of Rank-position Weights. Group Decisions and Negotiation 19, 441–456 (2010)
[15] García-Lapresta, J.L., Pérez-Román, D.: Measuring Consensus in Weak Orders. In: Herrera-Viedma, E., García-Lapresta, J.L., Kacprzyk, J., Fedrizzi, M., Nurmi, H., Zadrożny, S. (eds.) Consensual Processes. STUDFUZZ, vol. 267, pp. 213–234. Springer, Heidelberg (2011)
[16] Linstone, H.A., Turoff, M.: The Delphi Method: Techniques and Applications. Addison Wesley, New Jersey (1975)
[17] Ben-Arieh, D., Easton, T.: Multi-criteria Group Consensus under Linear Cost Opinion Elasticity. Decision Support Systems 43, 713–721 (2007)
[18] Ben-Arieh, D., et al.: Minimum Cost Consensus with Quadratic Cost Functions. IEEE Transactions on Systems, Man, and Cybernetics-Part A: Systems and Humans 39, 210–217 (2008)
[19] Zhang, G., et al.: Minimum Cost Consensus Models under Aggregation Operators. IEEE Transactions on Systems, Man, and Cybernetics-Part A: Systems and Humans 41, 1253–1261 (2011)
[20] Zadeh, L.A.: The concept of a linguistic variable and its application to approximate reasoning I-II-III. Information Sciences 357, 43–80 (1975)
[21] Zadeh, L.A.: A computational approach to fuzzy quantifiers in natural languages. Computer Mathematics with Applications 9, 149–183 (1983)
[22] Kacprzyk, J., Fedrizzi, M.: Soft' consensus measures for monitoring real consensus reaching processes under fuzzy preferences. Control and Cybernetics 15, 309–323 (1986)
[23] Kacprzyk, J., Fedrizzi, M.: A "soft" measure of consensus in the setting of partial (fuzzy) preferences. European Journal of Operational Research 34, 316–325 (1988)

[24] Kacprzyk, J., Fedrizzi, M.: A human–consistent degree of consensus based on fuzzy logic with linguistic quantifiers. Mathematical Social Sciences 18, 275–290 (1989)

[25] Kacprzyk, J., et al.: Group decision making and consensus under fuzzy preferences and fuzzy majority. Fuzzy Sets and Systems 49, 21–31 (1992)

[26] Fedrizzi, M., et al.: Consensus degrees under fuzzy majorities and fuzzy preferences using OWA (ordered weighted averaging) operators. Control and Cybernetics 22, 77–86 (1993)

[27] Kacprzyk, J., et al. (eds.): Consensus under Fuzziness (International Series in Intelligent Technologies. Kluwer Academic Publishers, Dordrecht (1997)

[28] Fedrizzi, M., et al.: An interactive multi-user decision support system for consensus reaching processes using fuzzy logic with linguistic quantifiers. Decision Support Systems 4, 313–327 (1988)

[29] Carlsson, C., et al.: Consensus in distributed soft environments. European Journal of Operational Research 185, 99–100 (1992)

[30] Fedrizzi, M., et al.: Consensual dynamics in group decision making. Presented at the Proc. 6th Intl. Fuzzy Systems Association World Congress, IFSA 1995, Sao Paolo, Brazil (1995)

[31] Fedrizzi, M., et al.: Consensual dynamics in group decision making. In: 6th Intl. Fuzzy Systems Association World Congress, IFSA 1995, Sao Paolo, Brazil, pp. 145–148 (1995)

[32] Fedrizzi, M., et al.: A dynamical model for reaching consensus in group decision making. In: ACM Symposium on Applied Computing, Nashville, Tennessee, USA, pp. 493–496 (1995)

[33] Fedrizzi, M., et al.: Soft consensus and network dynamics in group decision making. International Journal of Intelligent Systems 14, 63–77 (1999)

[34] Fedrizzi, M., Fedrizzi, M., Marques Pereira, R.A.: On the issue of consistency in dynamical consensual aggregation. In: Bouchon-Meunier, B., Gutiérrez-Ríos, J., Magdalena, L., Yager, R.R. (eds.) Technologies for Contructing Intelligent Systems. STUDFUZZ, vol. 89, pp. 129–137. Springer, Heidelberg (2002)

[35] Fedrizzi, M., et al.: Consensus modelling in group decision making: a dynamical approach based on fuzzy preferences. New Mathematics and Natural Computation 3, 219–237 (2007)

[36] Fedrizzi, M., et al.: Dynamics in Group Decision Making with Triangular Fuzzy Numbers. In: Presented at the HICSS, Hawaii, USA (2008)

[37] Fedrizzi, M., et al.: Consensual Dynamics in Group Decision Making with Triangular Fuzzy Numbers. In: HICSS 41, Hawaii, USA, p. 70 (2008)

[38] Fedrizzi, M., Fedrizzi, M., Pereira, R.A.M., Brunelli, M.: The dynamics of consensus in group decision making: Investigating the pairwise interactions between fuzzy preferences. In: Greco, S., Pereira, R.A.M., Squillante, M., Yager, R.R., Kacprzyk, J. (eds.) Preferences and Decisions. STUDFUZ, vol. 257, pp. 159–182. Springer, Heidelberg (2010)

[39] Cabrerizo, F.J., et al.: Analyzing consensus approaches in fuzzy group decison making: advantages and drawbacks. Soft Computing 14, 451–463 (2010)

[40] Herrera-Viedma, E., García-Lapresta, J.L., Kacprzyk, J., Fedrizzi, M., Nurmi, H., Zadrożny, S. (eds.): Consensual Processes. STUDFUZZ, vol. 267. Springer, Heidelberg (2011)

[41] Herrera-Viedma, E., et al.: A Review of Soft Consensus Models in a Fuzzy Environment. Information Fusion 17, 4–13 (2014)

[42] Pedrycz, W.: Why triangular membership functions? Fuzzy Sets and Systems 64, 21–30 (1994)

[43] Grzegorzewski, P., Mrówka, E.: Trapezoidal approximations of fuzzy numbers. Fuzzy Sets and Systems 153, 115–135 (2005)

[44] Ban, A.I., Corianu, L.: Nearest interval, triangular approximation of a fuzzy number preserving ambiguity. International Journal of Approximate Reasoning 53, 805–836 (2012)

[45] Zeng, W., Li, H.: Weighted triangular approximations of fuzzy numbers. International Journal of Approximate Reasoning 46, 137–150 (2007)

[46] Collan, M., et al.: Fuzzy pay-off method for real option valuation. Journal of Applied Mathematics and Decision Systems (2009)

[47] Collan, M.: Valuation of Industrial Giga-Investments: Theory and Practice. Fuzzy Economic Review XVI, 21–37 (2011)

[48] Grzegorzewski, P.: Metrics and orders in space of fuzzy numbers. Fuzzy Sets and Systems 97, 83–94 (1998)

[49] Kaufman, M., Gupta, M.: Introduction to fuzzy arithmetic. Van Nostrand Reinhold, New York (1991)

[50] Tran, L., Duckstein, L.: Comparison of fuzzy numbers using a fuzzy distance measure. Fuzzy Sets and Systems, 130 (2002)

[51] Voxman, W.: Some remarks on distances between fuzzy numbers. Fuzzy Sets and Systems 100, 353–365 (1997)

[52] Graham, J.L., et al.: Explorations of Negotiation Behaviors in Ten Foreign Cultures Using a Model Developed in the United States. Management Science 40, 72–95 (1994)

[53] Kumar, R., Worm, V.: Institutional dynamics and the negotiation process: comparing India to China. International Journal of Conflict Management 15, 304–334 (2004)

[54] Hindriks, K.V., Jonker, C.M., Tykhonov, D.: Analysis of Negotiation Dynamics. In: Klusch, M., Hindriks, K.V., Papazoglou, M.P., Sterling, L. (eds.) CIA 2007. LNCS (LNAI), vol. 4676, pp. 27–35. Springer, Heidelberg (2007)

[55] Lewicki, R.J., et al.: Negotiation 3rd ed. Irwin McGraw-Hill, Boston (1999)

[56] Kim, P.H., et al.: Power Dynamics in Negotiation. Academy of Management Journal 30, 799–822 (2005)

[57] Kaufmann, M., Gupta, M.: Fuzzy Mathematical Models in Engineering and Management Science. Elsevier Science Publishers B.V (1988)

Decision Making-Interactive
and Interactive Approaches

Weixia Li and Chengyi Zhang[*]

Abstract. In this chapter, firstly, according to the problem of the consistency of reciprocal judgment matrix, two kinds of consistency recursive iterative adjustment algorithms were given.Secondly, according to the consistency problem of the fuzzy complementary judgment matrix, the definition of the scale transition matrix of the fuzzy complementary judgment matrix was given, then one method of additive consistency recursive iterative adjustment algorithms about the fuzzy complementary judgment matrix was given.Thirdly, the definition of additive consistent intuitionistic fuzzy complementary judgement matrix was given, then the addition and subtraction algorithms of intuitionistic fuzzy value representing the relative importance degree in the matrix were given, and the definition of the scale transition matrix of intuitionistic fuzzy complementary judgement matrix was given, then additive consistency recursive iterative adjustment algorithms about the intuitionistic fuzzy complementary judgement matrix was given.Meanwhile, the priority vectors formula of intuitionistic fuzzy complementary judgment matrix was introduced in this paper.Lastly, based on additive consistency recursive iterative adjustment algorithms about the intuitionistic fuzzy complementary judgement matrix, the steps of intuitionistic fuzzy analytic hierarchy process were introduced, then the method was applied in actual examples, and the effectiveness was verified.

Keywords: Consistency recursive iterative adjustment algorithms, Intuitionistic fuzzy complementary judgement matrix, AHP, IFAHP.

Weixia Li
Department of Public Health,
Hainan Medical University, College   road no.3, Haikou, China
e-mail: liweixia851019@163.com

Chengyi Zhang
Department of Mathematics and Statistics,
Hainan Normal University, Long kun south road no.99, Haikou, China
e-mail: chengyizh@hainnu.edu.cn

[*] Corresponding author.

© Springer International Publishing Switzerland 2015   219
W. Pedrycz and S.-M. Chen (eds.), *Granular Computing and Decision-Making,*
Studies in Big Data 10, DOI: 10.1007/978-3-319-16829-6_9

1   Introduction

In the field of classic decision, the analytic hierarchy process (AHP) is a multiple attribute decision making method which is very widely used. With the fuzzy thought and methods emerging in the field of decision-making, especially, when the expert subjective judgment was given in the form of intuitionistic fuzzy sets, fuzzy analytic hierarchy process (FAHP) and intuitionistic fuzzy analytic hierarchy process (IFAHP) were proposed one by one. The core of AHP, FAHP and IFAHP is the problem of consistency for the judgment matrix, so the method of consistency test and correction for the judgment matrix is very important. The traditional consistency test and correction method has some deficiencies. This study firstly puts forward two methods of consistency recursive iteration adjustment algorithm for the judgment matrix in AHP problem. These two methods not only avoid large deviation with information of the original judgment matrix, but also have great effects in understanding the information of the judgment and improving the adjustment accuracy. On this basis, one consistency recursive iterative adjustment algorithm is given for the judgment matrix in FAHP problem. From the judgment matrix by orders, we adjust judgment matrix by electing the random value of the elements in each row vector, and get the consistency matrix by orders. Then we compare the deviation between the original judgment matrix and consistency adjustment matrix, and select random value of the element which satisfies the smallest deviation. Then the consistency adjustment matrix is complete consistency by orders.

In the end, according to characteristics of the information of the analytic hierarchy process under intuitionistic fuzzy environment, the consistency recursive iteration adjustment algorithm for intuitionistic fuzzy complementary judgment matrix is designed and developed in this study, which not only can apply core effect about hesitate degrees of intuitionistic fuzzy sets, but also has important theoretical significance and application value.

2   Consistency Adjustment Algorithm of the Reciprocal Judgment Matrix

Since 1980, analytic hierarchy process (AHP) has been widely applied to solve many significant practical problems. The key problem of AHP is the consistency of judgment matrix based on the comparison between the elements. The inconsistency of judgment matrix implies that the weight obtained from the elements is not in conformity with the actual situation, ultimately, it can not give the accurate sorting of each scheme. In recent years, there are many problems about the consistency checking and adjustment of the positive reciprocal judgment matrix: For example, some papers give several consistency adjustment methods based on the relationship between the weight vector derived from consistency positive reciprocal judgment matrix and eigenvectors of judgment matrix; some papers adjust the elements of positive reciprocal judgment matrix to be consistent based on probability theory and statistical knowledge; some papers introduce the concept of the perturbation matrix, and analyze relationship among the judgment matrix, the export matrix and measure matrix to adjust the judgment matrix; Based on the optimization point of view, some papers establish the optimization model to adjust consistency of the positive reciprocal judgment matrix; some papers propose interactive analysis method for the adjustment of the judgment matrix. However, the methods of the consistency judgment seem more complex, and also are lack of a theoretical basis, so the consistency adjustment method may be incomplete consistency, even the results appear seriously inconsistency with the information contained in the original judgment matrix. As to the consistency adjustment method of the reciprocal judgment matrix, two recursive iterations adjustment algorithm are introduced, which essence is to begin with reciprocal judgment matrix by order and fix elements values randomly of row vector to adjust other elements, then make positive reciprocal judgment matrix to be consistency by order, and elect random element corresponding the minimal deviation value and the corresponding consistency adjustment matrix of this order when compared the deviation value between the adjustment matrix and the original judgment matrix, so as to give the consistency adjustments of the reciprocal judgment matrix such by-order. The method avoids large deviation value between the adjusted consistency matrix and the original judgment matrix information, and adjusted positive reciprocal judgment matrix is complete consistency. It improves the awareness and understanding of judge information, as well as the accuracy of the adjustment. Then an example is given to adjust the reciprocal judgment matrix to be consistency by using the two kinds of recursive iterative adjustment algorithm.

2.1   Preliminaries

Definition 1. Let $A = (a_{ij})_{n \times n}$ be a judgment matrix, where $a_{ij} \in R$ ($i, j \in N$), if

$$a_{ij} > 0, \quad (i, j = 1, 2, \cdots, n) \tag{1}$$

$$a_{ji} = \frac{1}{a_{ij}}, \quad (i, j = 1, 2, \cdots, n) \tag{2}$$

then A is called the positive reciprocal judgment matrix.

Definition 2. Let $A = (a_{ij})_{n \times n}$ be a positive reciprocal judgment matrix for $(i, j = 1, 2, \cdots, n)$, if for each k, $a_{ij} = a_{ik}a_{kj}$, then A is called consistency positive reciprocal judgment matrix.

Theorem 1. Let $A = (a_{ij})_{n \times n}$ be a positive reciprocal judgment matrix where $(i, j = 1, 2, \cdots, n)$, and $w = (w_i)_{1 \times n}$ be the weight vector of $A = (a_{ij})_{n \times n}$, then for $\forall k = 1, 2, \cdots, n$,

$$w_i = a_{ik} / \sum_{i=1}^{n} a_{ik} \qquad (3)$$

2.2 Consistency Recursive Iterative Adjustment Algorithm of Positive Reciprocal Judgment Matrix

2.2.1 Basic Definition and Theorem

Symbols are as follows:

(1) Let $A = (a_{ij})_{n \times n}$ be a positive reciprocal judgment matrix, then $A^{(k)}$ signifies the leading principal submatrix of order K of A.

(2) Let $A^{(k)}$ be the leading principal submatrix of order K of A, then $A_k^{(s)}$ signifies the leading principal submatrix of order s of $A^{(k)}$ where $1 \leq s \leq k$.

(3) Let $A^{(k)}$ be the leading principal sub-matrix of order K of A, then $B^{(k)}$ signifies the consistency positive reciprocal judgment matrix of $A^{(k)}$.

(4) Let $A = (a_{ij})_{n \times n}$ be a positive reciprocal judgment matrix, then $C_k^{(k-1)} = (c_{ij}^{(k-1)})_{k \times k}$ signifies the leading principal sub-matrix of order K, in which the leading principal sub-matrix of order $k-1$ is consistency positive reciprocal judgment matrix $B^{(k-1)}$ and the elements in the kth row(column) are the same as A.

Definition 3. Let $A = (a_{ij})_{n \times n}$ be positive reciprocal judgment matrix, then $A_1 = (a'_{ij})_{n \times n}$ is called column normalized matrix of A, where $a'_{ij} = \dfrac{a_{ij}}{\sum\limits_{i=1}^{n} a_{ij}}$.

Definition 4. Let $A_1 = (a'_{ij})_{n \times n}$ and $B_1 = (b'_{ij})_{n \times n}$ be column normalized matrix of positive reciprocal judgment matrixes of $A = (a_{ij})_{n \times n}$ and $B = (b_{ij})_{n \times n}$, then $E = (e_{ij})_{n \times n}$ is called the deviation matrix between A and B, where $e_{ij} = a'_{ij} - b'_{ij}$.

Definition 5. Let $A = (a_{ij})_{n \times n}$ be positive reciprocal judgment matrix, C be the consistency judgment matrix of $A = (a_{ij})_{n \times n}$, and $W = (w_1, w_2, \cdots, w_n)$ be the weight vector of C, then $D = (d_{ij})_{n \times n}$ is called the derived matrix of $A = (a_{ij})_{n \times n}$, such that $d_{ij} = a_{ij} \dfrac{w_j}{w_i}$.

Theorem 2. Let $A = (a_{ij})_{n \times n}$ be positive reciprocal judgment matrix, if A is consistency positive reciprocal judgment matrix, then all the value of elements of its derived matrix D are one, and that is $D = \begin{pmatrix} 1 & 1 & \cdots & 1 \\ 1 & 1 & \cdots & 1 \\ \vdots & \vdots & \vdots & \vdots \\ 1 & 1 & \cdots & 1 \end{pmatrix}$.

2.2.2 Consistency Recursive Iterative Adjustment Algorithm of Positive Reciprocal Judgment Matrix 1

Let $A = (a_{ij})_{n \times n}$ be positive reciprocal judgment matrix, then the consistent recursive iterative adjustment algorithm is as follows:

Step 1: $A_1 = (1)$ and $A_2 = \begin{pmatrix} 1 & a_{12} \\ a_{21} & 1 \end{pmatrix}$ which are the leading principal sub-matrix of order one and order two of A respectively are consistency positive reciprocal judgment matrix.

Step 2: Suppose for each $k > 2$, $\forall 1 \le r \le k$, $C_k^{(k-1)} = (c_{ij}^{(k-1)})_{k \times k}$, let $t_{kr}^{(km)} = a_{km} c_{mr}^{(k-1)}$ and $t_{rk}^{(km)} = 1/t_{kr}^{(km)}$, that is $T_k^m = (t_{ij}^{(km)})_{k \times k}$ ($\forall 1 \le m < k$).

Step 3: Calculate the deviate value $E_k^m = (e_{ij}^{(km)})_{k \times k}$ of $A^{(k)}$ and T_k^m.

Step 4: Determine $J_k = \{l_k \mid s_k^{l_k} = \min\{s_k^m\}\}$ such that $s_k^m = \sum_{j=1}^{k} \sum_{i=1}^{k} \left| e_{ij}^{(km)} \right|$, and let $B^{(k)} = \{T_k^{l_k} \mid l_k = \min\{J_k\}\}$.

Step 5: Let $k = k + 1$. If $k \le n$, then go to step 2. Otherwise, continue to step 6.

Step 6: let $B = B^{(k)}$, then output B.

Step 7: End.

Theorem 3. Let $A = (a_{ij})_{n \times n}$ be positive reciprocal judgment matrix, and $A^{(k-1)}$ is adjusted to $B^{(k-1)}$. If the elements of $B^{(k)}$ were recursive iterations adjusted such that $b_{kj}^{(k)} = a_{kl_k} b_{l_k j}^{(k)}$, $b_{jk}^{(k)} = 1/b_{kj}^{(k)}$, then $B^{(k)}$ is consistency positive reciprocal judgment matrix.

Proof. Firstly, since the elements of the kth row of $B^{(k)}$ satisfy $b_{ks}^{(k)} = b_{kl_k}^{(k)} b_{l_k s}^{(k)}$, $b_{sj}^{(k)} = b_{sl_k}^{(k)} b_{l_k j}^{(k)}$ for $\forall 1 \le s \le k$ and $b_{kj}^{(k)} = b_{kl_k}^{(k)} b_{l_k j}^{(k)}$, then $b_{kj}^{(k)} = b_{ks}^{(k)} b_{sj}^{(k)}$ .Moreover, since $b_{kj}^{(k)} = b_{klk}^{(k)}b_{lkj}^{(k)}$, $b_{lkj}^{(k)} = b_{lki}^{(k)}b_{ij}^{(k)}$ for each $1 \le i, j \le k-1$, then $b_{ij}^{(k)} = b_{kj}^{(k)}b_{ik}^{(k)}$.

Hence, $b_{ij}^{(k)} = b_{is}^{(k)}b_{sj}^{(k)}$ where $\forall 1 \le s \le k$. Therefore, $B^{(h)}$ is consistency positive reciprocal judgment matrix.

2.2.3   Consistency Recursive Iterative Adjustment Algorithm of Positive Reciprocal Judgment Matrix 2

Let $A = (a_{ij})_{n \times n}$ be positive reciprocal judgment matrix, then the consistent recursive iterative adjustment algorithm is as follows:

Step 1: $A_1 = (1)$ and $A_2 = \begin{pmatrix} 1 & a_{12} \\ a_{21} & 1 \end{pmatrix}$ that are consistency positive reciprocal

judgment matrix are the leading principal sub-matrix of order one and order two of A respectively.

Step 2: Suppose for each $k > 2$, $\forall 1 \le r \le k$, $C_k^{(k-1)} = (c_{ij}^{(k-1)})_{k \times k}$, let

$t_{kr}^{(km)} = a_{km}c_{mr}^{(k-1)}$ and $t_{rk}^{(km)} = 1/t_{kr}^{(km)}$, that is $T_k^m = (t_{ij}^{(km)})_{k \times k}$ ($\forall 1 \le m < k$).

Step 3: Let $A^{(k)} = (a_1^{(k)}, a_2^{(k)}, \cdots, a_k^{(k)})$ and $T_k^m = (t_1^{km}, t_2^{km}, \cdots, t_k^{km})$ where $\forall 1 \le h \le k$,

$a_h^{(k)}$ is the line vector of $A^{(k)}$ and t_h^{km} is the line vector of T_k^m , then calculate

the value $\cos \theta_{mh}^{(k)} = \dfrac{(a_h^{(k)}, t_h^{km})}{\left|a_h^{(k)}\right| \left|t_h^{km}\right|}$.

Step   4:   Determine $J_k = \{l_k \mid \cos \theta_{l_k}^{(k)} = \max\{\cos \theta_m^{(k)}\}\}$ such   that

$\cos \theta_m^{(k)} = \displaystyle\sum_{h=1}^{k} \cos \theta_{mh}^{(k)}$, and let $B^{(k)} = \{T_k^{l_k} \mid l_k = \min\{J_k\}\}$.

Step 5: Let $k = k+1$; If $k \le n$,then go to step 2. Otherwise, continue to step 6.

Step 6: let $B = B^{(k)}$, then output B .

Step 7: End.

2.3   Case

Let $A = \begin{pmatrix} 1 & 1/9 & 3 & 1/5 \\ 9 & 1 & 5 & 2 \\ 1/3 & 1/5 & 1 & 1/2 \\ 5 & 1/2 & 2 & 1 \end{pmatrix}$, then adjust A to be consistency positive reciprocal

judgment matrix by the two kinds of algorithm above and give the sorting weight vector of A .

On the one hand, by using the recursive iterations adjustment algorithm 1, we can obtain the consistency positive reciprocal judgment matrix of A as follows:
$\begin{pmatrix} 1 & 1/9 & 5/9 & 5/18 \\ 9 & 1 & 5 & 5/2 \\ 1.8 & 0.2 & 1 & 0.5 \\ 3.6 & 0.4 & 2 & 1 \end{pmatrix}$. Then by the theorem 1, we get the sorting weight vector

$w = (0.0649, 0.5844, 0.1169, 0.2338)'$ and the results are in the table 1 as following.

On the other hand, by using the recursive iterations adjustment algorithm 2, we get the consistency positive reciprocal judgment matrix of A as follows:
$\begin{pmatrix} 1 & 1/9 & 5/9 & 5/18 \\ 9 & 1 & 5 & 5/2 \\ 1.8 & 0.2 & 1 & 0.5 \\ 3.6 & 0.4 & 2 & 1 \end{pmatrix}$. Then by the theorem 1, we get the sorting weight vector

$w = (0.0613, 0.5521, 0.1104, 0.2761)'$ and the results are in the table 2 as follows.

Table 1 The results from recursive iterations adjustment algorithm 1

k	m	T_k^m	S_k^m	l_k	$B^{(k)}$
3	1	$\begin{pmatrix} 1 & 1/9 & 3 \\ 9 & 1 & 27 \\ 1/3 & 1/27 & 1 \end{pmatrix}$	3.729	2	$\begin{pmatrix} 1 & 1/9 & 5/9 \\ 9 & 1 & 5 \\ 1.8 & 1/5 & 1 \end{pmatrix}$
	2	$\begin{pmatrix} 1 & 1/9 & 5/9 \\ 9 & 1 & 5 \\ 1.8 & 1/5 & 1 \end{pmatrix}$	2.933		
4	1	$\begin{pmatrix} 1 & 1/9 & 5/9 & 1/5 \\ 9 & 1 & 5 & 1.8 \\ 1.8 & 0.2 & 1 & 0.36 \\ 5 & 5/9 & 5/18 & 1 \end{pmatrix}$	3.582	3	
	2	$\begin{pmatrix} 1 & 1/9 & 5/9 & 2/9 \\ 9 & 1 & 5 & 2 \\ 1.8 & 0.2 & 1 & 0.4 \\ 4.5 & 1/2 & 5/2 & 1 \end{pmatrix}$	3.446		$\begin{pmatrix} 1 & 1/9 & 5/9 & 5/18 \\ 9 & 1 & 5 & 5/2 \\ 1.8 & 0.2 & 1 & 0.5 \\ 3.6 & 0.4 & 2 & 1 \end{pmatrix}$
	3	$\begin{pmatrix} 1 & 1/9 & 5/9 & 5/18 \\ 9 & 1 & 5 & 5/2 \\ 1.8 & 0.2 & 1 & 0.5 \\ 3.6 & 0.4 & 2 & 1 \end{pmatrix}$	3.2		

Table 2 The results from recursive iterations adjustment algorithm 2

k	m	T_k^m	$\cos\theta_m^{(k)}$	l_k	$B^{(k)}$
3	1	$\begin{pmatrix} 1 & 1/9 & 3 \\ 9 & 1 & 27 \\ 1/3 & 1/27 & 1 \end{pmatrix}$	2.889	1	$\begin{pmatrix} 1 & 1/9 & 5/9 \\ 9 & 1 & 5 \\ 1.8 & 1/5 & 1 \end{pmatrix}$
	2	$\begin{pmatrix} 1 & 1/9 & 5/9 \\ 9 & 1 & 5 \\ 1.8 & 1/5 & 1 \end{pmatrix}$	2.962		
4	1	$\begin{pmatrix} 1 & 1/9 & 5/9 & 1/5 \\ 9 & 1 & 5 & 1.8 \\ 1.8 & 0.2 & 1 & 0.36 \\ 5 & 5/9 & 5/18 & 1 \end{pmatrix}$	3.899	2	$\begin{pmatrix} 1 & 1/9 & 5/9 & 2/9 \\ 9 & 1 & 5 & 2 \\ 1.8 & 0.2 & 1 & 0.4 \\ 4.5 & 1/2 & 5/2 & 1 \end{pmatrix}$
	2	$\begin{pmatrix} 1 & 1/9 & 5/9 & 2/9 \\ 9 & 1 & 5 & 2 \\ 1.8 & 0.2 & 1 & 0.4 \\ 4.5 & 1/2 & 5/2 & 1 \end{pmatrix}$	3.905		
	3	$\begin{pmatrix} 1 & 1/9 & 5/9 & 5/18 \\ 9 & 1 & 5 & 5/2 \\ 1.8 & 0.2 & 1 & 0.5 \\ 3.6 & 0.4 & 2 & 1 \end{pmatrix}$	3.895		

2.4 Conclusion

As to the consistency of the reciprocal judgment matrix, two kinds of consistency recursive iterative adjustment algorithm are provided. These two kinds of algorithm are complete consistency recursive iterative adjustment algorithm satisfying people's need. In the end the example is given to verify the practicality of the methods.

3 Consistency Adjustment Algorithm of Fuzzy Complementary Judgment Matrix

Analytic hierarchy process (AHP) is an effective way widely used in many fields to solve multiple goals and attribute decision making. With fuzzy theory being brought into AHP, fuzzy analytic hierarchy process(FAHP) was introduced. The core problem of FAHP is to construct fuzzy complementary judgment matrix by comparing the relative importance of the two factors, so the consistency of the judgment matrix given by experts is important. It directly affects whether sorting weight vectors of this judgment matrix can reflect actual objective sorting of compared projects. Some of the consistency judgment methods in former literatures have shortage, which ignore consistency of original judgment matrix. The situation of big deviation between the original judgment matrix and consistency adjustment matrix may appear, and the credibility of the sorting weight vectors of the consistency adjustment matrix is weak. Then some paper give the concept of consistency index for fuzzy complementary judgement matrix based on the consistency index, and adjust the judgement matrix. Some literatures provide a test for consistency that will insure a rational ordering of the normalized weights when using the methods for normalization. Some literatures give the extension principle by which the fuzzy local and global weights are determined. Then some paper propose a chi-square method (CSM) for multiplicative and fuzzy preference relations to obtain a priority vector. Some literatures select the elements which need to be adjusted by constructing the deviation matrix between the complete consistent fuzzy complementary judgement matrix and original judgement matrix, so as to adjust the fuzzy complementary judgement matrix. Some paper based on the additive consistency give the additive consistency adjustment of fuzzy complementary judgment matrix. Based on the multiplicative consistency definition of the complementary judgment matrix, this paper gives method for identifying the consistency of the fuzzy judgment matrix.

According to the problem of the consistency of fuzzy complementary judgment matrix, one consistency recursive iterative adjustment algorithm is given. From the fuzzy complementary judgment matrix by orders, we adjust judgment matrix by electing the random value of the elements in each row vector, and we get the consistency matrix by orders. Then we compare the deviation between the original judgment matrix and consistency adjustment matrix, and select random value of the element which satisfies the smallest deviation. Then the consistency adjustment matrix is complete consistency, and the deviation value between the original judgment matrix and consistency adjustment matrix is smaller. In the meantime, we discuss the situation of the scale transition, and the definition of the scale transition matrix of the fuzzy complementary judgment matrix is given. Then the consistency adjustment matrix is complete consistency, and the deviation value between the original judgment matrix and consistency adjustment matrix is smaller. Hence, it plays great role for the understanding of the judge information and the accuracy of the adjustment. At last, we apply the consistency recursive iterative adjustment algorithm to an actual fuzzy complementary judgment matrix to consistency adjustment.

3.1   Basic Definitions and Theorems

Definition 6. Let $A = (a_{ij})_{n \times n}$ be judgment matrix. If $a_{ij} \in [0,1]$, $a_{ij} + a_{ji} = 1$, and $a_{ii} = 0.5$, then A is called fuzzy complementary judgment matrix.

Definition 7. Let $A = (a_{ij})_{n\times n}$ be fuzzy complementary judgment matrix. If for each $i, j, k \in \{1, 2, \cdots, n\}$, $a_{ij} = a_{ik} - a_{jk} + 0.5$, then A is called consistency fuzzy complementary judgment matrix.

Theorem 4. Let $A = (a_{ij})_{n\times n}$ be fuzzy complementary judgment matrix, $\omega = (\omega_1, \omega_2, \cdots, \omega_n)$ be the sorting weight vector of A. If $a_{ij} = \omega_i - \omega_j + 0.5$ where $i, j \in (1, 2, \cdots, n)$, then A is called additive consistent fuzzy complementary judgment matrix.

Definition 8. Let $A = (a_{ij})_{n\times n}$ be fuzzy complementary judgment matrix and $C_k^{(k-1)} = (c_{ij}^{(k-1)})_{k\times k}$. If $I_k = \{m \mid |c_{tm}^{(k-1)} - a_{km}| < 0.5, \forall t \in \{1, 2, \cdots, k\}\}$, then I_k is the consistency index of the consistency adjustment of $A^{(k)}$ according to the elements in the kth row.

Definition 9. Let $R = (r_{ij})_{k\times k}$ and $T = (t_{ij})_{k\times k}$ be consistency fuzzy complementary judgment matrix., then $e = (\sqrt{\sum_{j=1}^{k}\sum_{i=1}^{k}(r_{ij} - t_{ij})^2})/k$ is called deviate value of R and T.

Definition 10. Let $A = (a_{ij})_{n\times n}$ be fuzzy complementary judgment matrix, then

$P = (p_i)_{n\times 1}$ is called sum and normalized vector of line where $p_i = \dfrac{\sum_{j=1}^{n} a_{ij}}{\sum_{i=1}^{n}\sum_{j=1}^{n} a_{ij}}$.

Theorem 5. Let $A = (a_{ij})_{n\times n}$ be fuzzy complementary judgment matrix, P be sum and normalized vector of line. Then A is consistency fuzzy complementary judgment matrix if and only if P is the sorting weight vector of A.

Theorem 6. Let $A = (a_{ij})_{n\times n}$ be fuzzy complementary judgment matrix and w be the sorting weight vector of A. If A is additive consistent fuzzy complementary judgment matrix, then $a_{ij} = w_i - w_j + 0.5$.

3.2 The Consistency Adjustment Algorithms of Fuzzy Complementary Judgment Matrix

3.2.1 Basic Definition and Theorem

Symbols are as follows:

(1) Let $A = (a_{ij})_{n \times n}$ be fuzzy complementary judgment matrix, then $A^{(k)}$ signifies K order master array of A.

(2) Let $A^{(k)}$ be K order master array of A, then $A_k^{(s)}$ signifies s order master array of $A^{(k)}$ where $1 \le s \le k$.

(3) Let $A^{(k)}$ be K order master array of A, then $B^{(k)}$ signifies the additive consistent fuzzy complementary judgment matrix of $A^{(k)}$.

(4) Let $A = (a_{ij})_{n \times n}$ be fuzzy complementary judgment matrix, then $C_k^{(k-1)} = (c_{ij}^{(k-1)})_{k \times k}$ signifies K order master array, which satisfied $k-1$ order master array is additive consistent fuzzy complementary judgment matrix $B^{(k-1)}$ and the elements in the kth row(column) are the same as A.

3.2.2 Scale Transition Matrix of Fuzzy Complementary Judgment Matrix

Under the 0.1-0.9 nine scales, for the consistent judgment of fuzzy complementary judgment matrix: If for each k, adjust a_{ij} to a'_{ij} such that $a'_{ij} = a_{ik} - a_{jk} + \frac{1}{2}$. If the unconsistency degree of $A = (a_{ij})_{n \times n}$ is serious, then the scale of the elements may overflow, that is $a_{ij} < 0$ or $a_{ij} > 1$. Let the additive consistent fuzzy complementary judgment matrix $A' = (a'_{ij})_{n \times n}$ be the adjustment matrix, $b = -0.3$ and $a = 1.3$, then we give the "$b-a$" scales.

Considering it is inconvenient that the relative importance scale is negative, we unify scales into positive range. Hence, we transform the adjustment matrix of additive consistent fuzzy complementary judgment matrix A' into the scale transition matrix B whose elements take for 0.1-0.9 scales such that $b_{ij} = f(a'_{ij})$

$$= \frac{0.8a'_{ij}}{a-b} + (0.1 - \frac{0.8b}{a-b}).$$

Definition 11. Let $A' = (a'_{ij})_{n \times n}$ be additive consistent fuzzy complementary judgment matrix. If $\exists i, j$ such that $a'_{ij} < 0$, then transform all the elements of

A' into the elements of $B = (b_{ij})_{n \times n}$ by $b_{ij} = f(a'_{ij}) = \dfrac{0.8a'_{ij}}{a-b} + (0.1 - \dfrac{0.8b}{a-b})$,then

B is called the scale transition matrix of A'.

Theorem 7. Let $A' = (a'_{ij})_{n \times n}$ be additive consistent fuzzy complementary judgment matrix, and $B = (b_{ij})_{n \times n}$ is the scale transition matrix of A' such that

$b_{ij} = f(a'_{ij}) = \dfrac{0.8a'_{ij}}{a-b} + (0.1 - \dfrac{0.8b}{a-b})$ .Then $B = (b_{ij})_{n \times n}$ is fuzzy complementary judgment matrix.

Proof: Since $a + b = 1$ and $a'_{ji} + a'_{ij} = 1$,   then

$$b_{ij} + b_{ji} = [\dfrac{0.8a'_{ij}}{a-b} + (0.1 - \dfrac{0.8b}{a-b})] + [\dfrac{0.8a'_{ji}}{a-b} + (0.1 - \dfrac{0.8b}{a-b})]$$

$$= \dfrac{0.8(a'_{ji} + a'_{ij})}{a-b} + 0.2 - \dfrac{1.6b}{a-b}$$

$$= \dfrac{0.8}{a-b} + 0.2 - \dfrac{1.6b}{a-b} \quad = \dfrac{0.8(1-2b)}{1-2b} + 0.2 = 1 \quad . \quad \text{For} \quad a'_{ij} = 0.5 \quad , \quad \text{then}$$

$$b_{ii} = f(a'_{ii}) = \dfrac{0.8a'_{ii}}{a-b} + (0.1 - \dfrac{0.8b}{a-b}) = \dfrac{0.8 \times 0.5}{a-b} - \dfrac{0.8b}{a-b} + \dfrac{0.1a - 0.1b}{a-b}$$

$$= \dfrac{0.4(a+b)}{a-b} - \dfrac{0.8b}{a-b} + \dfrac{0.1a - 0.1b}{a-b} = 0.5 \ . \ \text{Hence,} \ B \ \text{is fuzzy complementary}$$

judgment matrix.

Theorem 8. Let $A' = (a'_{ij})_{n \times n}$ be additive consistent fuzzy complementary judgment matrix, and $B = (b_{ij})_{n \times n}$ is the scale transition matrix of A' such that

$b_{ij} = f(a'_{ij}) = \dfrac{0.8a'_{ij}}{a-b} + (0.1 - \dfrac{0.8b}{a-b})$ .Then $B = (b_{ij})_{n \times n}$ is additive consistent fuzzy complementary judgment matrix.

Proof: Since $a'_{ij} = a_{ik} - a_{jk} + \dfrac{1}{2}$,then $a_{ik} - a_{jk} = a'_{ij} - \dfrac{1}{2}$ and $a + b = 1$.Thus,

$$b_{ik} - b_{jk} + \dfrac{1}{2} = [\dfrac{0.8a_{ik}}{a-b} + (0.1 - \dfrac{0.8b}{a-b})] - [\dfrac{0.8a_{jk}}{a-b} + (0.1 - \dfrac{0.8b}{a-b})] + \dfrac{1}{2}$$

$$= \dfrac{0.8(a_{ik} - a_{jk})}{a-b} + \dfrac{1}{2} = \dfrac{0.8(a_{ij} - 0.5)}{a-b} + \dfrac{1}{2} = \dfrac{0.8a_{ij}}{a-b} + 0.1 + 0.4 - \dfrac{0.4}{a-b}$$

$$= \frac{0.8a_{ij}}{a-b} + 0.1 + \frac{0.8b}{a-b} = b_{ij} \quad , \quad \text{hence,} \quad B \text{ is additive consistent fuzzy}$$

complementary judgment matrix.

3.2.3 Consistency Recursive Iterative Adjustment Algorithm of Fuzzy Complementary Judgment Matrix

Let $A = (a_{ij})_{n \times n}$ be fuzzy complementary judgment matrix, then additive consistent adjustment algorithm is as follows:

Step 1: $A_1 = (1)$ and $A_2 = \begin{pmatrix} 1 & a_{12} \\ a_{21} & 1 \end{pmatrix}$ that are consistent fuzzy complementary judgment matrix are the one order master array and the two order master array of A respectively.

Step 2: Suppose for $k > 2$, $h = 1, 2, \cdots, k-1$, all the $A_k^{(h)}$ have been adjusted to additive consistent fuzzy complementary judgment matrix. Then we adjust the elements of the $k th$ row of $C_k^{(k-1)} = (c_{ij}^{(k-1)})_{k \times k}$ whose $k-1$ order master array equals $B^{(k)}$. If $I_i \neq \phi$ ($i = 1, 2, \cdots, k-1$), then $t_{kt}^{(km)} = a_{km} - c_{mt}^{(k-1)} + 0.5$, $t_{tk}^{(km)} = 1 - t_{kt}^{(km)}$ where $t \in \{1, 2, \cdots, k\}$, continue to Step 3. Otherwise, if $\exists 1 \leq h \leq k-1$, and $I_h = \phi$, then go to Step 7.

Step 3: If $I_k \neq \phi$, then let $T_k^m = (t_{ij}^{(km)})_{k \times k}$ where $m \in I_k$ and calculate deviate value e_k^m of $A^{(k)}$ and T_k^m, continue to Step 4. Otherwise, go to Step 5.

Step 4: Determine $J_k = \{l_k \mid l_k \in I_k\}$ such that $e_k^{l_k} = \min\{e_k^m\}$, then let $B^{(k)} = \{T_k^{l_k} \mid l_k = \min_{l_k \in J_k}(l_k)\}$ and go to Step 8.

Step 5: Let $T_k^m = (b_{ij}^{(km)})_{k \times k}$ where $m \in \{1, 2, \cdots, k\}$, then calculate deviate value e_k^m of $A^{(k)}$ and T_k^m.

Step 6: Determine $J_k = \{l_k \mid l_k \in I_k\}$ such that $e_k^{l_k} = \min\{e_k^m\}$, then get the scale transition matrix $S_k^{l_k}$ of $T_k^{l_k}$. Let $B^{(k)} = \{S_k^{l_k} \mid l_k = \min_{l_k \in J_k}(l_k)\}$, then go to Step 8.

Step 7: If $\exists 1 \leq h \leq k-1$ such that $I_h = \phi$, then get $b_{kt}^{(km)} = f(a_{km}) - c_{mt}^{(k-1)} + 0.5$, $b_{tk}^{(km)} = 1 - b_{kt}^{(km)}$ by the scale transition formula. Calculate the scale transition matrix $S^{(k)}$ of $A^{(k)}$, and let $A^{(k)} = S^{(k)}$, then go to Step 3.

Step 8: Let $k = k+1$; If $k \leq n$, then go to Step 2. Otherwise, continue to Step 9.

Step 9: let $B = B^{(k)}$, then output B.

Step 10: End.

Theorem 9. Let $A = (a_{ij})_{n \times n}$ be fuzzy complementary judgment matrix, and $B^{(h)}$ be the adjustment matrix by consistency recursive iterative adjustment algorithm such that $b_{hj}^{(h)} = a_{hr_h} - b_{jr_h}^{(j)} + \dfrac{1}{2}$ and $b_{jh}^{(j)} = 1 - b_{hj}^{(h)}$ where $\forall 2 \leq h \leq n$ and $\forall 2 \leq j \leq h-1$. Then $B^{(h)}$ is additive consistent fuzzy complementary judgment matrix.

Proof: According to consistency recursive iterative adjustment algorithm above, we get

$b_{ij}^{(i)} = a_{ir_i} - b_{jr_i}^{(j)} + \dfrac{1}{2}$ where $1 \leq k \leq n$. Since $b_{ik}^{(i)} = a_{ir_i} - b_{kr_i}^{(k)} + \dfrac{1}{2}$ for each $1 \leq k \leq n$,

then $a_{ir_i} = b_{ik}^{(i)} + b_{kr_i}^{(k)} - \dfrac{1}{2}$. Thus, $b_{ij}^{(i)} = (b_{ik}^{(i)} + b_{kr_i}^{(k)} - \dfrac{1}{2}) - b_{r_i j}^{(r_i)} + \dfrac{1}{2} = b_{ik}^{(i)} - b_{jk}^{(j)} + \dfrac{1}{2}$

where $1 \leq k \leq h$, hence, $B^{(h)}$ is additive consistent fuzzy complementary judgment matrix.

3.2.4 Case

Let the original judgment matrix $A = \begin{pmatrix} 0.5 & 0.1 & 0.6 & 0.7 \\ 0.9 & 0.5 & 0.8 & 0.4 \\ 0.4 & 0.2 & 0.5 & 0.9 \\ 0.3 & 0.6 & 0.1 & 0.5 \end{pmatrix}$, then we adjust it by

the consistency recursive iterative adjustment algorithm and get the results in table 3, then we get the sorting weight vectors.

Table 3 Adjust the fuzzy complementary judgment matrix by the algorithm

k	m	T_k^m	I_k	e_k^m	l_k	$B^{(k)}$
3	1	$\begin{pmatrix} 0.5 & 0.1 & 0.6 \\ 0.9 & 0.5 & 1 \\ 0.4 & 0 & 0.5 \end{pmatrix}$	$\{2\}$	—	2	$\begin{pmatrix} 0.5 & 0.1 & 0.4 \\ 0.9 & 0.5 & 0.8 \\ 0.6 & 0.2 & 0.5 \end{pmatrix}$
	2	$\begin{pmatrix} 0.5 & 0.1 & 0.4 \\ 0.9 & 0.5 & 0.8 \\ 0.6 & 0.2 & 0.5 \end{pmatrix}$		2		
4	1	$\begin{pmatrix} 0.5 & 0.1 & 0.4 & 0.7 \\ 0.9 & 0.5 & 0.8 & 1.1 \\ 0.6 & 0.2 & 0.5 & 0.8 \\ 0.3 & -0.1 & 0.2 & 0.5 \end{pmatrix}$	ϕ	0.26	1	$\begin{pmatrix} 0.5 & 0.3 & 0.45 & 0.6 \\ 0.7 & 0.5 & 0.65 & 0.8 \\ 0.55 & 0.35 & 0.5 & 0.65 \\ 0.4 & 0.2 & 0.35 & 0.5 \end{pmatrix}$
	2	$\begin{pmatrix} 0.5 & 0.1 & 0.4 & 0 \\ 9 & 0.5 & 0.8 & 0.4 \\ 0.6 & 0.2 & 0.5 & 0.1 \\ 1 & 0.6 & 0.9 & 0.5 \end{pmatrix}$		0.38		
	3	$\begin{pmatrix} 0.5 & 0.1 & 0.4 & 0.8 \\ 0.9 & 0.5 & 0.8 & 1.2 \\ 0.6 & 0.2 & 0.5 & 0.9 \\ 0.2 & -0.2 & 0.1 & 0.5 \end{pmatrix}$		0.29		

Hence, we get the additive consistent fuzzy complementary judgment

matrix $\begin{pmatrix} 0.5 & 0.3 & 0.45 & 0.6 \\ 0.7 & 0.5 & 0.65 & 0.8 \\ 0.55 & 0.35 & 0.5 & 0.65 \\ 0.4 & 0.2 & 0.35 & 0.5 \end{pmatrix}$, then we get the sorting weight vectors

$w = (0.2312 \quad 0.3313 \quad 0.2563 \quad 0.1812)'$.

3.2.5 Conclusion

One consistency recursive iterative adjustment algorithm which resolved consistency problem of the fuzzy complementary judgment matrix is given. From the fuzzy complementary judgment matrix by orders, we adjust judgment matrix by electing the random value of the elements in each row vector, and get the consistency matrix by orders. Then we select random value of the element which satisfies the smallest deviation. Then the consistency adjustment matrix is complete consistency, and the deviation value between the original judgment matrix and consistency adjustment matrix is smaller. In the meantime, we discuss the situation of the scale transition, and the definition of the scale transition matrix of the fuzzy complementary judgment matrix is given. At last, we use the actual example to verify the effectiveness of the adjustment algorithms.

4　Consistency Adjustment Algorithm of Intuitionistic Fuzzy Complementary Judgment Matrix

Since 1980, the decision-making theory of analytic hierarchy process(AHP) and fuzzy analytic hierarchy process (FAHP) have been more and more practically applied. As the generalization of fuzzy sets, intuitionistic fuzzy set (IFS) is more practical, scientific and reasonable than fuzzy value in solving the fuzziness and the uncertainty problems, but it makes calculation more complicated, so there are few research of the IFAHP. At present, there are few the research paper about the IFCJM. The definition of IFCJM is introduced in some paper, then ACIFCJM and multiplicative consistent intuitionistic fuzzy complementary judgement matrix (MCIFCJM) are given, but there are few research of consistency adjustment algorithms of IFCJM. In fact, the reason is that the elements of the IFCJM are intuitionistic fuzzy value, but the operation process of this kind of matrix is very complex. Thus, the method of consistency adjustment is difficult, then the theory about intuitionistic fuzzy analytic hierarchy process is few. One kind of consistency adjustment method has been given in some paper, and this method was based on the approximation theory in a paper and inversion theory between the IFS and the fuzzy set. Although a paper has resolved the consistency adjustment problem of IFCJM, this method isn't based on the proper of IFCJM itself.Hence, We consider directly dealing with strict consistency adjustment, then we apply it in IFAHP, and open a new way for the research and application of AHP under the intuitionistic fuzzy environment. the definition of additive consistent intuitionistic fuzzy complementary judgement matrix (ACIFCJM) is given; The addition and subtraction algorithms of intuitionistic fuzzy value representing the relative importance degree in the matrix are given, then the definition of the scale transition matrix of intuitionistic fuzzy complementary judgement matrix (IFCJM) is given; The additive consistency recursive iterative adjustment algorithm about the IFCJM is given, then priority vectors formula of IFCJM is introduced; At last, the steps of intuitionistic fuzzy analytic hierarchy process (IFAHP) are introduced, then the method is applied in actual examples, and its effectiveness is verified.

4.1 IFCJM and Its Propers

Definition 12. Let $A = (a_{ij})_{n \times n}$ be a judgement matrix, where $a_{ij} = (t_{ij}, f_{ij}, \pi_{ij})$

$(i, j \in N)$, if $t_{ij} \in [0,1]$, $f_{ij} \in [0,1]$, $t_{ji} = f_{ij}$, $\pi_{ji} = \pi_{ij}$, $t_{ii} = f_{ii} = 0.5$,

$t_{ij} + f_{ij} \leq 1$, then A is called IFCJM.

Note: The meaning of the elements $a_{ij} = (t_{ij}, f_{ij}, \pi_{ij})$ is as follows: t_{ij} represents importance degree of x_i relative to x_j , f_{ij} represents importance degree of x_j relative to x_i and π_{ij} represents the uncertainty importance degree of x_i relative to x_j. The scale of the elements t_{ij} and f_{ij} takes for 0.1-0.9 nine scales.

Definition 13. Let $A = (a_{ij})_{n \times n}$ be an IFCJM where $a_{ij} = (t_{ij}, f_{ij}, \pi_{ij})$, if for each k , $a_{ij} = a_{ik} - a_{jk}$, such that $t_{ij} = t_{ik} - t_{jk} + 0.5$, $f_{ij} = f_{ik} - f_{jk} + 0.5$ and $\pi_{ij} = \pi_{ik} - \pi_{jk}$, then A is called ACIFCJM.

Definition 14. Let $A = (a_{ij})_{n \times n}$ be an IFCJM, a_{kj} and a_{ij} be the elements of A , then $a_{ij} - a_{kj} = (t_{ij} - t_{kj} + 0.5, f_{ij} - f_{kj} + 0.5, \pi_{ij} - \pi_{kj})$ is called the subduction and $a_{ij} + a_{kj} = (t_{ij} + t_{kj} - 0.5, f_{ij} + f_{kj} - 0.5, \pi_{ij} + \pi_{kj})$ is called addition.

Note: let a_{ij} be adjusted for b_{ij} such that $b_{ij} = a_{ik} - a_{jk}$ where $b_{ij} = (tb_{ij}, fb_{ij})$. If the unconsistency degree of $A = (a_{ij})_{n \times n}$ is serious, then the scale of the elements may overflow 0.1-0.9 scales, so that $-0.3 \leq tb_{ij} \leq 1.3, -0.3 \leq fb_{ij} \leq 1.3$, $-0.8 \leq \pi b_{ij} \leq 0.8$.

Definition 15. Let $b_{ij} = (tb_{ij}, fb_{ij})$ be intuitionistic fuzzy value, then $a_{ij} = f(b_{ij})$ is called the conversion equation from b_{ij} to $a_{ij} = (t_{ij}, f_{ij}, \pi_{ij})$ such that

$$\begin{cases} t_{ij} = 0.1 + (tb_{ij} + 0.3)/3 \\ f_{ij} = 0.1 + (fb_{ij} + 0.3)/3 \\ \pi_{ij} = 1 - t_{ij} - f_{ij} \end{cases}$$

Definition 16. Let $B = (b_{ij})_{n \times n}$ be ACIFCJM where $b_{ij} = (tb_{ij}, fb_{ij})$. If $tb_{ij} < 0$ or $fb_{ij} < 0$ or $tb_{ij} > 1$ or $tb_{ij} > 1$, then all the elements of B are adjusted for

$$P_{ij} = (tp_{ij}, fp_{ij}, \pi p_{ij}) \text{ of } P = (p_{ij})_{n \times n} \text{ such that } \begin{cases} tp_{ij} = 0.1 + (tb_{ij} + 0.3)/3 \\ fp_{ij} = 0.1 + (fb_{ij} + 0.3)/3 \\ \pi p_{ij} = 1 - t_{ij} - f_{ij} \end{cases} \cdot \text{ Then}$$

P is called the scale transition matrix of B.

Theorem 10. Let $A = (a_{ij})_{n \times n}$ be an IFCJM where $a_{ij} = (t_{ij}, f_{ij}, \pi_{ij})$, B be the ACIFCJM of A, P be the scale transition matrix of B, then the equally important degree $e_b = (0.5, 0.5, 0)$ of x_i relatives to x_j is transformed into $e_p = (11/30, 11/30, 8/30)$ of x_i relatives to x_j.

Theorem 11. Let $A = (a_{ij})_{n \times n}$ be an IFCJM where $a_{ij} = (t_{ij}, f_{ij}, \pi_{ij})$, B be the ACIFCJM of A, P be the scale transition matrix of B, then P is an ACIFCJM.

4.2 The Consistency Adjustment Algorithm

4.2.1 Basic Definition and Theorem

Symbols are as follows:

(1) Let $A = (a_{ij})_{n \times n}$ be an IFCJM where $a_{ij} = (t_{ij}, f_{ij}, \pi_{ij})$, then $A^{(k)}$ signifies the leading principal submatrix of order K of A.

(2) Let $A^{(k)}$ be the leading principal submatrix of order K of A, then $A_k^{(s)}$ signifies the leading principal submatrix of order s of $A^{(k)}$ where $1 \le s \le k$.

(3) Let $A^{(k)}$ be the leading principal submatrix of order K of A, then $B^{(k)}$ signifies the ACIFCJM of $A^{(k)}$.

(4) Let $A = (a_{ij})_{n \times n}$ be an IFCJM, then $C_k^{(k-1)} = (c_{ij}^{(k-1)})_{k \times k}$ signifies the leading principal submatrix of order K, which satisfied the leading principal submatrix of order $k-1$ is ACIFCJM $B^{(k-1)}$ and the elements in the kth row(column) are the same as A.

Definition   17.   Let $A = (a_{ij})_{n \times n}$ be   an   IFCJM, $B^{(k-1)}$ be   the

ACIFCJM   of $A^{(k-1)}$.   let $C_k^{(k-1)} = ([tc_{ij}^{(k-1)}, fc_{ij}^{(k-1)}])_{k \times k}$,   if

$I_k = \{m \mid |tc_m^{(k-1)} - t_{km}| < 0.5, |fc_m^{(k-1)} - f_{km}| < 0.5, \text{ and } -0.9 \le (tc_m^{(k-1)} - t_{km}) + (fc_m^{(k-1)} - f_{km}) \le 0, \forall t \in \{1,2,\cdots,k\}\}$,   then

I_k is called the consistency index of the consistency adjustment of $A^{(k)}$

according to the elements in the kth row.

Definition 18. Let $R = ([tr_{ij}, fr_{ij}])_{k \times k}$ and $P = ([tp_{ij}, fp_{ij}])_{k \times k}$ be consistent IFCJM,

then $e = (\sum\limits_{i=1}^{k} \sum\limits_{j=1}^{k} (|tr_{ij} - tp_{ij}| + |fr_{ij} - fp_{ij}|))/k$ is called deviate value of R and P.

4.2.2   The Additive Consistent Adjustment Algorithm of IFCJM

Let $A = (a_{ij})_{n \times n}$ be IFCJM where $a_{ij} = (t_{ij}, f_{ij}, \pi_{ij})$, then the additive
consistent adjustment algorithm is as follows:

Step 1: $A_1 = (1)$ and $A_2 = \begin{pmatrix} 1 & a_{12} \\ a_{21} & 1 \end{pmatrix}$ that are ACIFCJM are the leading principal

submatrix of order one and order two of A respectively.

Step 2: Suppose for each $k > 2$, $h = 1, 2, \cdots, k-1$, all the $A_k^{(h)}$ have been
adjusted for ACIFCJM. Then we adjust the elements in the kth row of
$C_k^{(k-1)} = ([tc_{ij}^{(k-1)}, fc_{ij}^{(k-1)}])_{k \times k}$ whose leading principal submatrix of order $k-1$
equals $B^{(k)}$. If $I_i \neq \phi$ ($i = 1, 2, \cdots, k-1$) , then $tb_{kt}^{(km)} = t_{km} - tc_{mt}^{(k-1)} + 0.5$,
$fb_{kt}^{(km)} = f_{km} - fc_{mt}^{(k-1)} + 0.5$, $\pi b_{kt}^{(km)} = \pi_{km} - \pi c_{mt}^{(k-1)}$ and $tb_{tk}^{(km)} = fb_{kt}^{(km)}$,
$fb_{tk}^{(km)} = tb_{kt}^{(km)}$ where $t \in \{1, 2, \cdots, k\}$. Continue to step 3. Otherwise, if
$\exists 1 \leq h \leq k-1$, such that $I_h = \phi$, then go to step 7.

Step 3: If $I_k \neq \phi$, then let $T_k^m = (t_{ij}^{(km)})_{k \times k}$ where $m \in I_k$ and calculate deviate

value e_k^m of $A^{(k)}$ and T_k^m. Continue to step 4. Otherwise, go to step 5.

Step 4: Determine $J_k = \{l_k \mid l_k \in I_k\}$ such that $e_k^{l_k} = \min\{e_k^m\}$, then let
$B^{(k)} = \{T_k^{l_k} \mid l_k = \min\limits_{l_k \in J_k}(l_k)\}$ and go to step 8.

Step 5: Let $T_k^m = (b_{ij}^{(km)})_{k \times k}$ where $m \in \{1, 2, \cdots, k\}$, then calculate deviate value

e_k^m of $A^{(k)}$ and T_k^m .

Step 6: Determine $J_k = \{l_k \mid l_k \in I_k\}$ such that $e_k^{l_k} = \min\{e_k^m\}$, then get
the scale transition matrix $S_k^{l_k}$ of $T_k^{l_k}$. Let $B^{(k)} = \{S_k^{l_k} \mid l_k = \min\limits_{l_k \in J_k}(l_k)\}$, then go to
step 8.

Step 7: If $\exists 1 \leq h \leq k-1$ such that $I_h = \phi$, then get $b_{km} = f(a_{km})$, $tb_{kt}^{(km)} = ta_{km} - tc_{mt}^{(k-1)} + 0.5$, $fb_{kt}^{(km)} = fa_{km} - fc_{mt}^{(k-1)} + 0.5$, and $tb_{tk}^{(km)} = fb_{kt}^{(km)}$, $fb_{tk}^{(km)} = tb_{kt}^{(km)}$ by the scale transition formula. Calculate the scale transition matrix $S^{(k)}$ of $A^{(k)}$, and let $A^{(k)} = S^{(k)}$, then go to step 3.

Step 8: Let $k = k+1$; If $k \leq n$, then go to step 2. Otherwise, continue to step 9.

Step 9: let $B = B^{(k)}$, then output B.

Step 10: End.

4.2.3   Priority Vectors Formula of IFCJM

Definition 19. Let $B = (b_{ij})_{n \times n}$ be consistent IFCJM where $b_{ij} = (t_{ij}, f_{ij}, \pi_{ij})$, and $e = (t_e, f_e, \pi_e)$ represents the equal importance degree of x_i relative to x_j, then $b = (b_i)_{n \times 1}$ is called sum and normalized vector of line where $b_i = (t_i, f_i)$,

$$t_i = (\sum_{j=1}^{n}(t_{ij} - t_e))/n + t_e, \ f_i = (\sum_{j=1}^{n}(f_{ij} - f_e))/n + f_e, \ i = 1, 2, \cdots, n.$$

Theorem 12. Let $x = < u_A(x), v_A(x) >$ be intuitionistic fuzzy number of IFS A, then $E_\lambda(x) = u_A(x) - v_A(x) + (2\lambda - 1)\pi_A(x)$ which is based on the risk preference coefficient is sorting function of the intuitionistic fuzzy number, where λ is decision makers' risk preference coefficient and $\lambda \in [0,1]$.

Definition 20 let $B = (b_{ij})_{n \times n}$ be consistent IFCJM, $b = (b_i)_{n \times 1}$ be sum and normalized vector of line of B where $b_i = (t_i, f_i)$. Then $c = (c_i)_{n \times 1}$ which is based on the risk preference coefficient is called sorting of intuitionistic fuzzy number of B where $c_i = E_\lambda(b_i) = u_{b_i} - v_{b_i} + (2\lambda - 1)\pi_{b_i}$.

Definition 21 lets $B = (b_{ij})_{n \times n}$ be consistent IFCJM, $c = (c_i)_{n \times 1}$ be sorting of intuitionistic fuzzy number of $B = (b_{ij})_{n \times n}$, then $W = (w_i)_{n \times 1}$ is weight ordering vector of B and satisfies as follows:

(1)   If $\exists c_i \leq 0$, then determine $R = (r_i)_{n \times 1}$ such that
$r_i = 0.8 c_i / (\max_i(c_i) - \min_i(c_i)) + (0.1 - 0.8 \min_i(c_i) / (\max_i(c_i) - \min_i(c_i)))$

and $w_i = r_i / \sum_{i=1}^{n} r_i$.

(2)   If $\forall c_i > 0$, then $w_i = c_i / \sum_{i=1}^{n} c_i$.

4.3 Analytic Hierarchy Process under the Intuitionistic Fuzzy Environment and Its Application

4.3.1 Intuitionistic Fuzzy Analytic Hierarchy Process

Suppose $V = (v_{ij})_{n \times n}$ as IFCJM, it is usually given by decision maker's personal experience and practical conditions and other factors. However, it is worthy of discussion how to avoid the subjective arbitrary of decision-makers to confirm the value of v_{ij}. In order to ensure v_{ij} more reasonable, decision-makers can ask k experienced experts to compare the attributes each other in the same level. Suppose $v_{ij}^q = (t_{ij}^q, f_{ij}^q)$ is given by the expert q, we can get the arithmetic mean of all the t_{ij}^q and all the f_{ij}^q respectively to summarize the opinions of experts, that is

$$t_{ij} = (\sum_{q=1}^{k} t_{ij}^q)/k , \ f_{ij} = (\sum_{q=1}^{k} f_{ij}^q)/k , \ \text{then } v_{ij} = (t_{ij}, f_{ij}) \ \text{can be confirmed, where}$$

$i, j = 1, 2, \cdots, n$.

Step 1: By analyzing the relationship of different factors in the system, a systematic hierarchical structure can be established.

Step 2: By asking the experts to establish the IFCJM $V = (v_{ij})_{n \times n}$ which is the pair-wise comparisons of all the elements on the same level with respect to the element on the above level, where $v_{ij} = (t_{ij}, f_{ij})$, $i, j = 1, 2, \cdots, n$, and t_{ij}, f_{ij} are given quantity standard on the scale of 0.1 to 0.9.

Step 3: Do the consistency adjustment for the IFCJM V by additive consistency recursive iterative adjustment algorithm. Then obtain the ACIFCJM.

Step 4: Calculate the priority vector. Calculate the weights of the elements on every level, and then calculate the combined weight of all the elements on the bottom level with respect to the element on the top level.

4.3.2 The Example Analysis

Recently, people pay more attention on the evaluation of ecological architecture. Based on others' research, a comprehensive evaluation model of ecological architecture is constructed by using IFAHP.

Step 1: Construct an evaluation index system and establish a hierarchical structure in Table 4.

Table 4 The evaluation index system of the evaluation of ecological architecture

Target layer	Primary index	Secondary index
O : Ecological construction level	$A1$: Land-saving and outdoor environment	$B1$: Wasted land use
		$B2$: Surrounding air quality
		$B3$: Surrounding the daylighting
		$B4$: Surrounding greening
		$B5$: Public facilities
	$A2$: Indoor environment	$B6$: Indoor air quality
		$B7$: Indoor thermal environment
		$B8$: Indoor light environment
		$B9$: Indoor sound environment
	$A3$: Energy conservation and energy use	$B10$: Energy saving of main body Of building
		$B11$: Conventional optimal utilization of energy system
		$B12$: Renewable energy and energy recovery
		$B13$: Energy consumption and environmental
	$A4$: Operation management	$B14$: To the periphery environment influence
		$B15$: Waste treatment
		$B16$: Equipment monitoring system
		$B17$: Management measure and system
	$A5$: Water saving and water use	$B18$: Water supply and drainage system
		$B19$: Sewage treatment and recycling system
		$B20$: The rain recovery and utilization
		$B21$: Water saving appliances and facilities
	$A6$: Saving material and resource reuse	$B22$: Green building materials
		$B23$: Use local materials
		$B24$: Recycling
		$B25$: Indoor decoration

Step 2: We ask the senior experts to give the evaluation values and get the IFCJM which is the pair-wise comparisons for all the primary indexes (or secondary

indexes) with respect to the objective in the target layer (or primary indexes) as follows:

Table 5 IFCJM O which is the pair-wise comparison for all the primary indexes with respect to the objective in the target layer

O	A1	A2	A3	A4	A5	A6
A1	(0.5,0.5)	(0.6,0.4)	(0.7,0.2)	(0.7,0.1)	(0.8,0.2)	(0.9,0.1)
A2	(0.4,0.6)	(0.5,0.5)	(0.6,0.4)	(0.6,0.3)	(0.7,0.2)	(0.8,0.2)
A3	(0.2,0.7)	(0.4,0.6)	(0.5,0.5)	(0.5,0.4)	(0.6,0.3)	(0.7,0.2)
A4	(0.1,0.7)	(0.3,0.6)	(0.4,0.5)	(0.5,0.5)	(0.8,0.1)	(0.9,0.1)
A5	(0.2,0.8)	(0.2,0.7)	(0.3,0.6)	(0.1,0.8)	(0.5,0.5)	(0.6,0.3)
A6	(0.1,0.9)	(0.2,0.8)	(0.2,0.7)	(0.1,0.9)	(0.3,0.6)	(0.5,0.5)

Table 6 IFCJM $A1$ which is the pair-wise comparison for all the secondary indexes with respect to $A1$ in the primary indexes layer

A1	B1	B2	B3	B4	B5
B1	(0.5,0.5)	(0.1,0.7)	(0.1,0.7)	(0.3,0.6)	(0.2,0.8)
B2	(0.7,0.1)	(0.5,0.5)	(0.5,0.5)	(0.8,0.1)	(0.8,0.2)
B3	(0.7,0.1)	(0.5,0.5)	(0.5,0.5)	(0.8,0.1)	(0.8,0.2)
B4	(0.6,0.3)	(0.1,0.8)	(0.1,0.8)	(0.5,0.5)	(0.6,0.3)
B5	(0.8,0.2)	(0.2,0.8)	(0.2,0.8)	(0.3,0.6)	(0.5,0.5)

Similarly, we get the IFCJM $A2$, $A3$, $A4$, $A5, A6$.

Step 3: Do the consistency adjustment for the IFCJM, and get the weight vector.

(1) Do the consistency adjustment for the IFCJM, and get the ACIFCJM as follows in table 7:

Table 7 The consistency adjustment for the IFCJM

$$
O^{(1)} \begin{pmatrix}
0.5 & (0.6,0.4) & (0.7,0.3) & (0.7,0.2) & (0.8,0.1) & (0.9,0) \\
(0.4,0.6) & 0.5 & (0.6,0.4) & (0.6,0.3) & (0.7,0.2) & (0.8,0.1) \\
(0.3,0.7) & (0.4,0.6) & 0.5 & (0.5,0.4) & (0.6,0.3) & (0.7,0.2) \\
(0.2,0.7) & (0.3,0.6) & (0.4,0.5) & 0.5 & (0.6,0.4) & (0.7,0.3) \\
(0.1,0.8) & (0.2,0.7) & (0.3,0.6) & (0.4,0.6) & (0.5,0.5) & (0.6,0.4) \\
(0,0.9) & (0.1,0.8) & (0.2,0.7) & (0.3,0.7) & (0.4,0.6) & 0.5
\end{pmatrix}
$$

$$
A^{(1)} \begin{pmatrix}
(0.3667,0.3667) & (0.2333,0.4333) & (0.2333,0.4333) & (0.4,0.3667) & (0.5333,0.4) \\
(0.4333,0.2333) & (0.3667,0.3667) & (0.3667,0.3667) & (0.4667,0.2333) & (0.6667,0.3333) \\
(0.4333,0.2333) & (0.3667,0.3667) & (0.3667,0.3667) & (0.4667,0.2333) & (0.6667,0.3333) \\
(0.3667,0.4) & (0.2333,0.4667) & (0.2333,0.4667) & (0.3667,0.3667) & (0.5333,0.4333) \\
(0.4,0.5333) & (0.3333,0.6667) & (0.3333,0.6667) & (0.4333,0.5333) & (0.3667,0.3667)
\end{pmatrix}
$$

Similarly, we get the ACIFCJM $A^{(2)}$, $A^{(3)}$, $A^{(4)}$, $A^{(5)}$, $A^{(6)}$.

(2) Then we get the weight sorting vector of the ACIFCJM in the following table 8:

Table 8 The weight sorting vector

α	0.5
$W^{(1)}$	$(0.2937,0.2372,0.1808,0.1526,0.0961,0.0397)'$
$\omega^{(1)}$	$(0.1842,0.2519,0.2519,0.1729,0.1391)'$
$\omega^{(2)}$	$(0.4299,0.3452,0.1759,0.0489)'$
$\omega^{(3)}$	$(0.3664,0.3664,0.0701,0.1971)'$
$\omega^{(4)}$	$(0.3664,0.0701,0.1971,0.3664)'$
$\omega^{(5)}$	$(0.4405,0.1019,0.1442,0.3135)'$
$\omega^{(6)}$	$(0.377,0.123,0.2077,0.2923)'$

Step 4: Aggregate the global weight.

Let $\alpha = 0.5$, then we can get a weight matrix $W^{(1)}$ of all the secondary indexes on the third level with respect to all the primary indexes on the second level where $W^{(1)} = (\omega^{(1)}, \omega^{(2)}, \omega^{(3)}, \omega^{(4)}, \omega^{(5)})$. Aggregate the global weight. The combined weights of the elements in the secondary index level relative to the objective in the target layer is

$$W^{(2)} = W^{(1)} \cdot w^T$$

$= (0.0541, 0.074, 0.074, 0.0508, 0.0409, 0.102,\ 0.0819, 0.0417,\ 0.0116,$

$\quad 0.0662, 0.0662, 0.0127, 0.0356, 0.0559, 0.0107, 0.0301, 0.0559,$

$\quad 0.0423, 0.0098,\ 0.0139,\ 0.0301, 0.015, 0.0049, 0.0082, 0.0116)\,.$

5   Conclusion

This paper gives two kinds of consistency recursive iterative adjustment algorithms for judgment matrixes in AHP. Then one method of additive consistency recursive iterative adjustment algorithms about the fuzzy complementary judgment matrix was given. It is critical that additive consistency recursive iterative adjustment algorithms about the intuitionistic fuzzy complementary judgement matrix was given, and an actual case shows the effectiveness of the consistency recursive iterative adjustment algorithms.

Acknowledgments. This research is partly funded by the National Natural Science Foundation of China (NSFC) under Grant No. 71140008 & 71361008,and Special Application Technology Research and Development of Hainan province under Grant No. ZDXM2014107. Corresponding author: Chengyi Zhang. E-mail: chengyizh@hainnu.edu.cn.

References

1. Saaty, T.L.: The analytic hierarchy process. McGraw-Hill, New York (1980)
2. Saaty, T.L.: A scaling method for priorities in hierarchical structure. Math. Psychology 15(3), 234–281 (1977)
3. Saaty, T.L., Vargas, L.G.: Comparison of eigen value logarithmic least squares and least squares methods in estimating ratios. Mathematical Modelling 5, 309–324 (1984)
4. Xu, Z.S.: A practical method for improving consistency of judgement matrix. Systems Engineering 16(6), 61–63 (1998)
5. Liu, W.L.: A new method of rectifying judgement matrix. Systems Engineering-theory & Practice 9(9), 100–104 (1999)
6. Wu, Z.N., Zhang, W.G., Guan, X.J.: A statistical method to check and rectify the consistency of a judgement matrix. Systems Engineering 20(3), 67–71 (2002)
7. He, B., Meng, Q.: A new method of checking the consistency of a judgment matrix. Journal of Industrial Engineering and Engineering Management 16(4), 92–94 (2002)
8. Zhang, Q.H., Long, X.: An iterative algorithm for improving the consistency of judgement matrix in AHP. Mathematics In Practice and Theory 31(5), 565–568 (2001)
9. Zhan, T.S., Li, H., Wan, L.P.: A greedy algorithms to accelerating rectify judgement matrix on AHP through measure matrix. Mathematics In Practice and Theory 34(11), 94–97 (2004)
10. Wang, G.H., Liang, L.: A method of regulating judgement matrix according as consistent rule. Systems Engineering 19(4), 90–96 (2001)
11. Jin, J.L., Wei, Y.M., Pan, J.F.: Accelerating genetic algorithm for correcting judgement matrix consistency in analytic hierarchy process. Systems Engineering-Theory & Practice 24(1), 63–69 (2004)
12. Wang, X., Dong, Y.C., Chen, Y.H.: Consistency modification of judgment matrix based on a genetic simulated annealing algorithm. Journal of Systems Engineering 21(l), 107–111 (2006)
13. Luo, Z.Q.: A new method for adjusting inconsistency judgment matrix in AHP. Systems Engineering-Theory & Practice 6, 84–92 (2004)

14. Zang, J.J.: Fuzzy analytic hierarchy process. Fuzzy Systems and Math. 14, 80–88 (2000)
15. Kołodziejczyk, W.: Orlovsky's concept of decision-making with fuzzy preference relation—further results. Fuzzy Sets Systems 19(1), 11–20 (1986)
16. Gogus, O., Boucher, T.: A consistency test for rational weights in multi-criterion decision analysis with fuzzy pairwise comparisons. Fuzzy Sets and Systems 86(2), 129–138 (1997)
17. Leung, L.C., Cao, D.: On consistency and ranking of alternatives in fuzzy AHP. European Journal of Operational Research 124(1), 102–113 (2000)
18. Wang, Y.M., Fan, Z.P., Hua, Z.: A chi-square method for obtaining a priority vector from multiplicative and fuzzy preference relations. European Journal of Operational Research 182(1), 356–366 (2007)
19. Wu, X.H., Lv, Y.J., Yang, F.: The verifier and adjustment of consistency for fuzzy complementary judgement matrix. Fuzzy Systems and Mathematics 24(2), 105–111 (2010)
20. Zhang, X.X., Liu, J.X., Lei, J., Yang, B.A.: The approximate threshold of additive consistency index for fuzzy complementary matrix. Computer Simulation 27(1), 316–333 (2010)
21. Xu, G.L., Lin, L.: Method for identifying and improving the consistency of fuzzy complementary judgment matrix. Operations Research and Management Science 20(1), 93–97 (2011)
22. Orlovsky, S.A.: Decision making with a fuzzy preference relation. Fuzzy Sets and Systems 1, 155–167 (1978)
23. Xu, Z.S.: Uncertain multiple attribute decision making methods and applications. Tsinghua University Press, Beijing (2004)
24. Atanassov, K.: Intuitionistic fuzzy sets. Fuzzy Sets and Systems 20(1), 87–96 (1986)
25. Xu, Z.S.: A survey of preference relations. International Journal of General Systems 36(2), 179–203 (2007)
26. Xu, Z.S.: Approaches to multiple attribute decision making with intuitionistic fuzzy preference information. Systems Engineering-Theory & Practice 11, 62–71 (2007)
27. Wang, L.: The analytic hierarchy process based on vague sets and the application of the process. Hainan Normal University Master's Thesis (2011)
28. Zhang, C.Y., Zhou, H.Y.: The fuzzy approximation on vague sets and the similarity measure of fuzzy entropy. Computer Engineering and Applications 33, 20–21 (2006)
29. Chen, J.M., Li, H.X.: A ranking method and decision-making with consistent risk preference in intuitionistic fuzzy numbers. Fuzzy Systems and Math. 24(6), 85–91 (2010)
30. Xu, Z.S.: Study on the relation between two classes of scales in AHP. Systems Engineering-Theory & Practice 19(7), 97–101 (1999)
31. Lei, Y., Shi, H.X., Yang, X.J., Zhu, H.G., Pei, X.M.: A comprehensive assessment model of ecological architecture based on AHP and fuzzy evaluation theory. Architectural Journal S2, 50–54 (2010)

Collaborative Decision Making by Ensemble Rule Based Classification Systems

Han Liu and Alexander Gegov

Abstract. Rule based classification is a popular approach for decision making. It is also achievable that multiple rule based classifiers work together for group decision making by using ensemble learning approach. This kind of expert system is referred to as ensemble rule based classification system by means of a system of systems. In machine learning, an ensemble learning approach is usually adopted in order to improve overall predictive accuracy, which means to provide highly trusted decisions. This chapter introduces basic concepts of ensemble learning and reviews Random Prism to analyze its performance. This chapter also introduces an extended framework of ensemble learning, which is referred to as Collaborative and Competitive Random Decision Rules (CCRDR) and includes Information Entropy Based Rule Generation (IEBRG) and original Prism in addition to PrismTCS as base classifiers. This is in order to overcome the identified limitations of Random Prism. Each of the base classifiers mentioned above is also introduced with respects to its essence and applications. An experimental study is undertaken towards comparative validation between the CCRDR and Random Prism. Contributions and Ongoing and future works are also highlighted.

Keywords: Data Mining, Machine Learning, Rule Based Classification, Ensemble Learning, Collaborative Decision Making, Random Prism.

1 Introduction

Rule based classification is a common approach used for decision making. It is also feasible for multiple rule based classifiers to collaborate for group decision making by adopting ensemble learning approaches. This kind of expert system is

Han Liu · Alexander Gegov
University of Portsmouth, School of Computing, Buckingham Building,
Lion Terrace, PO1 3HE Portsmouth, United Kingdom
e-mail: {Han.Liu,Alexander.Gegov}@port.ac.uk

© Springer International Publishing Switzerland 2015

W. Pedrycz and S.-M. Chen (eds.), *Granular Computing and Decision-Making*,
Studies in Big Data 10, DOI: 10.1007/978-3-319-16829-6_10

245

referred to as ensemble rule based classification system by means of a system of systems. In this context, the ensemble rule based classification system is seen as a super system and consists of a number of single rule based classification systems, each of which is seen as a sub-system of the ensemble rule based classification system. In machine learning, an ensemble learning approach is usually adopted in order to improve overall predictive accuracy, which means to provide highly trusted decisions.

Ensemble learning can be done in parallel or sequentially. In the former way, there are no collaborations among different algorithms in training and only their predictions are combined for final decision making [1]. In this context, the final prediction is typically made by means of majority voting as part of the classification tasks. In the latter way of ensemble learning, the first algorithm learns a model from data and then the second algorithm learns to correct the former one etc [1]. In other words, the model built by the first algorithm is further corrected by the following algorithms sequentially. In parallel ensemble learning, a popular approach is to take sampling to a data set in order to get a set of samples. A classification algorithm is then used to train a classifier on each of these samples. The group of classifiers constructed will make predictions on test instances independently and final predictions on the test instances will be made based on majority voting. A commonly used sampling method is Bagging [2]. The Bagging method is useful especially when the base classifier is not stable due to high variance of data sample. This is because the method is robust and does not lead to overfitting as the number of generated hypothesizes is increased [1]. Some unstable classifiers include neural networks, decision trees and some other rule based methods [3].

In this chapter, all of the base classifiers used for ensemble learning tasks are rule based classification methods, namely original Prism [4], PrismTCS [5] and Information Entropy Based Rule Generation (IEBRG) [6]. All of the three methods follow 'separate and conquer' approach [7], which is one of the rule generation approaches. This is because each of the three methods generates if-then rules directly from training instances. The other approach of rule generation is referred to as 'divide and conquer' approach [8], which generates classification rules in the intermediate form of decision trees. As the generation aims to construct decision trees, the above approach is also referred to as Top-Down Induction of Decision Trees (TDIDT). A principal problem that usually arises with rule based classification methods is the overfitting of generated hypothesis to training data [9]. As mentioned earlier, the Bagging method is robust and helps avoid overfitting for rule based classifiers. It thus motivates the use of Bagging as a sampling method for ensemble learning tasks, especially when rule based methods are used as base classifiers.

The rest of this chapter is organized as follows. Section 2 introduces the three rule based classification methods, namely original Prism, PrismTCS and IEBRG. An existing ensemble learning method, called Random Prism [10, 11], is also introduced in the Section 2 in order to comparatively analyze the performance of the method. Section 3 introduces an extended framework of ensemble learning, which is referred to as Collaborative and Competitive Random Decision Rules (CCRDR) and includes the three base classifiers mentioned above. An experimental study is undertaken towards comparative validation between the CCRDR and Random Prism in Section 4. The contributions and further directions of this research area are also highlighted in Section 5.

2   Related Work

As mentioned in Section 1, this chapter investigates parallel ensemble learning approaches which use Bagging as the sampling method and rule based methods as base classifiers. Therefore, this section introduces three rule based methods, namely original Prism, PrismTCS and IEBRG, the Bagging method and Random Prism.

2.1   Original Prism

The original Prism method was introduced by Cendrowska in [4] and the basic procedure of the underlying Prism algorithm is illustrated in Fig. 1. This algorithm is primarily aimed at avoiding the generation of complex rules with many redundant terms [9] such as the 'replicated subtree problem' [4] that arises with decision trees as illustrated in Fig. 2.

Execute the following steps for each classification (*class= i*) in turn and on the original training data S:

1. $S'=S$.
2. Remove all instances from S' that are covered from the rules induced so far. If S' is empty then stop inducing further rules
3. Calculate the conditional probability from S' for *class=i* for each *attribute-value pair*.
4. Select the *attribute-value pair* that covers *class= i* with the highest probability and remove all instances from S' that comprise the selected *attribute-value pair*
5. Repeat 3 and 4 until a subset is reached that only covers instances of *class= i* in S'. The induced rule is then the conjunction of all the *attribute-value pairs* selected.

Repeat 1-5 until all instances of *class i* have been removed

*For each rule, no one attribute can be selected twice during rule generation

---

Fig. 1 Basic Prism algorithm [12]

The original Prism algorithm cannot directly handle continuous attributes as it is based on the assumption that all attributes in a training set are discrete. When continuous attributes are actually present in a dataset, these attributes should be discretized by preprocessing the dataset prior to generating classification rules [12, 13, 14]. In addition, Bramer's Inducer Software handles continuous attributes as described in [12, 13, 14].

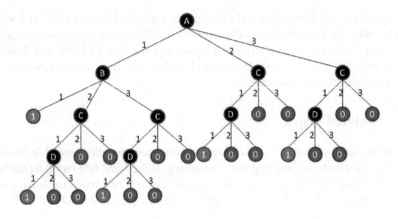

Fig. 2 Cendrowska's replicated sub-tree example [16, 17]

On the other hand, the original Prism algorithm does not take into account clashes, i.e. a set of instances in a subset of a training set that are identical apart from being assigned to different classes but cannot be separated further [12, 14]. Clashes usually occur in two principal ways:

1) One of the instances has at least one incorrect record for its attribute values or its classification [12].
2) The clash set has both (or all) instances correctly recorded but it is impossible to discriminate between them on the basis of the attributes recorded and thus it may be required to examine further attributes [12].

However, the Inducer software implementation [15] of Prism can handle clashes and the strategy of handling a clash is illustrated in Fig. 3. This way of dealing with clashes would result in underfitting of generated hypothesis to training data. This is because there would be a large number of instances that are not covered by the generated rule set if the rules that cover the instances are discarded. In testing stage, the way of clash handling would also make a large number of unseen instances left unclassified. This is because the algorithm does not generate a default rule that assigns a default classification (usually majority class) [15] to those instances that the generated rule set does not cover.

Another problem that arises with Prism is tie-breaking, i.e. if there are two or more attribute-value pairs which have equal highest probability in a subset (see step 3 in Fig.1). The original Prism algorithm makes an arbitrary choice in step 4 as illustrated in Fig. 1 whereas the Inducer software makes the choice using the highest total target class frequency [12].

If a clash occurs while generating rules for class i:
1. Determine the majority class for the subset of instances in the clash set.
2. If this majority class is target *class i*, then compute the induced rule by assigning all instances in the clash set to class i. If it is not, discard the whole rule.
3. If the induced rule is discarded, then all instances that match the target class should be deleted from the training set before the start of the next rule induction. If the rule is kept, then all instances in the clash set should be deleted from the training data.

Fig. 3 Dealing with clashes in Prism [12, 16, 17]

Also, the original Prism may generate a rule set which may result in a classification conflict in predicting unseen instances. This can be illustrated by the example below:

Rule 1: If x=1 and y=1 then class= a
Rule 2: If z=1 then class= b

What should the classification be for an instance with x=1, y=1 and z=1? One rule gives *class a*, the other one gives *class b*. A method is required to choose only one classification to classify the unseen instance [12]. Such a method is known as a conflict resolution strategy. Bramer mentioned in [12] that Prism uses the 'take the first rule that fires' strategy in dealing with the conflict problem and therefore it is required to generate the most important rules first. However, the original Prism cannot actually introduce an order to a rule according to its importance as each of those rules with a different target class is independent of each other. As mentioned in [5, 13, 14], this version of Prism would restore the training set to its original size after the completion of rule generation for class i and before the start for class $i+1$. This indicates the rule generation for each class may be done in parallel so the algorithm cannot directly rank the importance among rules with different target classes. Thus the 'take the first rule that fires' strategy may not deal with the classification confliction well.

2.2 PrismTCS

Bramer pointed out that the original Prism algorithm always deletes instances covered by those rules generated so far and then restores the training set to its original size after the completion of rule generation for class i and before the start for class $i+1$. This results in a high number of iterations resulting in high computational cost [5] when the training data is very large. For the purpose of increasing the computational efficiency, a modified version of Prism, called PrismTCS, was developed by Bramer [5]. PrismTCS always chooses the minority class as the target class pre-assigned to a rule being generated as its consequence. Besides this, it does not reset the dataset to its original state and thus introduces an order to each rule according to its importance [5, 13, 14]. Therefore, PrismTCS is not only faster in generating rules compared with the original Prism, but also provides a similar level of classification accuracy [5, 13, 14].

As mentioned in Section 2.1, original Prism has some disadvantages in dealing with continuous attributes, tie-breaking, clashes and classification conflict. For each of these issues, Bramer introduces a corresponding solution in [12]. Each of the solutions is also applied to PrismTCS for each of the corresponding issues. In comparison to original Prism, PrismTCS can deal with conflict of classification better. This is because PrismTCS generates a set of ordered rules as mentioned earlier in this section. However, similar with original Prism, the way of dealing with clashes also results in underfitting of training data. As mentioned earlier, PrismTCS always chooses the minority class in the current training set as the target class of the rule being generated. Since the training set is never restored to its original size as mentioned above, it can be proven that one class could always be selected as target class until all instances of this class have been deleted from the training set because the instances of this minority class covered by the current rule generated should be removed prior to generating the next rule. This case may result in that the majority class in the training set may not be necessarily selected as target class to generate a list of rules until the termination of the whole generation process. In this case, there is not even a single rule having the majority class as its consequence (right hand side of this rule).

Although PrismTCS can generate a rule set which includes a default rule as introduced in [15] and thus leads to the decrease of number of unclassified instances, the default rule is likely to give a wrong classification to those unseen instances that are not covered by the generated rule set. This is because the assumption needs to be guaranteed that the training set covers complete patterns in a domain, which is in order to make the default rule unlikely to give wrong classifications. Otherwise, the rule set could still underfit the training set as the conditions of classifying instances to the other classes are probably not strong enough.

2.3   *Information Entropy Based Rule Generation*

IEBRG is developed in [6] in order to overcome the limitations of both original Prism and PrismTCS. This method is attribute-value-oriented like Prism but it uses the 'from cause to effect' approach. In other words, it does not have a target class pre-assigned to the rule being generated. The main difference from Prism is that IEBRG focuses mainly on minimizing the uncertainty for each rule being generated no matter what the target class is. A popular technique used to measure the uncertainty is information entropy introduced by Shannon in [18]. The basic idea of IEBRG is illustrated in Fig.4 as below:

1.	Calculate the conditional entropy of each attribute-value pair in the current subset
2.	Select the attribute-value pair with the smallest entropy to spilt on, i.e. remove all other instances that do not comprise the attribute-value pair.
3.	Repeat step 1 and 2 until the current subset contains only instances of one class (the entropy of the resulting subset is zero).
4.	Remove all instances covered by this rule.

Repeat 1-4 until there are no instances remaining in the training set.

* For each rule, no one attribute can be selected more than once during generation.

Fig. 4 IEBRG algorithm

As mentioned in Section 2.1, all versions of Prism need to have a target class pre-assigned to the rule being generated. In addition, an attribute might not be relevant to each particular classification and sometimes only one value of an attribute is relevant [19]. Therefore, the Prism method chooses to pay more attention to the relationship between attribute-value pair and a particular class. However, the class to which the attribute-value pair is highly relevant is probably unknown, as can be seen from the example in Table 1 below with reference to the lens 24 dataset reconstructed by Bramer in [12]. This dataset shows that P (class=3|tears=1) =1 illustrated by the frequency table for attribute "tears". The best rule generated first would be *if tears=1 then class=3*.

Table 1 Lens 24 dataset example

Class Label	Tears=1	Tears=2
Class=1	0	4
Class=2	0	5
Class=3	12	3
total	12	12

This indicates that the attribute-value "tears=1" is only relevant to class 3. However, this is actually not known before the rule generation. According to PrismTCS strategy, the first rule being generated would select "class =1" as target class as it is the minority class (Frequency=4). Original Prism may select class 1 as well because it is in a smaller index. As described in [12], the first rule generated by Original Prism is "if astig=2 and tears=2 and age=1 then class=1". It indicates that the computational efficiency is slightly worse than expected and the resulting rule is more complex. When a large data set is used for training, the Prism method would be even likely to generate an incomplete rule covering a clash set as mentioned in Section 2.2 if the target class assigned is not a good fit to some of those attribute-value pairs in the current training set. Then the whole rule would be discarded resulting in underfitting and redundant computational effort.

In order to find a better strategy for reducing the computational cost, the IEBRG method is developed in [6]. In this method, the first iteration of the rule generation process for the "lens 24" dataset can make the resulting subset's entropy reach 0. Thus the first rule generation is complete and its rule is represented by "if tears=1 then class=3".

In dealing with continuous attributes, IEBRG takes the same way as applied to the Prism family, which includes original Prism and PrismTCS in the Inducer software implementation. With regard to tie-breaking, IEBRG deals with this issue in the way similar to that Prism family does, which means that when two or more attribute-value pairs have the same smallest entropy value the one with the highest total frequency is selected as introduced by Bramer in [12]. IEBRG can also deal with conflict of classification well because the method also generates a set of ordered rules like PrismTCS. In dealing with clashes, majority voting, which assigns the most common classification of the instances in the clash set to the current rule [12], is usually used for IEBRG, especially when the objective is to validate this method and to find its potential in improving accuracy and computational efficiency.

In comparison with the Prism family, this algorithm would reduce significantly the computational cost when the training set is large. In addition, in contrast to Prism, the IEBRG method deals with clashes by assigning a majority class in the clash set to the current rule. This would potentially reduce the underfiting of rule set thus reducing the number of unclassified instances although it may increase the number of misclassified instances. As mentioned in [12], Prism prefers to discard a rule rather than to give a wrong classification when a clash occurs and thus is more noise tolerant than TDIDT. However, if the reason that a clash occurs is not due to noise and the training set covers a large amount of data, then it would result in serious underfitting of the rule set by discarding rules as it would leave many unseen instances unclassified at prediction stage. The fact that Prism would decide to discard the rules in some cases is probably because it uses the so-called 'from effect to cause' approach. As mentioned in Section 2.1, each rule being generated should be pre-assigned a target class and then the conditions should be searched by adding terms (antecedents) until the adequacy conditions are met. Sometimes, it may not necessarily receive adequacy conditions even after all attributes have been examined. This indicates the current rule covers a clash set that contains instances of more than one class. If the target class is not the majority class, this indicates the search of causes is not successful so the algorithm decides to withdraw the task by discarding the incomplete rule and deleting all those instances that match the target class in order to avoid the same case to happen all over again [13, 14]. This actually not only increases the irrelevant computation cost but also results in underfitting of the rule set. On the other hand, the IEBRG would also have the potential to avoid occurring clashes better compared with Prism. This is due to the strategy of rule generation from IEBRG as mentioned earlier in this section.

2.4   Bagging

As mentioned in Section 1, Bagging is a popular method of data sampling for ensemble learning tasks due to its robustness in avoiding overfitting. The term Bagging stands for bootstrap aggregating which is a method for sampling of data with replacement [1]. In detail, the Bagging method is to take a sample with a size as same as the original data set and to randomly select an instance from the original data set to be put into the sample set. This means that some instances in the original set may appear more than once in the sample set and some other instances may not even appear once in the sample set. According to the principle of statistics, the bagging method would produce a sample that is expected to contain 63.2% of the original data instances [1, 2, 10, 11]. Therefore, the Bagging method is useful especially when the base classifier is not stable due to high variance of data sample as mentioned in Section 1 and thus helpful to rule based classification methods in avoiding overfitting. For example, the method is successfully applied with PrismTCS into Random Prism for construction of ensemble learners [10, 11], which is further introduced in Section 2.5.

2.5   Random Prism

Random Prism, an existing ensemble learning method [10, 11], follows the parallel ensemble learning approach and uses Bagging for sampling as illustrated in Fig.5. It has been proven in [10, 11] that Random Prism is a noise-tolerant method alternative to Random Forests [20]. However, the Random Prism has two aspects in which can be improved in training and testing stages respectively. The above two aspects are also mentioned with suggestions for further improvements in [10, 11].

The first aspect is that there is only a single base classifier, PrismTCS, involved in training stage for Random Prism, which cannot always generate strong hypothesis (robust models). In fact, it is highly possible that a single algorithm performs well on some samples but poorly on the others. From this point of view, it is motivated to extend the ensemble learning framework by including multiple base classifiers involved in training stage. This is in order to achieve that on each data sample the learner created is much stronger.

On the other hand, Random Prism uses weighted majority voting to determine the final prediction on test instances. In other words, each model is assigned a weight, which is equal to the overall accuracy checked by validation data from the sample. In prediction stage, each model is used to predict unseen instances and give an individual classification. The ensemble learning system then makes the final classification based on weighted majority voting instead of traditional majority voting. For example, there are three base classifiers: A, B and C. A predicts the classification X with the weight 0.8 and both B and C predicts classification Y with the weights 0.55 and 0.2 respectively so the final classification is X if using weighted majority voting (weight for X: 0.8> 0.55+0.2=0.75) but is Y if using traditional majority voting (frequency for Y: 2>1). However, for the weighted majority voting mentioned above, the strategy in determining the weight is not reliable enough especially for unbalanced data sets. This is because it is highly possible that a classifier performs better on predicting positive instances but worse on negative instances if it is a two class classification task. The similar case can also happen in multi-class classification tasks. Therefore, it is more reasonable to use the individual accuracy for a single classification (e.g. true positive rate) as the weight.

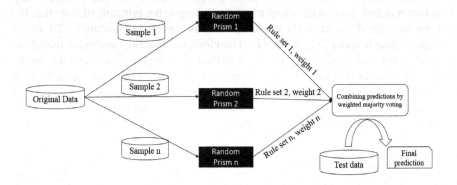

Fig. 5 Random Prism Framework with Bagging [10, 11]

Therefore, an extended framework of ensemble learning, referred to as Collaborative and Competitive Random Decision Rules (CCRDR), is developed in order to overcome the limitations and introduced in Section 3.

3 Collaborative and Competitive Random Decision Rules

As mentioned in Section 2.5, Random Prism is a noise tolerant ensemble learning algorithm alternative to Random Forests [20]. However, it has two weak points in training and testing stages respectively and thus has space for improvement. This section introduces an advanced ensemble learning framework extended from Random Prism with the aim to overcome the two weak points which are mentioned above and described in Section 2.5. This section introduces a new framework that addresses the two weak points.

The framework developed in the authors' recent research is referred to as Collaborative and Competitive Random Decision Rules (CCRDR) and illustrated in Fig.6, which indicates that the ensemble learning framework includes both cooperative learning and competitive learning involved.

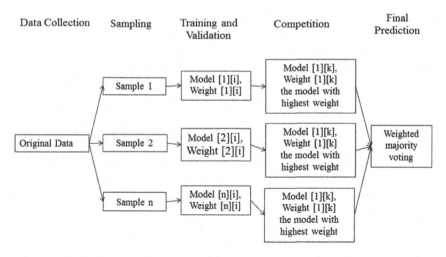

Fig. 6 Procedures of Proposed Ensemble Learning

The first weak point of Random Prism is that there is only a single base classifier involved in training stage, which cannot always generate robust models as mentioned in Section 2.5. In order to overcome the limitation, the ensemble learning framework is modified in the way that the framework can include multiple base classifiers for training. Due to this modification, there is competition involved among the classifiers constructed on a same sample of training data. In other words, there are multiple learning algorithms applied to each sample of training data, which implies that multiple classifiers are constructed on each sample. In this context, it becomes achievable to find better classifiers to be involved in testing stage and worse classifiers to be absent through competition among the classifiers. The competition is based upon the weight (confidence) of each of the classifiers by means of overall accuracy estimated by validation data. In the extended framework, only the classifier with the highest weight (confidence) is eligible to be involved in testing stage. The modification with regard to the first weak point is also reflected from the second part of the name of the method namely 'Competitive Random Decision Rules'. The name of the method indicates that any rule based classification methods are eligible for being involved in training stage as base classifiers. This modification theoretically contributes to that on each sample of data the learners constructed become much stronger.

The second weak point is regarding the way of determining the weight of a classifier for weighted majority voting as mentioned in Section 2.5. In order to overcome the limitation, confusion matrix, which reflects the individual accuracy for each single class such as true positive rate and true negative rate, is recommended in [10, 11]. However, the individual accuracy for a single classification reflected from confusion matrix is not effective in some special cases. In contrast, precision for a particular classification would be more reliable in determining the weight of a classifier. For example, there are 5 positive instances out of 20 in a test set and a classifier correctly predicts the 5 instances as positive but incorrectly predicts other 5 instances as positive as well. In this case, the recall/true positive rate is 100% as all of the five positive instances are correctly classified. However, the precision on positive class is only 50%. This is because the classifier predicts 10 instances as positive and only five of them are correct. This case indicates the possibility that high recall could result from low frequency of a particular classification. Therefore, precision is sometimes more reliable in determining the weight of a classifier on a particular prediction from this point of view. Overall, both precision and recall would usually be more reliable than overall accuracy in determining weight of a classifier especially for unbalanced data sets but it is important to determine which one of the two metrics is used in resolving special issues.

The modifications to Random Prism with regard to its two weak points generally aim to improve the robustness of models built in training stage and to more accurately measure the confidence of each single model in making a particular prediction. In this chapter, original Prism, PrismTCS and IEBRG are used as base classifiers in the CCRDR framework due to the better noise tolerance of Prism family in comparison with TDIDT as well as the advantages of IEBRG listed in Section 2.3 in comparison with Prism family. However, in general, this framework could incorporate any type of rule based classification methods or even other type of machine learning methods such as Neural Networks [36] and Support Vector Machine [37]. With regard to the way to choose machine learning methods that are incorporated into the framework, it is typically based on theoretical analysis on the suitability of a particular method to a particular dataset. For example, some methods cannot directly deal with continuous attributes such as some rule based methods. In this case, it is required to discretize continuous attributes by preprocessing the dataset prior to training stage. One of popular approaches is ChiMerge [38]. There are also some methods that cannot effectively discrete attributes such as Neural Networks and Support Vector Machine. In this case, it needs to split the discrete attributes into n binary attributes, while n is the number of values for the attribute, and each of the n binary attributes corresponds to a value of the original attribute. For example, gender is a discrete attribute with two values (male and female) and can be divided into two binary attributes named male and female respectively. Each of the binary attributes is judged either yes or no. If a dataset contains a large number of discrete attributes and each of them has a large number of possible values, it would significantly increase the number of attributes for the dataset resulting in the curse of dimensionality [39]. On the basis of above description, one way to decide which methods are chosen for training could be based on the type of attributes as part of data characteristic. On the other hand, as mentioned in Section 2.4, the training instances are randomly selected from original dataset and different methods may demonstrate different level of robustness with respect to the change of sample. Therefore, the decision on choosing methods could also be based on the robustness of a particular method validated in experimental studies. Appropriate selection of algorithms would obviously help increase the overall performance of using the CCRDR framework with respects to both predictive accuracy and computational efficiency. The empirical validation of CCRDR framework against Random Prism is introduced in Section 4.

The authors also define a novel way of understanding ensemble learning in the context of system theory by referring an ensemble classifier to as an ensemble rule based classification system. This is because an ensemble classifier actually consists of a number of single base classifiers as mentioned in Section 1. Therefore, in the context of system theory, an ensemble rule based classification system consists of a group of single rule based classification systems as mentioned in Section 1, each of which is a subsystem of the ensemble system. In other words, it is a system of systems like a set of sets in set theory. In addition, an ensemble rule based classification system can also be a subsystem of another ensemble system in theory. In other words, a super ensemble rule based classification system contains a number of clusters, each of which represents a subsystem that consists of a group of single rule based systems.

4   Comparative Validation

The validation of CCRDR framework against Random Prism is in terms of classification accuracy. The experimental study is undertaken by splitting a data set into a training set and a test set in the ratio of 80:20. For each data set, the experiment is repeated five times and the average of the corresponding accuracies is used for comparative validation. The reason is that ensemble learning is usually computationally more expensive because the size of data set dealt with by ensemble learning is as same as n times the size of the original data set when using Bagging. In other words, a data set should be pre-processed to get n samples, each of which has the same size of original data set. In addition, the proposed ensemble learning method includes two or more base classifiers in general (three base classifiers in this experiment) used for each of the n samples. Therefore, in comparison with single learning such as use of IEBRG or Prism, the computational efforts would be the same as $3*n$ times that conducted by a single learning task. In this situation, the experimental environment would be computationally quite constrained on a single computer if cross validation is used to measure the accuracy. On the other hand, instances in each sample are randomly selected with replacement from the original data set. Thus the classification results are not deterministic and the experiment is setup in the way mentioned above to make the results more convincing. Besides, the accuracy performed by random guess is also calculated and compared with that performed by each chosen algorithm. This is in order to check whether a chosen algorithm really works on a particular data set as mentioned earlier. The validation of the proposed ensemble learning method does not include this measure of efficiency. This is because, on the basis of above descriptions, the computation conducted using the proposed method is theoretically much more complex if it is done on a single computer. However, the efficiency can be easily improved in practice by adopting parallel data processing techniques and is thus not a critical issue.

In addition, the comparison is also against the random classifier, which predicts classification by random guess. The corresponding accuracy depends on number of classifications and distribution of these classifications. For example, if the objective function is a two class classification problem and the distribution is 50:50, then the accuracy performed by random guess would be 50%. Otherwise, the accuracy must be higher than 50% in all other cases. This setup of experimental study is in order to indicate the lower bound of accuracy to judge if an algorithm really works on a particular data set.

All of the data sets used in this evaluation are retrieved from UCI repository [21], some of which contain missing values in input attributes or class attributes. This is usually a far large issue that needs to be dealt with effectively as it would result in infinite loops for rule based methods in training stage. In machine learning tasks, there are typically two ways of dealing with missing values [12]:

1) Replace all missing values by the most frequent occurring value for each attribute.
2) Discard all instances with missing values.

In this experimental study, the first way is adopted because all of the chosen data sets are relatively small. It indicates that if the second way is adopted both training and test sets would be too small to be representative samples. Under this kind of situation, the model generated is likely to introduce biased patterns with low confidence especially if the model overfits the training data. However, this way of dealing with missing values also potentially introduces noise to the data set. Thus such an experimental setup would also provide the validation with respect to the noise tolerance of an algorithm in the meantime. On the other hand, if missing values are in the class attribute, then the best approach would be by adopting the second way mentioned above. This is because the first way mentioned above is likely to introduce noises to the data sets and thus incorrect patterns and predictive accuracies would be introduced. It is also mentioned in [12] that the first way is unlikely to prove successful in most cases and thus the second way would be the best approach in these cases. In practice, the two ways of dealing with missing values can easily be achieved by using the implementations in some popular machine learning software such as Weka [22, 23].

The validation is divided into two parts of comparison. The first part is to prove empirically that combination of multiple learning algorithms would usually outperforms a single algorithm as a base classier for ensemble learning with respect to accuracy. The second part is to prove that the use of precision instead of overall accuracy or recall as the weight of a classifier would be more reliable in making final prediction. In Table 2, CCRDR I represents that the weight of a classifier is determined by the overall accuracy of the classifier. In addition, the CCRDR II and III represent the weight determined using precision for the former and using recall for the latter.

Table 2 Ensemble learning results

Dataset	Random Prism	CCRDR I	CCRDR II	CCRDR III	Random classifier
anneal	71%	78%	79%	**80%**	60%
balance-scale	44%	56%	**68%**	64%	43%
diabetes	66%	68%	**73%**	68%	54%
heart-statlog	68%	71%	**74%**	63%	50%
ionosphere	65%	68%	**69%**	65%	54%
lympth	68%	60%	**89%**	65%	47%
car	69%	68%	**71%**	70%	33%
breast-cancer	70%	72%	**74%**	73%	58%
tic-tac-toe	63%	65%	66%	**67%**	55%
breast-w	**85%**	75%	81%	75%	55%
hepatitis	81%	84%	**87%**	82%	66%
heart-c	70%	74%	**83%**	65%	50%
lung-cancer	75%	79%	**88%**	75%	56%
vote	67%	82%	**95%**	80%	52%
page-blocks	**90%**	**90%**	**90%**	89%	80%

The results in Table 2 show that all of the chosen methods outperform the random classifier in classification accuracy. This indicates that all of the methods really work on the chosen data sets. In the comparison between Random Prism and CCRDR I, the results show that the latter method outperforms the former method in 12 out of 15 cases. This indicates empirically that combination of multiple learning algorithms usually helps generate a stronger hypothesis in making classifications. This is because the combination of multiple algorithms could achieve both collaboration and competition. The competition among these classifiers, each of which is built by one of the chosen algorithms, would make it achievable that for each sample of training data the learner constructed is much stronger. All of the stronger learners then effectively collaborate on making classifications so that the predictions would be more accurate.

As mentioned earlier, the second part of comparison is to validate that precision would usually be a more reliable measure than overall accuracy and recall for the weight of a classifier. The results in Table 2 indicate that in 12 out of 15 cases CCRDR II outperforms CCRDR I and III. This is because in prediction stage each individual classifier would first make classifications independently and their predictions are then combined in making a final classification. For the final prediction, each individual classifier's prediction would be assigned a weight to server for final weighted majority voting. The weight is actually used to reflect how reliable the individual classification is. The heuristic answer would be based on the historical record on how many times the classifier has recommended this classification and how correct it is. This could be effectively measured by precision. The weakness of overall accuracy is that this measure can only reflect the reliability of a classifier on average rather than in making a particular classification as mentioned in Section 2.5. Thus overall accuracy cannot satisfy this goal mentioned above. In addition, although recall can effectively reflect the reliability of a classifier in making a particular classification, the reliability is affected by the frequency of a particular classification and thus cheats the final decision maker, especially when the frequency of the classification is quite low as mentioned in Section 3. Therefore, the results prove empirically that precision would be more reliable in determining the weight of a classifier for weighted majority voting.

The basis of above description with regard to CCRDR validates that combination of multiple learning algorithms would be more effective in improving the overall accuracy of classification and that precision would be a more reliable measure in determining the weight of a classifier to successfully serve for weighted majority voting, especially on unbalanced data sets.

5   Conclusion and Future Work

This chapter reviews an existing ensemble learning method called Random Prism and three rule based classification methods, namely original Prism, PrismTCS and IEBRG. An extended framework of ensemble learning is developed, which is referred to as CCRDR and includes the three methods mentioned above as the base classifiers. The experimental study reports that CCRDR outperforms Random Prism in terms of classification accuracy while the overall accuracy measured by validation data is used as the weight of a particular base classifier. In addition, the study also reports that precision would usually be more reliable than recall and overall accuracy in measuring the confidence of a classification made by a classifier. However, all of the data sets used in the validation introduced in Section 4 are noise free and include well representative samples. Each method may have a particular level of noise tolerance and stability with regard to change of sample. Therefore, the authors will further check the tendency with respect to the change of level of predictive accuracy as the change of noise level. The authors will also further check the variance of the accuracy when the sample of training and test data is changed. These are in order to validate the noise tolerance and stability of the CCRDR against Random Prism.

On the other hand, as mentioned in Section 4, ensemble learning methods are usually computationally more expensive than single learning methods such as IEBRG and Prism family. This is because the size of data set dealt with by ensemble learning is as same as n times the size of the original data set. In the CCRDR framework, the size would be $m \times n$ times the size of the original data set, where m is the number of learning algorithms involved in training stage. However, as mentioned in Section 1, this type of ensemble learning tasks belongs to parallel ensemble learning, which indicates the tasks can be parallelized to improve the computational efficiency in both training and testing stages. In practice, each company or organization may have branches in different cities or countries so the databases for the companies or organizations are actually distributed over the world. As the existence of high performance cloud and mobile computing technologies, the ensemble learning framework can also be easily transplant into distributed or mobile computing environments such as multi-agent systems [29].

However, the theoretical framework introduced in this chapter still has space for extension. The ensemble learning concepts introduced in the chapter focus on parallel learning, which means that the building of each classifier is totally parallel to the others without collaborations in training stage and only their predictions in testing stage are combined for final decision making. However, the ensemble learning could also be done in sequential ways with collaborations in training stage. For example, there are two learning algorithms involved; the first one learns a model and the second one learns to correct the former as mentioned in Section 1. This is a direction to extend the theoretical framework further.

So far, ensemble learning concepts introduced in the machine learning literature lie in single learning tasks. In other words, all algorithms involved in ensemble learning need to achieve the same learning outcomes in different strategies. This is defined as local learning by the authors in the chapter. In this context, the further direction would be to extend the ensemble learning framework to achieve global learning by means of different learning outcomes. The different learning outcomes are actually not independent of each other but have interconnections. For example, the first learning outcome is a prerequisite for achieving the second learning outcome. This direction of extension is towards evolving machine learning approach in a universal vision. To fulfil this objective, the networked rule bases as illustrated in Fig.7 can actually provide this kind of environment for discovering and resolving problems in a global way.

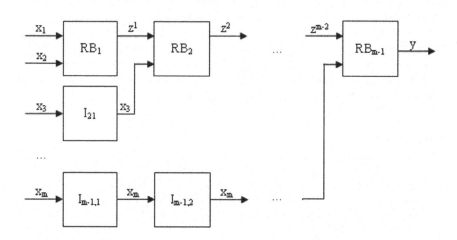

Fig. 7 Rule Based Network (modular rule bases) from [24, 25, 26, 27, 28]

In this network, each node represents a single rule base. The nodes can be connected sequentially or in parallel. In detail, each variable labelled x_{m-1}, while m represents the number of layer in which the node locates, represents an input and y represents the output. In addition, each of these labels labelled z^{m-2} represents an intermediate variable, which means this kind of variable is used as output for a former rule base and then again as inputs for a latter rule base as illustrated in Fig.7. On the other hand, there are two kinds of nodes representing rule bases as illustrated in Fig.7, one of which is a type of standard rule bases and labelled RB_{m-1}. This kind of nodes is used to transform the input(s) to output(s). The other type of nodes, in addition to the standard type, represents identities. It can be seen from the Fig.7 that this type of nodes does not make changes between inputs and outputs. This indicates the functionality of an identity is just like an email transmission, which means the inputs are exactly the same as the outputs.

In practice, a complex problem could be subdivided into a number of smaller subproblems. The sub-problems may need to be solved sequentially in some cases. They can also be solved in parallel in other cases. In connection to machine learning context, each sub-problem could be solved by using a machine learning approach. In other words, the solver to each particular sub-problem could be a single machine learner or an ensemble learner of which a single rule base can consist.

In military process modelling and simulation, each networked rule base can be seen as a chain of command (chained rule bases [24]) with radio transmissions (identities). In a large scale raid, there may be more than one chain of command. From this point of view, the networked topology should have more than one networked rule bases parallel each other. All these networked rule bases should finally connect to a single rule base which represents the Centre of command.

The basis of above descriptions highlights the further directions of this research area. The extensions with respects to both sequential ensemble learning and networked rule bases would improve the intractability among different algorithms or models during the process of collaborative decision making. In addition, this also improves towards reduction of complexity in problem solving by dividing a complex problem into a set of simple problems. Therefore, the contributions would also be to complexity management [30], systems engineering [31, 32, 33, 34] and Big Data processing [35] in addition to machine learning.

References

1. Knoneko, I., Kukar, M.: Machine Learning and Data Mining: Introduction to Principles and Algorithms. Horwood Publishing, Chichester (2007)
2. Breiman, L.: Bagging Predictors. Machine Learning 2(24), 123–140 (1996)
3. Tan, P.N., Steinbach, M., Kumar, V.: Introduction to Data Mining. Pearson Education, Inc., New Jersey (2006)
4. Cendrowska, J.: PRISM: an algorithm for inducing modular rules. International Journal of Man-Machine Studies 27, 349–370 (1987)
5. Bramer, M.A.: Automatic induction of classification rules from examples using N-Prism. In: Research and Development in Intelligent Systems, vol. XVI, pp. 99–121. Springer, Cambridge (2000)

6. Liu, H., Gegov, A.: Induction of modular classification rules by Information Entropy Based Rule Generation. In: Sgurev, V., Yager, R., Kacprzyk, J. (eds.) Innovative Issues in Intelligent Systems. Springer (in press)

7. Michalski, R.S.: On the Quasi-Minimal solution of the general covering problem. In: Proceedings of the Fifth International Symposium on Information Processing, Bled, Yugoslavia, pp. 125–128 (1969)

8. Quinlan, J.R.: C4.5: Programs for Machine Learning. Morgan Kaufman (1993)

9. Bramer, M.A.: Using J-Pruning to Reduce Overfitting of Classification Rules in Noisy Domains. In: Hameurlain, A., Cicchetti, R., Traunmüller, R. (eds.) DEXA 2002. LNCS, vol. 2453, p. 433. Springer, Heidelberg (2002)

10. Stahl, F., Bramer, M.A.: Random Prism: a noise-tolerant alternative to Random Forests. Expert Systems, Wiley Online Library (2013)

11. Stahl, F., Bramer, M.A.: Random Prism: an alternative to Random Forests. In: Research and Development in Intelligent Systems XXVIII, pp. 5–18. Springer (2011)

12. Bramer, M.A.: Principles of Data Mining. Springer, London (2007)

13. Stahl, F., Bramer, M.A.: Jmax-pruning: A facility for the information theoretic pruning of modular classification rules. Knowledge-Based Systems 29, 12–19 (2012)

14. Stahl, F., Bramer, M.A.: Induction of modular classification rules: using Jmax-pruning. In: Thirtieth SGAI International Conference on Innovative Techniques and Applications of Artificial Intelligence, Cambridge, December 14-16 (2011)

15. Bramer, M.A.: Inducer: a public domain workbench for data mining. International Journal of Systems Science 36(14), 909–919 (2005)

16. Liu, H., Gegov, A., Stahl, F.: J-measure based hybrid pruning for complexity reduction in classification rules. WSEAS Transaction on Systems 12(9), 433–446 (2013)

17. Liu, H., Gegov, A., Stahl, F.: Unified Framework for Construction of Rule Based Classification Systems. In: Pedrycz, W., Chen, S.M. (eds.) Information Granularity, Big Data and Computational Intelligence. SBD, vol. 8, pp. 209–230. Springer, Heidelberg (2015)

18. Shannon, C.: A mathematical theory of communication. Bell System Technical Journal 27(3), 379–423 (1948)

19. Deng, X.: A Covering-based Algorithm for Classification: PRISM. CS831: Knowledge Discover in Databases (2012)

20. Breiman, L.: Random Forests. Machine Learning 45(1), 5–32 (2001)

21. Bache, K., Lichman, M.: UCI Machine Learning Repository. University of California, School of Information and Computer Science, Irvine (2013), http://archive.ics.uci.edu/ml

22. Hall, M., Frank, E., Holmes, G., Pfahringer, B., Reutemann, P., Witten, I.H.: The WEKA Data Mining Software: An Update; SIGKDD Explorations 11(1) (2009)

23. Machine Learning Open Source Software, http://jmlr.org/mloss/ (accessed on June 15, 2014)

24. Gegov, A.: Fuzzy Networks for Complex Systems: A Modular Rule Base Approach. Springer, Berlin (2010)

25. Gegov, A., Petrov, N., Vatchova, B.: Advanced Modelling of Complex Processes by Rule Based Networks. In: 5th IEEE International Conference on Intelligent Systems, London, July 7-9, pp. 197–202 (2010)

26. Gegov, A., Petrov, N., Vatchova, B., Sanders, D.: Advanced Modelling of Complex Processes by Fuzzy Networks. WSEAS Transactions on Circuits and Systems 10(10), 319–330 (2011)

27. Gegov, A.: Rule Based Networks: Theory and Applications. Plenary Lecture in IEEE International Conference on Intelligent Systems, London, UK (2010)
28. Gegov, A.: Rule Based Network Models for Complex Systems. Tutorial in ENNS International Conference on Neural Networks, Sofia, Bulgaria (2013)
29. Wooldridge, M.: An Introduction to Multi-Agent Systems. John Wiley & Sons (2002)
30. Gegov, A.: Complexity Management in Fuzzy Systems. Springer, Berlin (2007)
31. Schlager, J.: Systems engineering: key to modern development. IRE Transactions EM-3(3), 64–66 (1956)
32. Sage, A.P.: Systems Engineering. Wiley IEEE (1992)
33. Hall, A.D.: A Methodology for Systems Engineering. Van Nostrand Reinhold (1962)
34. Goode, H.H., Machol, R.E.: System Engineering: An Introduction to the Design of Large-scale Systems. McGraw-Hill (1957)
35. Gaber, M.M., Stahl, F., Gomes, J.B.: Pocket Data Mining. SBD, vol. 2. Springer, Heidelberg (2014)
36. Stahl, F., Jordanov, I.: An Overview of Use of Neural Networks for Data Mining Tasks. WIREs: Data Mining and Knowledge Discovery 3(2), 193–208 (2012)
37. Press, W.H., Teukolsky, S.A., Vetterling, W.T., Flannery, B.P.: Section 16.5. Support Vector Machines. In: Numerical Recipes: The Art of Scientific Computing, 3rd edn. Cambridge University Press, New York (2007)
38. Kerber, R.: ChiMerge: Discretization of Numeric Attributes. In: Proceedings of the 10th National Conference on Artificial Intelligence, pp. 123–128. AAAI Press (1992)
39. Houle, M.E., Kriegel, H.-P., Kröger, P., Schubert, E., Zimek, A.: Can Shared-Neighbor Distances Defeat the Curse of Dimensionality? In: Gertz, M., Ludäscher, B. (eds.) SSDBM 2010. LNCS, vol. 6187, pp. 482–500. Springer, Heidelberg (2010)

A GDM Method Based on Granular Computing for Academic Library Management

Francisco Javier Cabrerizo*, Raquel Ureña, Juan Antonio Morente-Molinera, and Enrique Herrera-Viedma

Abstract. An academic library, as a service organization, has to maintain a level of service quality that, satisfying its users, will assure funding for its existence and development. To do so, the general manager, which is in charge of distributing the funding, asks to the staff of the library about their opinions in the allocation of the budget. An important issue here is the level of agreement achieved among the staff before making a decision. In this paper, we propose a group decision making method based on granular computing aiding to the general manager to decide about funding distribution according to the staff's opinions. Assuming fuzzy preference relations to represent the preferences of the staff, a concept of a granular fuzzy preference relation is developed, where each pairwise comparison is formed as an information granule instead of a single numeric value. It offers the required flexibility to increase the level of agreement within the staff, using the fact that the most suitable numeric representative of the fuzzy preference relation is selected.

Keywords: Group decision making, Academic library, Granular computing, Consensus.

Francisco Javier Cabrerizo
Department of Software Engineering and Computer Systems,
Universidad Nacional de Educación a Distancia (UNED),
C/ Juan del Rosal 16, 28040 - Madrid, Spain
e-mail: cabrerizo@issi.uned.es

Raquel Ureña · Juan Antonio Morente-Molinera · Enrique Herrera-Viedma
Department of Computer Science and Artificial Intelligence,
University of Granada,
C/ Periodista Daniel Saucedo Aranda s/n, 18071 - Granada, Spain
e-mail: {raquel,jamoren,viedma}@decsai.ugr.es

* Corresponding author.

© Springer International Publishing Switzerland 2015
W. Pedrycz and S.-M. Chen (eds.), *Granular Computing and Decision-Making*
Studies in Big Data 10, DOI: 10.1007/978-3-319-16829-6_11

1 Introduction

Since the appearance of the first libraries in Academic Institutions, until today, the main role of academic libraries, and at the same time the fundamental reason of their existence, has been the support of the educational and research work performed within an Academic Institution [20, 27]. It is such an integral, functional part of a University, that it is very difficult to imagine a University without a library [5].

By the reason of academic libraries play an important role in the educational progress, it is vital to improve their services. However, since funding for higher education and universities has been reduced year after year [3, 46], the improvement of the quality of service in academic libraries according to the available funding is a critical and important task. Therefore, there is a need to determine the value and measure the performance of the academic library to distribute the funds among the different services which are offered to the users.

The funding distribution is a complex task because it is necessary to adapt the distribution to the users' needs, which are different in each case. For instance, it is not the same to manage an academic library whose users are students of Engineering that one which is frequented by specialists in History, because resources and information are different.

Usually, the person in charge of distributing the funds, called general manager, asks to the staff of the academic library about their impressions because they deal directly with the users and know their needs and worries about the library services. In addition, the general manager has confidence in the staff to consider their criteria about the distribution of the budget.

Considering the above factors, the main problem for the general manager is to rank the different library services in order to distribute the funds among them according to the staff's opinions. This situation can be seen as a Group Decision Making (GDM) problem, as it includes all the required conditions for this kind of problems.

A GDM problem is a situation where there is a set of possible alternatives to solve the problem and a group of decision makers who express their preferences about the alternatives [10, 18, 39]. Each decision maker may approach the decision process from a different angle, but they have a common interest in reaching an agreement on choosing the best alternative to solve the problem. In the case of the funding distribution among the library services, the alternatives are these services, and the set of decision makers of the problem is the staff of the academic library. Furthermore, here, the objective is to classify the library services from best to worst in order to distribute the funds.

An important question here is the level of agreement achieved among the staff of the academic library before making the decision. When decisions are made by a group of decision makers, it is recommendable that they are engaged in a consensus process [6, 28], in which all the decision makers discuss their reasons for making decisions in order to arrive at a sufficient level of agreement that is acceptable (to the highest possible extent) to all. Otherwise, it could be obtained solutions that are not well accepted by some decision makers in the group [6, 41], because they could consider that the solution achieved does not reflect their preferences, and hence, they might reject it. Therefore, GDM problems are usually faced by applying a consensus process before obtaining a final solution [9, 22, 23, 31, 33, 43, 47, 49, 57]. In any case, consensus needs that each decision maker has to allow a certain degree of flexibility and be ready to make modifications of his/her first opinions and, here, information granularity [36, 37, 38] may come into play.

Information granularity is an important design asset and may offer to each member of the staff a real level of flexibility using some initial preferences which can be adjusted with the intent to obtain a higher level of consensus. There exist several different representation formats in which decision makers can express their preferences [12, 13]. However, fuzzy preference relations [12, 26] have been widely used because they have proved to offer a very expressive representation and also because they present good properties allowing to operate with them easily [12, 26]. Hence, assuming that each member of the staff provides his/her preferences using a fuzzy preference relation, this required flexibility is brought into the fuzzy preference relations by allowing them to be granular rather than numeric. We consider that the entries of the fuzzy preference relations are not plain numbers but information granules, say intervals [4], fuzzy sets [51, 52, 53, 54], rough sets [42], probability density functions [56], and so on. In summary, information granularity that is present here serves as an important modeling asset, offering an ability of the member of the staff to exercise some flexibility to be used in modifying his/her initial opinion when becoming aware of the opinions of the other members of the staff. To do so, the fuzzy preference relation is elevated to its granular format.

The aim of this study is to present a GDM method based on granular computing to improve the academic library management. To do so, an allocation of information granularity as a key component to facilitate the achievement of consensus is proposed. We introduce a certain level of granularity in the realization of the granular representation of the fuzzy preference relations. It supplies the required flexibility to increase the level of consensus among the opinions expressed by the members of the staff. This proposed concept of granular fuzzy preference relation is used to optimize a performance index, which quantifies the level of consensus within the staff. Given the nature of the required optimization, the ensuing optimization problem is solved by engaging a machinery of population-based optimization, namely Particle Swarm Optimization (PSO) [32]. We should point out that the granulation formalism being considered concerns intervals (sets). However, it applies equally well to any other formal scheme of information granulation.

This study is outlined as follows. In Section 2, we start with the GDM scenario considered in this study and we describe the method which is utilized to obtain the level of consensus reached among the members of the staff. Section 3 deals with the GDM method based on granular computing for academic library management proposed in this study. Furthermore, we describe the use of PSO as the underlying optimization tool, giving strong attention to the content of the particles used in the method and a way in which the information granularity component is utilized in the adjustment of the single numeric values of the original fuzzy preference relations.

An example of application of the proposed method is reported in Section 4. Finally, we point out some conclusions and future works in Section 5.

2 Group Decision Making

In a classical GDM situation [10, 18, 29], there is a problem to solve, a solution set of possible alternatives, $X = \{x_1, x_2, \ldots, x_n\}$ ($n \geq 2$), and a group of two or more decision makers, $E = \{e_1, e_2, \ldots, e_m\}$ ($m \geq 2$), characterized by their background and knowledge, who express their opinions about the alternatives to achieve a common solution. In a fuzzy context, the objective is to classify the alternatives from best to worst, associating with them some degrees of preference expressed in the $[0, 1]$ interval.

As aforementioned, among the different representation formats that decision makers may use to express their opinions, fuzzy preference relations [29, 34] are one of the most used because of their effectiveness as a tool for modelling decision processes and their utility and easiness of use when we want to aggregate decision makers' preferences into group ones [29, 44].

Definition 1. A fuzzy preference relation PR on a set of alternatives X is a fuzzy set on the Cartesian product $X \times X$, i.e., it is characterized by a membership function $\mu_{PR} : X \times X \rightarrow [0, 1]$.

A fuzzy preference relation PR may be represented by the $n \times n$ matrix $PR = (pr_{ij})$, being $pr_{ij} = \mu_{PR}(x_i, x_j)$ ($\forall i, j \in \{1, \ldots, n\}$) interpreted as the preference degree or intensity of the alternative x_i over x_j: $pr_{ij} = 0.5$ indicates indifference between x_i and x_j ($x_i \sim x_j$), $pr_{ij} = 1$ indicates that x_i is absolutely preferred to x_j, and $pr_{ij} > 0.5$ indicates that x_i is preferred to x_j ($x_i \succ x_j$). Based on this interpretation we have that $pr_{ii} = 0.5$ $\forall i \in \{1, \ldots, n\}$ ($x_i \sim x_i$). Since pr_{ii}'s (as well as the corresponding elements on the main diagonal in some other matrices) do not matter, we will write them as '–' instead of 0.5 [25, 29].

Usually, GDM problems are faced by applying two different processes before a final solution can be given [2, 30]:

- *Consensus process.* It refers to how to obtain the maximum degree of agreement within the group of decision makers.
- *Selection process.* It obtains the final solution according to the preferences given by the decision makers.

The selection process is composed of two steps [7, 40]: *aggregation* of preferences provided by the decision makers and *exploitation* of the aggregated preference obtained previously. Clearly, it is recommendable that decision makers discuss and negotiate in order to achieve a sufficient agreement. Once the consensus level is higher than a specified threshold, the selection process is applied. For this reason, consensus has become a major area of research in GDM [6, 22, 28, 31, 33, 43, 57].

In order to evaluate the agreement achieved among the decision makers, we need to compute coincidence existing among them [24]. Usual consensus approaches determine consensus degrees, which are used to measure the current level of consensus in the decision process, given at three different levels of a preference relation [8, 22]: pairs of alternatives, alternatives, and relation.

In such a way, once the fuzzy preference relations have been provided by the decision makers, the computation of the consensus degrees is carried out as follows:

1. For each pair of decision makers (e_k, e_l) $(k = 1, \ldots, m-1, l = k+1, \ldots, m)$ a similarity matrix, $SM^{kl} = (sm_{ij}^{kl})$, is defined as:

$$sm_{ij}^{kl} = 1 - |pr_{ij}^k - pr_{ij}^l| \tag{1}$$

2. Then, a consensus matrix, $CM = (cm_{ij})$, is calculated by aggregating all the $(m-1) \times (m-2)$ similarity matrices using the arithmetic mean as the aggregation function, ϕ:

$$cm_{ij} = \phi(sm_{ij}^{kl}), \ k = 1, \ldots, m-1, \ l = k+1, \ldots, m \tag{2}$$

In this case, the arithmetic mean is utilized as aggregation function, although different aggregation operators could be utilized according to the particular properties that we want to implement.

3. Once the consensus matrix has been computed, the consensus degrees are obtained at three different levels:

   a. *Consensus degree on pairs of alternatives.* The consensus degree on a pair of alternatives (x_i, x_j), called cp_{ij}, is defined to measure the consensus degree among all the decision makers on that pair of alternatives. In this case, this is expressed by the element of the collective similarity matrix CM:

$$cp_{ij} = cm_{ij} \tag{3}$$

   The closer cp_{ij} to 1, the greater the agreement among all the decision makers on the pair of alternatives (x_i, x_j).

   b. *Consensus degree on alternatives.* The consensus degree on the alternative x_i, called ca_i, is defined to measure the consensus degree among all the decision makers on that alternative:

$$ca_i = \frac{\sum_{j=1; j \neq i}^{n}(cp_{ij} + cp_{ji})}{2(n-1)} \tag{4}$$

   c. *Consensus degree on the relation.* The consensus degree on the relation, called cr, expresses the global consensus degree among all the decision makers' opinions. It is computed as the average of all the consensus degree for the alternatives:

$$cr = \frac{\sum_{i=1}^{n} ca_i}{n} \tag{5}$$

The consensus degree of the relation, cr, is the value used to control the consensus situation. The closer cr is to 1, the greater the agreement among all the decision makers' opinions.

3 Applying a GDM Method Based on Granular Computing in Academic Library Management

Academic libraries have changed from the storehouses of books to the powerhouses of knowledge and information since the middle of the 20th century. The information and communication technology, which is responsible for this revolution, has drastically changed the organization, management and operation mode of modern academic libraries.

Nowadays, the existence of an academic library is fully dependent on the satisfaction of its users. Therefore, academic libraries are now more concerned about their users, and their satisfaction, the quality of their services, and their proper marketing. A user is satisfied when the academic library is able to rise to his/her expectations or actual needs. Thus, the academic library and the information professionals have to properly understand the users, what they want, how they want it and when they want the documents or information from the academic library.

It is needless to say that it is important to have a system through which the user needs are taken into account and it has to be used to improve the quality of the library services by distributing the funds according to these needs. There are several methods, tools or techniques to measure, control and improve the service quality of an academic library. In particular, the LibQUAL+ survey model [15] is a popular method to evaluate the quality of the academic libraries according to the user satisfaction.

We propose a GDM method in which the members of the staff are who give their opinions about the needs of funding in each library service with the aim of distributing the funds according to the user needs. Taking it into account, the general manager will distribute the funds among the library services in order to improve their quality.

To establish the library services to be assessed by the members of the staff, we follow the LibQUAL+ survey model [15], where three library services are considered:

- *Affect of service*. This library service assesses empathy, responsiveness, assurance, and reliability of library employees. In this case, new funds could contribute to improve the staff knowledge by means of courses, to hire new staff, and so on.
- *Information control*. This library service measures how users want to interact with the modern library and include scope, timeliness and convenience, ease of navigation, modern equipment, and self-reliance. In this case, new funds could contribute to buy new books, to subscribe new journals, to actualize old computers, and so on.

- *Library as place*. This library service measures the usefulness of space, the symbolic value of the library, and the library as a refuge for work of study. In this case, new funds could contribute to buy new furniture, to build new rooms, and so on.

Furthermore, by considering the advance of use of new technologies in traditional academic libraries [21], we identify other library services related to the development of the new 2.0 functionality in the library activities:

- *New 2.0 functionality*. This library service measures the usefulness of the web 2.0 services. In this case, new funds could contribute to actualize the web page, to make new web services (wikis, blogs, podcast, etc.), and so on.

In such a way, we consider four library services which could be potential receptors of funds depending on the general manager's decision which is taken according to the opinions provided by the members of the staff. To do so, we present a GDM model that collects the individual staff's opinions about the funding needs of the library services and shows to the general manager the computed ranking of these services. With this kind of model, if the staff members reach a consensual collective solution, the problem can have a quickly and precise solution. As it is known, the consensus calls for some flexibility exhibited by all members of the group, who in the name of cooperative pursuits give up their initial opinions and show a certain level of elasticity. Therefore, here, information granularity [36, 37, 38] can be used.

In the following subsections, the two steps of the GDM model proposed in this study, that is, the consensus process and the selection process, are described in detail.

3.1 Improving Consensus through an Allocation of Information Granularity

The improvement of consensus becomes a very important aspect in order to arrive a solution that each member of the staff is comfortable with. It is not necessary to say that it calls for some flexibility exhibited by them.

The changes in the opinions provided by the members of the staff are articulated through alterations of the entries of the fuzzy preference relations. In such a way, if the pairwise comparisons of the fuzzy preference relations are not managed as single numeric values, which are inflexible, but rather as information granules, it will bring the essential factor of flexibility.

This means that the fuzzy preference relation is abstracted to its granular format. The notation $G(PR)$ is used to emphasize the fact that we are interested in granular fuzzy preference relations, where $G(.)$ represents a specific granular formalism being used here (for instance, fuzzy sets, intervals, probability density functions, rough sets, and alike). We propose the concept of granular fuzzy preference relation and accentuate a role of information granularity being regarded here as an important conceptual and computational resource which can be exploited as a means to increase the level of consensus achieved among the members of the staff. In short, the level of granularity is treated as synonymous of the level of flexibility, which makes easy the improvement of consensus.

The higher level of granularity is offered to the members of the staff, the higher the possibility of arriving at decisions accepted by all them. In this contribution, the granularity of information is articulated through intervals and, hence, the length of such intervals can be sought as a level of granularity α. As here we are using interval-valued fuzzy preference relations, $G(PR) = P(PR)$, where $P(.)$ denotes a family of intervals.

The flexibility given by the level of granularity can be effectively used to optimize a certain optimization criterion to capture the essence of the reconciliation of the individual opinions. In what follows, we give the details both the optimization criterion to be optimized and its optimization using the PSO framework.

3.1.1 The Optimization Criterion

We suppose that each member of the staff feels equally comfortable when choosing any fuzzy preference relation whose values are located within the bounds established by the fixed level of granularity α. This level of granularity is employed to increase the level of consensus within the staff by bringing all opinions close to each other.

This goal is realized by maximizing the global consensus degree among all the preferences of the staff members, which is quantified in terms of the consensus degree on the relation described in Sect. 2:

$$Q = cr \tag{6}$$

Therefore, the optimization problem reads as follows:

$$\text{Max}_{PR^1, PR^2, \ldots, PR^m \in P(PR)} Q \tag{7}$$

The aforementioned maximization problem is carried out for all interval-valued fuzzy preference relations admissible because of the introduced level of information granularity α. This fact is underlined by including a granular form of the fuzzy preference relations allowed in the problem, i.e., $PR^1, PR^2, PR^3, PR^4, \ldots, PR^m$, are elements of the family of interval-valued fuzzy preference relations, specifically, $P(PR)$.

This optimization task is not an easy one. Because of the nature of the indirect relationship between optimized fuzzy preference relations, which are selected from a quite large search space formed by $P(PR)$, it calls for the use of advanced techniques of global optimization, as for instance: evolutionary optimization, genetic algorithms, PSO, ant colonies, simulated annealing, and so forth.

Here the optimization of the fuzzy preference relations, coming from the space of interval-valued fuzzy preference relations, is realized by means of the PSO, which is a viable optimization alternative for this problem, as it offers a substantial level of optimization flexibility and does not come with a prohibitively high level of computational overhead as this is the case of other techniques of global optimization (for example, genetic algorithms). Of course, some other optimization mechanisms could be use as well.

3.1.2 PSO Environment in Optimization of Fuzzy Preference Relations

The optimization of the fuzzy preference relations coming from the space of granular preference relations (more precisely, interval-valued fuzzy preference relations) is realized by means of the PSO, which occurs to a viable optimization alternative for this problem.

It is a population-based stochastic optimization technique developed by Kennedy and Eberhart [32], which is inspired by social behavior of bird flocking or fish schooling. A particle swarm is a population of particles, which are possible solutions to an optimization problem located in the multidimensional search space. The PSO is well documented in the existing literature with numerous modifications and augmentations [17, 19, 32, 45, 59].

One the one hand, what is important in this setting is finding a suitable mapping between problem solution and the particle's representation. Here, each particle represents a vector whose entries are located in the interval $[0,1]$. Basically, if there is a group of m members of the staff and a set of n alternatives, the number of entries of the particle is $m \cdot n(n-1)$.

Starting with the initial fuzzy preference relation provided by the member of the staff and assuming a given level of granularity α (located in the unit interval), let us consider an entry pr_{ij}. The interval of admissible values of this entry of $P(PR)$ implied by the level of granularity is equal to:

$$[a,b] = [\max(0, pr_{ij} - \alpha/2), \min(1, pr_{ij} + \alpha/2)] \tag{8}$$

Let assume that the entry of interest of the particle is x. It is transformed linearly according to the expression $z = a + (b-a)x$. For example, consider that pr_{ij} is equal to 0.7, the admissible level of granularity $\alpha = 0.1$, and the corresponding entry of the particle is $x = 0.4$. Then, the corresponding interval of the granular fuzzy preference relation computed as given by Eq. (8) becomes equal to $[a,b] = [0.65, 0.75]$. Subsequently, $z = 0.69$, and, therefore, the modified value of pr_{ij} becomes equal to 0.69.

The overall particle is composed of the individual segments, where each of them is concerned with the optimization of the parameters of the fuzzy preference relations.

On the other hand, the performance of each particle during its movement is assessed by means of some performance index (fitness function). Here, the aim of the PSO is the maximization of the consensus achieved among the members of the staff. Therefore, the fitness function, f, associated with the particle is defined as:

$$f = Q \tag{9}$$

being Q the optimization criterion presented previously. The higher the value of f, the better the particle is.

Finally, it is important to note that, in this contribution, the generic form of the PSO algorithm is used. Here, the updates of the velocity of a particle are realized in the form $\mathbf{v}(t+1) = w \times \mathbf{v}(t) + c_1 \mathbf{a} \cdot (\mathbf{z}_p - \mathbf{z}) + c_2 \mathbf{b} \cdot (\mathbf{z}_g - \mathbf{z})$ where "t" is an index of the generation and $\cdot$ denotes a vector multiplication realized coordinatewise. $\mathbf{z}_p$ denotes the best position reported so far for the particle under discussion while $\mathbf{z}_g$ is the best position overall and developed so far across the entire population. The current velocity $\mathbf{v}(t)$ is scaled by the inertia weight (w) which emphasizes some effect of resistance to change the current velocity. The value of the inertia weight is kept constant through the entire optimization process and equal to 0.2 (this value is commonly encountered in the existing literature [35]). By using the inertia component, we form the memory effect of the particle. The two other parameters of the PSO, that is $\mathbf{a}$ and $\mathbf{b}$, are vectors of random numbers drawn from the uniform distribution over the $[0,1]$ interval. These two update components help form a proper mix of the components of the velocity. The second expression governing the change in the velocity of the particle is particularly interesting as it nicely captures the relationships between the particle and its history as well as the history of overall population in terms of their performance reported so far. The next position (in iteration step "t+1") of the particle is computed in a straightforward manner: $\mathbf{z}(t+1) = \mathbf{z}(t) + \mathbf{v}(t+1)$.

When it comes to the representation of solutions, the particle $\mathbf{z}$ consists of "$m \cdot n(n-1)$" entries positioned in the $[0,1]$ interval that corresponds to the search space. One should note that while PSO optimizes the fitness function, there is no guarantee that the result is optimal, rather than that we can refer to the solution as the best one being formed by the PSO.

3.2 Selection Process

When all members of the staff have provided their fuzzy preference relations about the library services and the above procedure has been applied in order to increase the level of consensus, a ranking of library services can be obtained applying a selection process [7, 25]. This selection process is carried out in two sequential phases:

- *Aggregation.* It defines a collective fuzzy preference relation indicating the global preference between every pair of alternatives.
- *Exploitation.* It transforms the global information about the alternatives into a global ranking of them, from which a set of alternatives is derived.

In what follows, we describe in detail both the aggregation phase and the exploitation phase.

3.2.1 Aggregation: The Collective Fuzzy Preference Relation

In this phase, a collective fuzzy preference relation, $PR^c = (pr_{ij}^c)$, is obtained by aggregating all individual fuzzy preference relations, $\{PR^1, \ldots, PR^m\}$. Each value pr_{ij}^c represents the preference of the alternative x_i over the alternative x_j according to the majority of the staff' opinions. To do that, an OWA operator is used [50].

Definition 2. An OWA operator of dimension n is a function $\phi : [0,1]^n \longrightarrow [0,1]$, that has a weighting vector associated with it, $W = (w_1, \ldots, w_n)$, with $w_i \in [0,1]$, $\sum_{i=1}^n w_i = 1$, and it is defined according to the following expression:

$$\phi_W(a_1, \ldots, a_n) = W \cdot B^T = \sum_{i=1}^m w_i \cdot a_{\sigma(i)} \tag{10}$$

being $\sigma : \{1, \ldots, n\} \longrightarrow \{1, \ldots, n\}$ a permutation such that $p_{\sigma(i)} \geq a_{\sigma(i+1)}$, $\forall i = 1, \ldots, n-1$, i.e., $a_{\sigma(i)}$ is the i-highest value in the set $\{a_1, \ldots, a_n\}$.

The OWA operators fill the gap between the operators Min and Max. It can be immediately verified that OWA operators are commutative, increasing monotonous and idempotent, but in general not associative.

In order to classify OWA aggregation operators with regards to their localization between "or" and "and", Yager [50] introduced the measure of *orness* associated with any vector W expressed as:

$$\text{orness}(W) = \frac{1}{n-1} \sum_{i=1}^n (n-i)w_i \tag{11}$$

This measure, which lies in the unit interval, characterizes the degree to which the aggregation is like an "or" (Max) operation. Note that the nearer W is to an "or", the closer its measure is to one; while the nearer it is to an "and", the closer is to zero. As we move weight up the vector we increase the orness(W), while moving weight down causes us to decrease orness(W). Therefore, an OWA operator with much of nonzero weights near the top will be an "orlike" operator (orness(W) ≥ 0.5), and when much of the weights are nonzero near the bottom, the OWA operator will be "andlike" (orness(W) < 0.5).

A natural question in the definition of the OWA operator is how to obtain the associated weighting vector. In [50], it was defined an expression to obtain W that allows to represent the concept of fuzzy majority [29] by means of a fuzzy linguistic non-decreasing quantifier Q [55]:

$$w_i = Q\left(\frac{i}{n}\right) - Q\left(\frac{i-1}{n}\right), \, i = 1, \ldots, n \tag{12}$$

The membership function of Q is given by Eq. (13), with $a, b, r \in [0,1]$. Some examples of non-decreasing proportional fuzzy linguistic quantifiers are: "most" (0.3, 0.8), "at least half" (0, 0.5), and "as many as possible" (0.5, 1).

$$Q(r) = \begin{cases} 0 & \text{if } r < a \\ \frac{r-a}{b-a} & \text{if } a \leq r \leq b \\ 1 & \text{if } r > a \end{cases} \qquad (13)$$

When a fuzzy quantifier Q is used to compute the weights of the OWA operator ϕ, it is symbolized by ϕ_Q.

3.2.2 Exploitation: Ranking the Library Services

Using again the OWA operator and the concept of fuzzy majority (of alternatives), two choice degrees of alternatives may be used: the *quantifier-guided dominance degree (QGDD)* and the *quantifier-guided non-dominance degree (QGNDD)* [7, 12, 25]. These choice degrees will act over the collective preference relation resulting in a global ranking of the alternatives (library services), from which the solution will be obtained.

- $QGDD_i$: This quantifier guided dominance degree quantifies the dominance that one alternative has over all the others in a fuzzy majority sense. It is defined as follows:

$$QGDD_i = \phi_Q(pr^c_{i1}, pr^c_{i2}, \ldots, pr^c_{i(i-1)}, pr^c_{i(i+1)}, \ldots, pr^c_{in}) \qquad (14)$$

- $QGNDD_i$: This quantifier guided non-dominance degree gives the degree in which each alternative is not dominated by a fuzzy majority of the remaining alternatives. It is defined as follows:

$$QGNDD_i = \phi_Q(1 - p^s_{1i}, 1 - p^s_{2i}, \ldots, 1 - p^s_{(i-1)i}, 1 - p^s_{(i+1)i}, \ldots, 1 - p^s_{ni}) \qquad (15)$$

where $p^s_{ji} = max\{pr^c_{ji} - pr^c_{ij}, 0\}$ represents the degree in which x_i is strictly dominated by x_j. When the fuzzy quantifier represents the statement "all", whose algebraic aggregation corresponds to the conjunction operator Min, this non-dominance degree coincides with Orlovski's non-dominated alternative concept [34].

The application of the above choice degrees of alternatives over X may be carried out according to two different policies: *sequential policy* and *conjunctive policy* [11, 25]. On the one hand, in the sequential policy, one of the choice degrees is selected and applied to X according to the preference of the members of the staff, obtaining a selection set of alternatives. If there is more than one alternative in this selection set, then, the other choice degree is applied to select the alternative of this set with the best second choice degree. One the other hand, in the conjunctive policy, both choice degrees are applied to X, obtaining two selection sets of alternatives. The final selection set of alternatives is obtained as the intersection of these two selection sets of alternatives.

The latter conjunction selection process is more restrictive than the former sequential selection process because it is possible to obtain an empty selection set.

4 Example of Application

In this section, we show an example of application which helps quantifying the performance of the GDM method proposed in this study. In particular, we highlight the advantages, which are brought by an effective allocation of information granularity in the improvement of consensus.

Proceeding with the details of the optimization environment, PSO was used with the following values of the parameters which are selected as a result of intensive experimentation:

- The size of the swarm consisted of 50 particles. This size of the population was found to produce "stable" results meaning that very similar or identical results were reported in successive runs of the PSO. Due to the research space, this particular size of the population was suitable to realize a search process.
- The number of iterations (or generations) was set to 200. It was observed that after 200 iterations, there were no further changes of the values of the fitness function.
- The parameters in the update equation for the velocity of the particle were set as $c_1 = c_2 = 2$. These values are commonly encountered in the existing literature.

Let us suppose that the manager of an academic library wants to invest a sum of money in improving the following library services:

- x_1: Affect of service.
- x_2: Information control.
- x_3: Library as place.
- x_4: New 2.0 functionality.

To do so, the general manager asks to the four members of the staff, $\{e_1, e_2, e_3, e_4\}$, about their opinions on what library services need more funds in order to improve them. According to opinions of the members of the staff, the general manager will distribute the funds among the different library services.

The four members of the staff provide the following fuzzy preference relations:

$$PR^1 = \begin{pmatrix} - & 0.40 & 0.30 & 0.30 \\ 0.40 & - & 0.70 & 0.70 \\ 0.40 & 0.10 & - & 0.20 \\ 0.20 & 0.30 & 0.70 & - \end{pmatrix} \quad PR^2 = \begin{pmatrix} - & 0.20 & 0.60 & 0.30 \\ 0.60 & - & 0.50 & 0.30 \\ 0.20 & 0.30 & - & 0.50 \\ 0.10 & 0.30 & 0.90 & - \end{pmatrix}$$

$$PR^3 = \begin{pmatrix} - & 0.20 & 0.50 & 0.10 \\ 0.40 & - & 0.20 & 0.80 \\ 0.50 & 0.40 & - & 0.90 \\ 0.90 & 0.10 & 0.40 & - \end{pmatrix} \quad PR^4 = \begin{pmatrix} - & 0.60 & 0.20 & 0.60 \\ 0.40 & - & 0.60 & 0.20 \\ 0.80 & 0.60 & - & 0.50 \\ 0.40 & 0.60 & 0.60 & - \end{pmatrix}$$

Furthermore, the general manager has decided that the minimum consensus threshold is 0.75. Therefore, the consensus level achieved among the members of the staff has to be higher than 0.75 in order to apply the selection process.

4.1 Consensus Process

Considering a given level of granularity α, Fig. 1 illustrates the performance of the PSO quantified in terms of the fitness function (optimization criterion) obtained in successive generations. The most notable improvement is noted at the very beginning of the optimization, and afterwards, there is a clearly visible stabilization, where the values of the fitness function remain constant.

To put the obtained optimization results in a certain context, we report the performance obtained when no granularity is allowed ($\alpha = 0$), that is, when considering the entries of the fuzzy preference relations are single numeric values. In such a case, the corresponding consensus level achieved among the members of the staff is 0.722, which is lower than the minimum consensus threshold.

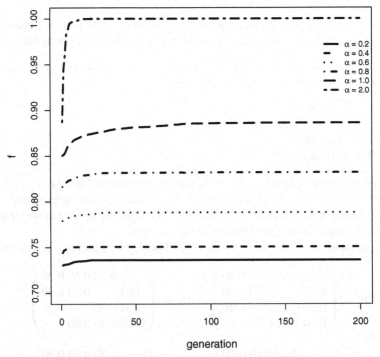

Fig. 1 Fitness function f in successive PSO generations for selected values of α

Comparing with the values obtained by the PSO, the level of agreement or consensus takes on now lower values. As we can see in Fig. 1, the higher the admitted level of granularity α, the higher the values obtained by the fitness function f. It is not surprising because the higher the level of granularity α, the higher the level of flexibility introduced in the fuzzy preference relations and, therefore, the possibility of achieving a higher level of consensus among the members of the staff. In particular, when each entry of the granular preference relation is treated as the whole $[0, 1]$ interval (it occurs when $\alpha = 2.0$), the value of the fitness function is the maximum one, which is 1. However, when the level of granularity is very high, the values of the entries of the fuzzy preference relation could be very different in comparison with the original values provided by the member of the staff and, therefore, he/she could reject them.

Comparing this consensus approach with the majority of the existing ones, where several consensus rounds are carried out, we can see that here the consensus is built in a single step. It reduces the amount of time required for building consensus. However, as negotiations among the decision makers are not included, the decision makers influencing each other are not considered.

4.2 Selection Process

Once the consensus among the members of the staff is higher than the minimum threshold, the selection process can be applied in order to rank the library services according to the opinions provided by the members of the staff.

For example, with a level of granularity $\alpha = 0.6$, the consensus level achieved among the members of the staff is 0.788, which is higher than 0.75.

In such a way, using this level of granularity, the new fuzzy preference relations obtained using the PSO are:

$$PR^1 = \begin{pmatrix} - & 0.22 & 0.12 & 0.12 \\ 0.22 & - & 0.52 & 0.52 \\ 0.22 & 0.08 & - & 0.10 \\ 0.10 & 0.12 & 0.52 & - \end{pmatrix} \quad PR^2 = \begin{pmatrix} - & 0.10 & 0.42 & 0.12 \\ 0.42 & - & 0.52 & 0.12 \\ 0.10 & 0.12 & - & 0.32 \\ 0.08 & 0.12 & 0.68 & - \end{pmatrix}$$

$$PR^3 = \begin{pmatrix} - & 0.10 & 0.32 & 0.08 \\ 0.22 & - & 0.10 & 0.60 \\ 0.32 & 0.22 & - & 0.68 \\ 0.68 & 0.08 & 0.22 & - \end{pmatrix} \quad PR^4 = \begin{pmatrix} - & 0.30 & 0.00 & 0.30 \\ 0.10 & - & 0.30 & 0.00 \\ 0.50 & 0.30 & - & 0.20 \\ 0.10 & 0.30 & 0.30 & - \end{pmatrix}$$

4.2.1 Aggregation

Once the fuzzy preference relations have been optimized in order to increase the consensus level, we aggregate them by means of OWA operator. We make use of the linguistic quantifier "most", defined in Sect. 3.2.1, which, applying Eq. (12), generates a weighting vector of four values to obtain each collective preference value pr_{ij}^c. As example, the collective preference value pr_{12}^c is calculated in the following way:

$$w_1 = Q(1/4) - Q(0) = 0 - 0 = 0$$
$$w_2 = Q(2/4) - Q(1/4) = 0.4 - 0 = 0.4$$
$$w_3 = Q(3/4) - Q(2/4) = 0.9 - 0.4 = 0.5$$
$$w_4 = Q(1) - Q(3/4) = 1 - 0.9 = 0.1$$
$$pr_{12}^c = w_1 \cdot pr_{12}^4 + w_2 \cdot pr_{12}^1 + w_3 \cdot pr_{12}^2 + w_4 \cdot pr_{12}^3 = 0.15$$

Then, the collective fuzzy preference relation is:

$$PR^c = \begin{pmatrix} - & 0.15 & 0.19 & 0.12 \\ 0.21 & - & 0.37 & 0.27 \\ 0.25 & 0.16 & - & 0.24 \\ 0.10 & 0.12 & 0.38 & - \end{pmatrix}$$

4.2.2 Exploitation

Using again the same linguistic quantifier "most" and Eq. (12), we obtain the following weighting vector $W = (w_1, w_2, w_3)$:

$$w_1 = Q(1/3) - Q(0) = 0.07 - 0 = 0.07$$
$$w_2 = Q(2/3) - Q(1/3) = 0.73 - 0.07 = 0.66$$
$$w_3 = Q(1) - Q(2/3) = 1 - 0.73 = 0.27$$

Using, for example, the quantifier guided dominance degree, $QGDD_i$, we obtain the following values:

$$QGDD_1 = 0.15$$
$$QGDD_2 = 0.26$$
$$QGDD_3 = 0.22$$
$$QGDD_4 = 0.13$$

Finally, applying the sequential policy with the quantifier guided dominance degree, the following ranking of alternatives is obtained:

$$x_2 \succ x_3 \succ x_1 \succ x_4$$

Using this information, the general manager will distribute the available funds according to this ranking. In such a way, information control will be the library service which will receive more funds whereas new 2.0 functionality will be the library service that will receive less funds.

5 Concluding Remarks and Future Works

Policies, budget plans, and other organizational tasks frequently involve group discussions or meetings due to their effectiveness in making decisions. Academic library management is one of these situations in which several decision makers have to interact to reach a decision. It is important in order to distribute the available funds to improve the library services. An important issue here is the level of agreement achieved among the group members before making the decision.

Here, we have proposed a GDM method based on granular computing for academic library management. To do so, we have proposed a method based on an allocation of information granularity as an important asset to increase the level of consensus achieved among the members of the staff. Here, information granularity is an important and useful asset that support to improve consensus among the preferences given by the member of the staff. It offers a badly needed flexibility so that the granular fuzzy preference relations can produce numeric realizations so that the level of consensus is increased. To do so, the PSO environment has been shown to serve a suitable optimization framework.

In the future, it is worth continuing this research in several directions:

- The granular representation of fuzzy preference relations discussed in this contribution was the one using intervals. However, any other formalism of granular computing, especially fuzzy sets, could be equally applicable here.
- In the scenario analyzed in this contribution, a uniform allocation of granularity has been discussed, where the same level of granularity α has been allocated across all the fuzzy preference relations. However, a non-uniform distribution of granularity could be considered, where these levels are also optimized so that each member of the staff might have an individual value of α becoming available to his/her disposal.
- PSO has been used an optimization framework because it offers a great deal of flexibility. Different fitness functions could be easily accommodated and a multi-objective optimization can be sought. The need for the two-objective becomes apparent in case of a GDM problem where, in addition to the criterion of consensus, one can consider a maximization of the consistency achieved by each decision maker in his/her opinions. Due to the complexity of most GDM problem, decision makers' opinions may not satisfy formal properties that fuzzy preference relations are required to verify. Consistency is one of them, and it is associated with the transitivity property. Definition 1 dealing with a preference relation does not imply any kind of consistency property. In fact, preference values of a fuzzy preference relation can be contradictory. However, the study of consistency is crucial for avoiding misleading solutions in GDM [1, 14, 16, 48, 58]. To make a rational choice, properties to be satisfied by such fuzzy preference relations have been suggested [26]. For instance, the additive transitivity property could be used as it facilitates the verification of consistency in the case of fuzzy preference relations.

Acknowledgements. The authors would like to acknowledge FEDER financial support from the Projects FUZZYLING-II Project TIN2010-17876 and TIN2013-40658-P, and also the financial support from the Andalusian Excellence Projects TIC-05299 and TIC-5991.

References

1. Alonso, S., Cabrerizo, F.J., Chiclana, F., Herrera, F., Herrera-Viedma, E.: An interactive decision support system based on consistency criteria. Journal of Multiple-Valued Logic & Soft Computing 14(3-5), 371–385 (2008)
2. Alonso, S., Pérez, I.J., Cabrerizo, F.J., Herrera-Viedma, E.: A linguistic consensus model for web 2.0 communities. Applied Soft Computing 13(1), 149–157 (2013)
3. Baalhavaeji, F., Isfandyari-Moghaddam, A., Aqili, S.V., Shakooii, A.: Quality assessment of academic libraries' performance with a special reference to information technology-based services: Suggesting an evaluation checklist. The Electronic Library 28(4), 592–621 (2010)
4. Bargiela, A.: Interval and ellipsoidal uncertainty models. In: Pedrycz, W. (ed.) Granular Computing: An Emerging Paradigm, pp. 23–57. Physica-Verlag (2001)
5. Bokos, G.D.: From the "diffusion" of functions to the "recomposition" of the role: The future of the academic libraries in the context of educational and research. In: Proceedings of the 8th Panhellenic Conference of Academic Libraries, pp. 46–56 (1999)
6. Butler, C.T., Rothstein, A.: On conflict and consensus: A handbook on formal consensus decision making. Tahoma Partk (2006)
7. Cabrerizo, F.J., Heradio, R., Pérez, I.J., Herrera-Viedma, E.: A selection process based on additive consistency to deal with incomplete fuzzy linguistic information. Journal of Universal Computer Science 16(1), 62–81 (2010)
8. Cabrerizo, F.J., Moreno, J.M., Pérez, I.J., Herrera-Viedma, E.: Analyzing consensus approaches in fuzzy group decision making: Advantages and drawbacks. Soft Computing 14(5), 451–463 (2010)
9. Cabrerizo, F.J., Pérez, I.J., Herrera-Viedma, E.: Managing the consensus in group decision making in an unbalanced fuzzy linguistic context with incomplete information. Knowledge-Based Systems 23(2), 169–181 (2010)
10. Chen, S.J., Hwang, C.L.: Fuzzy multiple attributive decision making: Theory and its applications. Springer, Berlin (1992)
11. Chiclana, F., Herrera, F., Herrera-Viedma, E., Poyatos, M.C.: A classification method of alternatives of multiple preference ordering criteria based on fuzzy majority. The Journal of Fuzzy Mathematics 4(4), 801–813 (1996)
12. Chiclana, F., Herrera, F., Herrera-Viedma, E.: Integrating three representation models in fuzzy multipurpose decision making based on fuzzy preference relations. Fuzzy Sets and Systems 97(1), 33–48 (1998)
13. Chiclana, F., Herrera, F., Herrera-Viedma, E.: A note on the internal consistency of various preference representations. Fuzzy Sets and Systems 131(1), 75–78 (2002)
14. Chiclana, F., Mata, F., Martínez, L., Herrera-Viedma, E., Alonso, S.: Integration of a consistency control module within a consensus model. International Journal of Uncertainty Fuzziness and Knowledge-Based Systems 16(1), 35–53 (2008)
15. Cook, C., Heath, F.M.: Users' perception of library service quality: A LibQUAL+ qualitative study. Library Trends 49(4), 548–584 (2001)
16. Cutello, V., Montero, J.: Fuzzy rationality measures. Fuzzy Sets and Systems 62(1), 39–54 (1994)
17. Daneshyari, M., Yen, G.G.: Constrained multiple-swarm particle swarm optimization within a cultural framework. IEEE Transactions on Systems, Man and Cybernetics, Part A: Systems and Humans 42(2), 475–490 (2012)
18. Fodor, J., Roubens, M.: Fuzzy preference modelling and multicriteria decision support. Kluwer, Dordrecht (1994)

19. Fu, Y., Ding, M., Zhou, C.: Angle-encoded and quantum-behaved particle swarm optimization applied to three-dimensional route planning for UAV. IEEE Transactions on Systems, Man and Cybernetics, Part A: Systems and Humans 42(2), 511–526 (2012)
20. Grigoriadou, G., Kipourou, A., Mouratidis, E., Theodoridou, M.: Digital academic libraries: An important tool in engineering education. In: Proceedings of the 7th Baltic Region Seminar on Engineering Education, pp. 41–44 (2003)
21. Heradio, R., Cabrerizo, F.J., Fernández-Amorós, D., Herrera, M., Herrera-Viedma, E.: A fuzzy linguistic model to evaluate the quality of library 2.0 functionalities. International Journal of Information Management 33(4), 642–654 (2013)
22. Herrera, F., Herrera-Viedma, E., Verdegay, J.L.: A model of consensus in group decision making under linguistic assessments. Fuzzy Sets and Systems 78(1), 73–87 (1996)
23. Herrera, F., Herrera-Viedma, E., Verdegay, J.L.: A rational consensus model in group decision making using linguistic assessments. Fuzzy Sets and Systems 88(1), 31–49 (1997)
24. Herrera, F., Herrera-Viedma, E., Verdegay, J.L.: Linguistic measures based on fuzzy coincidence for reaching consensus in group decision making. International Journal of Approximate Reasoning 16(3-4), 309–334 (1997)
25. Herrera-Viedma, E., Herrera, F., Alonso, S.: Group decision-making model with incomplete fuzzy preference relations based on additive consistency. IEEE Transactions on Systems, Man and Cybernetics - Part B: Cybernetics 37(1), 176–189 (2007)
26. Herrera-Viedma, E., Herrera, F., Chiclana, F., Luque, M.: Some issues on consistency of fuzzy preference relations. European Journal of Operational Research 154(1), 98–109 (2004)
27. Herrera-Viedma, E., López-Gijón, J.: Libraries' social role in the information age. Science 339(6126), 1382 (2013)
28. Herrera-Viedma, E., Cabrerizo, F.J., Kacprzyk, J., Pedrycz, W.: A review of soft consensus models in a fuzzy environment. Information Fusion 17, 4–13 (2014)
29. Kacprzyk, J.: Group decision making with a fuzzy linguistic majority. Fuzzy Sets and Systems 18(2), 105–118 (1986)
30. Kacprzyk, J., Fedrizzi, M., Nurmi, H.: Group decision making and consensus under fuzzy preferences and fuzzy majority. Fuzzy Sets and Systems 49(1), 21–31 (1992)
31. Kacprzyk, J., Zadrozny, S.: Soft computing and web intelligence for supporting consensus reaching. Soft Computing 14(8), 833–846 (2010)
32. Kennedy, J., Eberhart, R.C.: Particle swarm optimization. In: Proceedings of the IEEE International Conference on Neural Networks, pp. 1942–1948 (1995)
33. Mata, F., Martínez, F., Herrera-Viedma, E.: An adaptive consensus support model for group decision making problems in a multi-granular fuzzy linguistic context. IEEE Transactions on Fuzzy Systems 17(2), 279–290 (2009)
34. Orlovski, S.A.: Decision-making with a fuzzy preference relation. Fuzzy Sets and Systems 1(3), 155–167 (1978)
35. Pedrycz, A., Hirota, K., Pedrycz, W., Dong, F.: Granular representation and granular computing with fuzzy sets. Fuzzy Sets and Systems 203, 17–32 (2012)
36. Pedrycz, W.: The principle of justifiable granularity and an optimization of information granularity allocation as fundamentals of granular computing. Journal of Information Processing Systems 7(3), 397–412 (2011)
37. Pedrycz, W.: Granular computing: Analysis and design of intelligent systems. CRC Press/Francis Taylor, Boca Raton (2013)
38. Pedrycz, W.: Knowledge management and semantic modeling: A role of information granularity. International Journal of Software Engineering and Knowledge 23(1), 5–12 (2013)

39. Pérez, I.J., Cabrerizo, F.J., Alonso, S., Herrera-Viedma, E.: A new consensus model for group decision making problems with non homogeneous experts. IEEE Transactions on Systems, Man, and Cybernetics: Systems 44(4), 494–498 (2014)

40. Roubens, M.: Fuzzy sets and decision analysis. Fuzzy Sets and Systems 90(2), 199–206 (1997)

41. Saint, S., Lawson, J.R.: Rules for reaching consensus: A moderm approach to decision making. Jossey-Bass (1994)

42. Słowiński, R., Greco, S., Matarazzo, B.: Rough set analysis of preference-ordered data. In: Alpigini, J.J., Peters, J.F., Skowron, A., Zhong, N. (eds.) RSCTC 2002. LNCS (LNAI), vol. 2475, pp. 44–59. Springer, Heidelberg (2002)

43. Szmidt, E., Kacprzyk, J.: A consensus-reaching process under intuitionistic fuzzy preference relations. International Journal of Intelligent Systems 18(7), 837–852 (2003)

44. Tanino, T.: Fuzzy preference orderings in group decision making. Fuzzy Sets and Systems 12(2), 117–131 (1984)

45. Tsekouras, G.E., Tsimikas, J.: On training RBF neural networks using input-output fuzzy clustering and particle swarm optimization. Fuzzy Sets and Systems 221, 65–89 (2013)

46. Wiener, S.A.: Library quality and impact: Is there a relationship between new measures and traditional measures? Journal of Academic Librarianship 31(5), 432–437 (2005)

47. Xu, Z.S.: An automatic approach to reaching consensus in multiple attribute group decision making. Computers & Industrial Engineering 56(4), 1369–1374 (2009)

48. Xu, Z.S.: Consistency of interval fuzzy preference relations in group decision making. Applied Soft Computing 11(5), 3898–3909 (2011)

49. Xu, Z.S., Cai, X.: Group consensus algorithms based on preference relations. Information Sciences 181(1), 150–162 (2011)

50. Yager, R.R.: On ordered weighted averaging aggregation operators in multicriteria decision making. IEEE Transactions on Systems Man and Cybernetics 18(1), 183–190 (1988)

51. Zadeh, L.A.: Fuzzy sets. Information and Control 8(3), 338–353 (1965)

52. Zadeh, L.A.: The concept of a linguistic variable and its applications to approximate reasoning. Part I. Information Sciences 8(3), 199–243 (1975)

53. Zadeh, L.A.: The concept of a linguistic variable and its applications to approximate reasoning. Part II. Information Sciences 8(4), 301–357 (1975)

54. Zadeh, L.A.: The concept of a linguistic variable and its applications to approximate reasoning. Part III. Information Sciences 9(1), 43–80 (1975)

55. Zadeh, L.A.: A computational approach to fuzzy quantifiers in natural languages. Computers & Mathematics with Applications 9(1), 149–184 (1983)

56. Zadeh, L.A.: Toward a perception-based theory of probabilistic reasoning with imprecise probabilities. Journal of Statistical Planning and Inference 105(1), 233–264 (2002)

57. Zhang, G., Dong, Y., Xu, Y., Li, H.: Minimum-cost consensus models under aggregation operators. IEEE Transactions on Systems, Man and Cybernetics, Part A: Systems and Humans 41(6), 1253–1261 (2011)

58. Zhang, G., Dong, Y., Xu, Y.: Linear optimization modeling of consistency issues in group decision making based on fuzzy preference relations. Expert Systems with Applications 39(3), 2415–2420 (2012)

59. Zhang, Y., Wang, S., Phillips, P., Ji, G.: Binary PSO with mutation operator for feature selection using decision tree applied to spam detection. Knowledge-Based Systems 64, 22–31 (2014)

Spatial-Taxon Information Granules as Used in Iterative Fuzzy-Decision-Making for Image Segmentation

Lauren Barghout

Abstract. An image conveys multiple meanings depending on the viewing context and the level of granularity at which the viewer perceptually organizes the scene. In image processing, an image can be similarly organized by means of a standardized natural-scene-taxonomy, borrowed from the study of human visual taxometrics. Such a method yields a three-dimensional representation comprised of a hierarchy of nested spatial-taxons. Spatial-taxons are information granules composed of pixel regions that are stationed at abstraction levels within hierarchically-nested scene-architecture. They are similar to the Gestalt psychological designation of figure-ground, but are extended to include foreground, object groups, objects and salient object parts. By using user interaction to determine scene scale and taxonomy structure, image segmentation can be operationalized into a series of iterative two-class fuzzy inferences. Spatial-taxons are segmented from a natural image via a three-step process. This chapter provides a gentle introduction to analogous human language and vision information-granules; and decision systems, modeled on fuzzy natural vision-based reasoning, that exploit techniques for measuring human consensus about spatial-taxon structure. A system based on natural vision-based reasoning is highly non-linear and dynamical. It arrives at an end-point spatial-taxon by adjusting to human input as it iterates. Human input determines the granularity of the query and consensus regarding spatial-taxon regions. The methods of concept algebra developed for computing with words [42] [48] are applied to spatial-taxons. Tools from the study of chaotic systems, such as tools for avoiding iteration problems, are explained in the context of fuzzy inference.

Lauren Barghout
Berkeley Initiative for Soft Computing (BISC),
University of California at Berkeley, United States
e-mail: lauren.barghout@gmail.com

© Springer International Publishing Switzerland 2015   285
W. Pedrycz and S.-M. Chen (eds.), *Granular Computing and Decision-Making*
Studies in Big Data 10, DOI: 10.1007/978-3-319-16829-6_12

Keywords: image segmentation, visual taxometrics, spatial-taxon, fuzzy logic, granular computing, decision-making, natural vision processing, scene perception, computer vision, artificial intelligence, general artificial intelligence, machine learning, Ljapunov exponent, non-linear dynamical systems.

1   Introduction

Computer vision is a subset of the field of Artificial Intelligence (AI). AI researchers fall into two camps: those trying to build machines that think and/or act like people; or those trying to build machines that think and/or act rationally. The methods described in this chapter live squarely in the camp attempting to build machines that see and/or act as though they see like people.

In his IPMU 2014 talk titled "Unifying Logic and Probability: A New Dawn for AI?", Stuart Russell predicts that we are currently at a new dawn for AI [33]. Historically, he points out, classical AI noticed that the world had things in it and used first-order logic to model knowledge about those things. Modern AI noticed that the world had lots of uncertainty and used tools, such as probability, to model uncertain knowledge about the world. As he tells it:

> "What happened next, of course, is that classical AI researchers noticed the pervasive uncertainty, while modern AI researchers noticed, or remembered, that the world has things in it. Both traditions arrived at the same place: the world is uncertain and it has things in it." -*Stuart Russell [33]*

Researchers in granular computing and fuzzy set theory also noticed that the world had things in it. They noticed that there was uncertainty as to the probability of things, set-membership of things, the possibility of these things and, most relevant to this chapter, the meaning of these things [44], [45], [51]. They created computational methods that enabled precise variables to be replaced with granular variables,[1] bins of values defined by general constraints [51]. Thus fuzzy logical inference and control systems were able to handle information at the scale most appropriate to the task at hand. This is contrary to what happened in the field of computer vision, where computational precision was so subordinate in scale to the tasks at hand that they lost the ability to apply first-order logic to these variables.

As defined by Bargiela & Pedrycs, information granules are conceptual entities that compactly encapsulate information at a specific level of abstraction (scale). Their properties result from the aggregation of even smaller information granules. Information granules give rise to hierarchies of cognitive entities. [51] Bargiela & Pedrycs note that the techniques associated with information granules are particularly useful for images.

---

[1] I.e. a variable X when replaced with X is R W, where W is A, constrains the precision of X. See Zadeh 2006 for a full description.

"At the higher end of abstraction, we are interested in image description and interpretation. Here the level of detail (or level of abstraction) depends on the task we have to handle. Images perceived by humans are full of information granules. An image of any landscape consists of trees, houses, roads, lakes, shrubs, etc. They are spatially distributed and this distribution is an important factor in describing the content of the image. Interestingly, all these objects are generic information granules. " -*Bargiela & Pedrycs [1]*

The information granules we'll be looking at in this chapter are called spatial-taxons. They are regions of pixels that are organized in a nested hierarchy of information. As with human perceptual organization, the organization of spatial-taxons depends on context and granularity.

Human vision is typically thought of as an open universe problem because every possible outcome is unknown. However, in this chapter, we start by framing image segmentation as a closed universe problem defined by a classical binary tree composed of spatial-taxons. We can frame it this way because, by our method, every pixel in the image must belong to a spatial taxon or its complement.

Defined in this way, the image segmentation problem can be operationalized into a series of iterative two-class fuzzy inferences. This framing of image segmentation is the only classical set theory used in this chapter. I choose to use crisp sets because image segments, as used by most application programming interfaces (APIs), require crisp sets. The rest of the chapter is devoted to methods for using fuzzy decision making to generate these 'de-fuzzified'image segments.

Definition 1. Hierarchy of nested spatial-taxons:

Let X be the universe of discourse consisting of all pixels within the rectangular (or square) pixel array of an image, such that $X_{1,1}$ is located at the upper left corner, and pixel $X_{I,J}$ at the lower left corner. Let ST_0 be a nonempty set that contains all pixels in the universe of discourse (the image). ST_0 has two mutually exclusive children ST_1 and ST_0 - ST_1 such that $ST_1 \wedge (ST_0 - ST_1) = \emptyset$ and $ST_1 \vee (ST_0 - ST_1) = ST_0$ (the parent). We have now defined abstraction level 0 and level 1. The most abstract information granule is the whole image and the second most abstract level contains two mutually exclusive children subsets.

Let's next define the set ST_1 as having two children subsets: ST_2, $(ST_1 - ST_2)$. As before, these children are mutually exclusive, such that $ST_2 \wedge (ST_1 - ST_2) = \emptyset$ and $ST_1 \vee (ST_0 - ST_1) = ST_0$ (the parent). This is the third most abstract level in nested spatial-taxon hierarchy.[2]

Though the hierarchy of nested spatial-taxons comprise crisp sets, not all the common properties of crisp sets apply. For one thing, a parent-child

---

[2] I could have used subset $(ST_1 - ST_2)$ as a root for a new child subset. However, to make this chapter readable, I limit the definition and all the examples to spatial-taxon children stemming from the initial image root

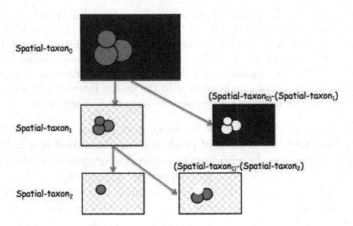

Fig. 1 A classical nested binary tree forms the architecture of a natural-scene-taxonomy. It's comprised of a hierarchy of nested spatial-taxons. Because each image (set) is a de-fuzzified output, it is represented as a crisp set. However, this frame can be made fuzzy and used in decision-making algorithms that require larger image information granules. By assuming the taxonomy prior to segmentation, segmentation becomes a series of two-class fuzzy inferences.

structure means that any one pixel may belong to many sets (though each at a different level of abstraction). This structure is not monotonic, i.e., *entailment does not always increase information.* This is especially true because the decision-making process described later may, over the course of its iteration, change its decision regarding pixel classification.

In the next section, I will introduce spatial-taxons within the framework of visual taxometrics. An example fuzzy-inference algorithm will be explained and applied to two simplified example images. Prior to explaining visual taxometrics, however, I will discuss an analogous linguistics problem. In this analogy, spatial-taxons correspond to basic-level words. Thus the isR relationships in visual taxometrics correspond to the isR relationships between basic-level words and their superordinate and subordinate word categories. My hope is that the framework for decision-making in the linguistics problem serves as a helpful analogue, clarifying and reinforcing the usefulness of such a framework for visual taxometrics.

2   Visual Taxometrics

2.1   *Taxometric Observation from an Analogous System: Language*

Before delving into the parsing of images into spatial-taxons, let's examine an analogous problem: the parsing of text documents for interpretation. To interpret a document, it must be subdivided into its relevant components, such as characters, words, sentences and paragraphs. Text documents (unlike images) have a standardized architecture with components designated by punctuation. Traditional punctuation and modern innovations, such as hypertext mark-up language (HTML), minimize uncertainty in the process of selecting which characters to include or exclude within a subdivision. [4]

Standardized architecture for written documents provides an example of a complex system that has proven to be stable across history, culture, and technical innovation. As pointed out by Nobel Laureate Herbert Simon (1962), "hierarchy is one of the central structural schemes that the architect of complexity uses." He further observes that hierarchic systems have some common properties that are independent of their specific content "and he roughly defines a complex system as a system in which "the whole is more than the sum of the parts... in the pragmatic sense that given the properties of the parts and the laws of their interaction, it is not a trivial matter to infer the properties of the whole."

Text-document architecture succeeds because its structure is independent of content semantics. Letters, words, sentences, and paragraphs follow the same structure - regardless of whether they belong to a document discussing fashion, religion or nature, or any other topic.

Now, let's shift our focus for a moment, and consider certain language restrictions or rules and how children learn them. In this case, we'll look at how children learn the relative granule restrictions between words. After all, the approach described in this chapter is squarely in the cognitive AI camp, a system built to think, see, or act like people. To build a decision-making system that infers spatial-taxons, we can borrow linguistic rules or restrictions from natural systems. In this case, the natural system we look to borrow from is the system parents use to teach language to children.

Parents, when teaching their children, chose words a child is most likely to grasp, i.e., words at the most basic level of abstraction. Abstraction levels in a word taxonomy range from the most basic term to the most specific, or specialized term, as well as ranging in the opposite direction from the most basic to the most generalized term.

> "Children do not try to guess what it is that the adult intends; rather they have certain concepts of these aspects of the world they find interesting and in successful cases of word acquisition it is the adult (at least in Western middle-class society) who guesses what the child is focused on and applies an appropriate word. " -*Katherine Nelson [26]*

Linguist Paul Bloom [14] describes the above quote as an example of word learning as an inductive process that stems from children's ability to form associations.[3]. The adult (or *trainer* in a machine learning context) provides the granular constraints by indicating the word's level of abstraction. In machine learning, the trainer provides this information per labeled data,

---

[3] Variable binding and the ability form associations about never before encountered things or experiences continues to be a challenge within the A.I. community.

and expects the machine learning systems to infer by example (as the child
infers by example). Bloom explains that children learn this relevancy principle directly from what we, in the fuzzy logic community, would call linguistic
constraints; those in the linguistic community call it linguistic support:

> Words are learned when they are relevant to what the child has in mind
> ...parents tailor their use of words to accord with their children's mental
> states. When interacting with young children, they tend to talk in the here
> and now, adjusting their conversational patterns to the fit the situation. They
> engage in *follow-in* labeling, in which they notice what their babies are
> looking at and name it. They even seem to have an implicit understanding
> that children assume that new words referring to objects will be basic-level
> names, such as 'dog 'or 'shoe ', so when adults present children with words
> that are not basic-level names, they use linguistic cues to make it clear that
> the words have a different status. For instance, when adults present part
> names to children they hardly ever point and say 'Look at the ears '. Instead
> they typically begin by talking about the whole object. 'This is a rabbit 'and
> then introduce the part name with a possessive construction 'and these are
> his ears. 'Similar linguistic support occurs for subordinates 'A pug is a kind
> of dog 'and superordinates 'These are animals'. 'Dogs and cats are kinds of
> animals'. " -*Lois Bloom [26]*

The analogy between linguistic and visual organization is powerful enough
to give us a very useful model for the structure of visual information. For
example, in visual hierarchies, the granule level is equivalent to the basic level
of words. And the role of the expert designing the fuzzy inference is somewhat
like the role of the parent teaching the child. The expert, in both cases, needs
to indicate (and in the case of the fuzzy inference expert, designate), by means
of constraints, the hierarchy status of any given word or spatial-taxon. The
child in this analogy is the homunculus/observer.

To establish an image-specific knowledge base, we use spatial-taxons, and
the domain-specific natural vision-processing rules used to derive the spatial-
taxons, in the role of the parent/teacher.[4]

If you accept this premise, then spatial-taxons, as defined in Definition 1,
are the appropriate level for granule computing within the visual taxometric
model. These information granules are also at the appropriate level to *elicit
human interaction and feedback*. Just as an adult provides linguistic support
to children learning non-basic level words, users can provide taxometric designation support regarding primary, subordinate and superordinate image
granules.[5]

---

[4] As put by Russell and Norvig "one might say that to solve a hard problem, you
have to almost know the answer already. "[34]

[5] Latin term means little man inside the head. Strong A.I. attempts to model
reasoning - hence the cartoon illustrates a privileged teacher correcting miscon-
ceptions. Note, Nobel Laureate Vladimir Vapnik suggests the use of the privileged
teacher, as an augmentation to support vector machines and Bayesian models.
However, they can also be used to inform a system of mistakes in abstraction
designation.

2.2 Spatial-Taxons: Basic Level Categories in Vision

Spatial-taxons are the granular level best suited to making decisions about the relevant way to slice-and-dice an image.

Prevailing theories on image segmentation are contrary to this approach. The majority of image segmentation systems work at the level of granularity of their feature extractors. Decision-making revolves around building larger information-granules comprised of pixel with similar features. Support-vector machines and/or machine learning techniques try to match these pixel regions with known reconstructions of objects. Though this approach works for specific image types, it still views image segmentation as an open universe problem - which when framed this way is likely to be intractable. I view trying to segment images at granules extracted by lower level visual attributes as

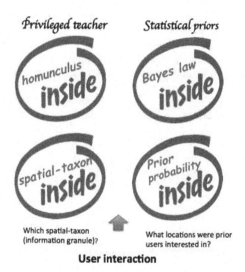

Fig. 2 Alternative knowledge models used in artificial intelligence. *Left Column* We try to solve the homunculus problem (little man inside the head, which becomes, in our context, the nested spatial taxonomy inside the head), by positing the following: that it is only by informing and constraining knowledge, that the homunculus can infer relationships between different levels of knowledge. In the case of language acquisition, relationships between taxonomy hierarchies are provided by a privileged teacher, such as the parent in the quote by Lois Bloom. *Right Column* Alternative A.I. models try to solve computer vision problems by measuring, at a high precision, statistical co-occurrences within very large data sets. They use these statistics to infer probability priors of the relations between data as mined from their training data. These methods fall short when the correct associations requires the A.I. system to extrapolate relations not already existing in the training data.

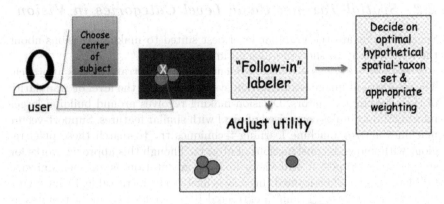

Adjust utility

Set of hypothetical spatial-taxons

Fig. 3 Graphical user interface helps identify spatial-taxon abstraction. Graphical user interface asks user to indicate the center of the subject. The system infers the abstraction level of the spatial-taxon hierarchy and increases the utility of hypothetical spatial-taxons that provide visual support. The graphical user interface mimics 'follow-in'behavior, enabled the system to notice where the user is looking.

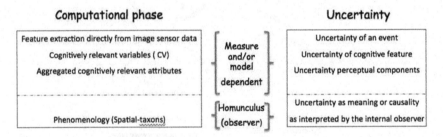

Fig. 4 Computational phase and corresponding uncertainty as per the visual-taxometric model. Unlike the visual taxometric approach, prevailing theories such as support-vector machines, machine learning and convolution networks attempt to infer relations at the feature extraction phase. At this phase uncertainty in the occurrence of a sensor event, or feature is typically managed with Bayesian statistics. In contrast, visual taxometrics uses fuzzy inference to manage uncertainty at low levels and makes decisions at the spatial-taxon level whose primary uncertainty is with respect to meaning and causality

analogous to trying to sub-divide property by trying to connect trees, bushes and rocks without the prior structure of a map or aerial view.

Images convey multiple meanings that depend on the context in which a viewer perceptually organizes the scene. Visual taxometrics provides tools for investigating context by distinguishing categorical scene structures from continuous percepts. Using these tools, Barghout [6] [11][6] showed that scene architecture is comprised of spatially distinct taxons that are characterized by the dichotomy of foreground/background sets, where the foreground is referred to as a spatial-taxon.

Spatial-taxons are regions (pixel-sets) that are figure-like, in that they are perceived as having a contour, are either 'thing-like', or a 'group of things', that draw our attention. Backgrounds, the complements of spatial-taxons, are shapeless regions that are perceived as the space existing around or behind the spatial-taxons. Notice that spatial-taxons are defined solely by their intrinsic properties, independent of the extrinsic properties of the particular objects of which they are composed.[7]

Spatial-taxon examples include foreground, figure, object groups or objects. [6] They are the 'building blocks', of scenes. In essence, they serve as a proxy for the figural status of a region. When human subjects are asked to mark the center of the subject of the image, they tend to choose the center of a spatial taxon, with little variance. They rarely choose locations defined solely by continuous visual percepts [6] [11]. This enables graphical user interfaces to query users directly, by asking them to locate what they are interested in (see figure 3). Graphical user interfaces can also query users indirectly, by using the reverse search queries to infer objects of interest. Because consensus location is centralized at the center of spatial taxons, this type of user query scales to large number of people. Furthermore, evidence suggests that the frequency at which people choose spatial-taxons at a particular abstraction level, regardless of image content follows a Zipfian distribution associated with Zipfs law [11]. This is consistent with the law of least effort demonstrated in other cognitive systems, and consistent with Simon's observations of complex systems [36].

The spatial-taxon view of scene perception assumes that humans parse scenes not between regions of similar features that vary continuously, but by categorically discrete spatial scene configurations. Theories of visual attention make a similar distinction. The 'spotlight theory'[38] assumes that attention regions vary continuously. Theories of 'object based'attention assume that attended spatial regions vary discretely to accommodate attended objects.

If humans are parsing scenes by inferring categories, then quantifying pixel-regions as to their aggregate 'trueness'relative to the category prototype is

---

[6] Jolicoeur, Gluck & Kosslyn[24] showed that entry level categories, similar to primary level words, exist in human vision

[7] This distinction between the intrinsic and extrinsic properties is what enables image segmentation without object recognition. Object recognition requires some sort of description of what the object looks like. If an object is subdivided, the object-part may not retain the 'looks'of the whole object. Suppose for example, we have a spatial-taxon comprised of a ladybug. If we cut the ladybug's head off - the head no longer fits the description of a ladybug. The region segmentation that comprises of the ladybug head, however, retains the properties of being a spatial-taxon.

necessary. As discussed earlier, fuzzy-logic provides tools for handling the partial or relative truth of meaning [53], and this allows us to make inferences based on visual percepts. [9]. I use the phrase "natural-vision-processing"to refer to the parsing of images into psychological variables whose relative truth (fuzzy membership) corresponds to human phenomenological interpretation.

The scene taxonomy enables us to side step the chicken/egg problem. Its spatial-taxons are independent of scene content semantics, providing the computational homunculus much more information as to where to look for 'visual-grammar conflicts'[9]. Adaptive filters can be chosen to optimize the spatial taxon cut. The optimal spatial taxon cut has been shown to be the point at which utility is maximized and use of attentional resources is minimized. Barghout [4] provides examples of how to calculate utility and attentional resource metrics. As in earlier models [5] [9] , the properties of local filters are changed to favor good grammar, i.e., transducer functions for contrast are shifting left or right. However, the taxonomy structure provides a theoretical basis on which to hang the concepts of good and bad visual grammar.

Furthermore, the Zipfian relationship between spatial-taxon hierarchies means that user feedback provided regarding a spatial-taxon at one abstraction can be used to infer information at all levels in the hierarchy.

In order to understand and use these ideas, we need concrete examples. I'll be using fuzzy cognitively relevant attributes, from the domain of vision science. I'm going to introduce spatial-taxon rules, adapted from Gestalt psychology, into sentential logic. As we continue through this section, we will apply these rules examples that increase in sophistication.

2.3   Spatial-Taxons: Gestalt Phenomenology

I'd like to take a moment to distinguish the the term "phenomenology "as used in this chapter. As a thought experiment image yourself as the person illustrated in figure 5. You are wearing stereoscopic 3D glasses that show different images to each eye. Your experience, when viewing the through the 3D glasses is of a unified image in depth. This experience (phenomenology/spatial-taxon) is in the bottom row of the chart in figure 4. It is a Gestalt experience that can not be subdivided without losing the phenomenology.

To demonstrate this to yourself, imagine that a light was flashed in the image shown to the left eye, as illustrated in the figure. Your experience will not enable you to determine which eye the flash occurred.

This distinction between sensory feature and the phenomenology experienced by people is important to keep in mind. Note that much of knowledge of inferred by the cognitively relevant variables used to infer the fused spatial-taxon are lost once the observer experiences a single unitary visual reality.

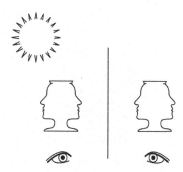

Fig. 5 Thought experiment: phenomenology of spatial-taxon. This thought experiment reveals the difference between the phenomenology of a Gestalt spatial-taxon and a group of features. Two pictures are presented separately to each eye. We experience the fused image of the vase (spatial-taxon) in depth. We do not experience a separate right eye vase, a separate left eye vase and a fused whole image of a vase. To demonstrate this to yourself, imagine a flash of light appears in the left image (as illustrated in the figure). Psychologists have found that humans can't tell which eye the flash of light is shown to, but they do experience a light flash in the whole image.

Fig. 6 Thought experiment: unitary experience of a fused object in depth. The fused image experienced when presented as in the previous figure. Note phenomenology as used in this chapter references the experience of the human observer within his or her head (also known as the homunculus)

2.4 Color Antecedent to Infer Spatial-Taxon Consequence Using Domain Specific Knowledge from Human Vision

As discussed earlier, the nested-spatial-taxon structure does not assume monotonicity. A monotonic logic-knowledge base grows as each predicate and consequence is verified to be true enough. In the natural-vision-processing system described however, the predicate-consequence rules defining a spatial-taxon are assumed by default. The default is circumscribed, once human interaction or iteration (box c, figure 5) provides a clue indicating that a non-default case considered, when decision system (box c, figure 5) needs to resolve conflicts. Thus the knowledge base can both expand and shrink.

Fig. 7 Photograph of a red apple on a white background

A real world example in which default logic might be circumscribed, is what vision scientists call color constancy: the perception of object colors as stable - despite variable color illumination. [21] A red apple, for example, looks red regardless of whether it is within a low-lit cupboard or on a windowsill in full sunshine. Color constancy is an example of visual grammar, that may cause the system to change its mind regarding the pixel regions designated to a particular spatial-taxon due to its spatial configuration. Another reason we don't assume monotonicity is because the belief set that defines a spatial-taxon does not grow as more evidence is accumulated.

In Wang's work on concept algebra [42] defines objects as follows:

Definition 2. Definition of Objects taken from Wang's paper on concept algebra and CWW (computing with words)

Let ϑ denote a finite or infinite nonempty set of objects, and A be a finite or infinite set of attributes, then a semantic environment or universal context θ is denoted by a triple: i.e.

$\theta \eqsim (\vartheta, A, R) = R: \vartheta \to \vartheta$ *alternatively*

$\vartheta \to A$ *alternatively*

$A \to \vartheta$ *alternatively*

$A \to A$

where R is a set of alternative relations between ϑ (i.e. spatial-taxons) and A (i.e. cognitively relevant attributes).

Notice that the universal context θ in which the objects ϑ , spatial-taxons, lives is the nested spatial-taxon hierarchy.

Definition 3. Spatial-taxons K:

Let $CV_a, ..., z$ be a set of information granules, subordinate to spatial-taxon information granules, defined as the aggregated fuzzy membership of all pixels within the universe of discourse X for which possibilistic (x isCVR) where CVR is a set of possibility distributions on cognitively relevant variable $CV_a, ..., z$

Let $CRA_a, ..., z$ be a set of information granules, defined as the aggregated fuzzy attributes $A(X) = GTU(x)$, where A are cognitively relevant attributes defined by the generalized-theory-of-uncertainty (GTU) restraint as defined in Zadeh (2006)

Knowledge:

If Poss(CV_a) andif Poss(CV_b) ... andif Poss(CV_z) then Spatial-taxon K

If CRA_a is true enough, then Spatial-taxon K

If CRA_b is true enough, then Spatial-taxon K

... If CRA_z is true enough, then Spatial-taxon K

Facts: CRA_a is μ_a CRA_b is μ_b ... CRA_z is μ_z

Conclusion:

then let Spatial-Taxon K belong to the set of hypothetical taxons under consideration.

'True enough' encompasses linguistic constraints designed by the domain specialist.

To clarify the definition of a spatial-taxon, let's take a trivial example. In this example, we define a spatial-taxon as being the color red and apply it to a photograph of an apple on a white background. Figure 8 shows the fuzzy partitioning of red, which will be defined shortly. in definition 4 and applied to the photograph of an apple. Note that all colors, including the white background and gray shadow have the possibility of being Red. Because Red and Green are mutually exclusive possible partitions of X is constrained by possibility functions as defined in 4. All pixels not red are the background, $\neg(spatialtaxon)$ Figure 9 shows the generalized constraints for red applied to the pixels in the universe of discourse X

Fig. 8 Graphical representation of generalized constraints for primary colors as defined in definition 3. The x-y plane of the image is shown in skew to enable height (fuzzy membership) to be vertical. The figure also shows the color opponency of the green leaf, which requires that red and green cannot be simultaneously perceived in the same location. This forces the green left to have zero possibility and zero membership of red.

Example 1. Let ST_0 be a non-empty set that contains all pixels in the universe of discourse in the photograph of the apple. ST_0 has two mutually exclusive children ST_{red} and ST_0-ST_{red} called the background, which is equal to $\neg$ Red.

If $Poss_{red}(x) > 1 \vee X_k$ is μ_{red} then $X \to ST_{red}$

X_k is μ_{red}.

$\therefore X_k$ is ST_k

In this trivial example, our decision has one hypothetical spatial-taxon to consider. The background in this example, $\neg$ red does not overlap our hypothetical spatial taxon. However, this is rarely the case in real vision segmentation problems. Consider that under this definition the leaf was designated as background. However, most people tend to group the apple and leaf in a superordinate spatial-taxon, referred to in lay terms as the 'foreground of the image '. Human interaction and the Zipfian relationship between spatial-taxon hierarchies enable the natural-vision-processing AI system shown in Figure 10 to combine information from many - sometimes conflicting hypothetical spatial-taxons.

The fuzzy-natural-vision-processing model begins by partitioning an image into a set of cognitively relevant fuzzy sets. In the model described I directly implement the fuzzy-logical-color-naming model introduced by Berlin & Kay in1969 [13] This yields the 11 interval-valued color antecedents as defined in definition 4. Many cultures also include a fuzzy partition for light blue. In this model I used English color names, in which there is no primary level word for light blue. This fuzzy partition enables us to answer many color related queries, such as is it possible for this color patch to be green? To what degree of truth does this color patch is green imply?

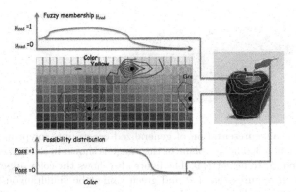

Fig. 9 Graphical representation of the constraint of Red, as adapted from the World Color Survey (see next section). Note that fuzzy membership extends deep into the yellow colors. The possibility function, is adapted from the color opponency, which requires that red and green can not be simultaneously perceived in the same location. Thus the green leaf is excluded from the spatial-taxon because it does not overlap green. Other spatial-taxon rules adapted from field perceptual organization would be required to group the green leaf with the red apple.

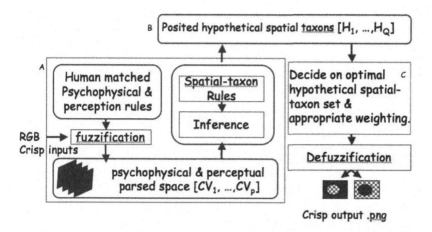

Fig. 10 Figure 4: Natural Vision Processing System. Natural vision processing system implements image segmentation as a nested two-class fuzzy inference system. Box A show domain specific knowledge that is used to design the fuzzy partitioning of cognitively relevant variables, such as color. Hypothetical spatial-taxon inference rules, also designed using domain specific knowledge, partition the image into fuzzy hypothetical spatial-taxon regions. Box C contains a decision making system that decides on the subset of spatial-taxons which the weights of the fuzzy rules. This section combines the spatial-taxons implication functions, such as the Max-min rule (Zadeh), which results in the defuzzified spatial taxon.

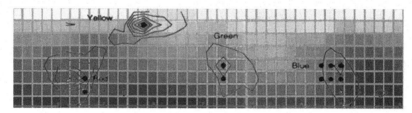

Fig. 11 World Color Survey Berlin and Kay 1968. Colors by Berlin and Kay in the 1969 world color naming survey. They discovered that all humans, regardless of culture, used a naming hierarchy that referred to the same color prototypes.

Definition 4. World Color Survey as Universe of Discourse *Let X be the universe of discourse consisting of all pixels within the rectangular (or square) pixel array of an image, as defined in definition 1. Let WCS be the universe of discourse first described in the world color survey by Berlin & Kay 1969. [13]*

For each pixel $X_{i,j}$ their exists one-to-many mappings with generalized constraints defined by the world color survey [13] such that GTC_{color} entails $X_{i,j}$.[8]

$GTC_{red} \rightarrow X_{i,j}$
$GTC_{green} \rightarrow X_{i,j}$
$GTC_{yellow} \rightarrow X_{i,j}$
$GTC_{blue} \rightarrow X_{i,j}$
$GTC_{white} \rightarrow X_{i,j}$
$GTC_{black} \rightarrow X_{i,j}$

and by opponent color theory

if $X_{i,j}$ is Green then $\neg Poss_{red}(x)$
if $X_{i,j}$ is Red then $\neg Poss_{Green}(x)$
if $X_{i,j}$ is Yellow then $\neg Poss_{Blue}(x)$
if $X_{i,j}$ is Blue then $\neg Poss_{Yellow}(x)$

and by derived color theory

if $X_{i,j}$ is (Red $\wedge$ Blue) then $GTC_{Purple} \rightarrow X_{i,j}$
if $X_{i,j}$ is (Red $\wedge$ Yellow) then $GTC_{Orange} \rightarrow X_{i,j}$
if $X_{i,j}$ is (Red $\wedge$ White) then $GTC_{Pink} \rightarrow X_{i,j}$
if $X_{i,j}$ is (Yellow $\wedge$ Black) then $GTC_{Brown} \rightarrow X_{i,j}$
if $X_{i,j}$ is (White $\vee$ Black) then $GTC_{Gray} \rightarrow X_{i,j}$

Note

It is beyond the scope of this paper to detail granularization of these colors. However, other works propose various linguistic constraints such as Barghout(2003) which uses Gaussian distributions for "a little bit of colorname"and "some colorname"and a saturating sigmoid constraint "very color ". Figure 6 provides a graphical description and figure 7 shows the fuzzy red partition within the universe of discourse defined by World Color Survey.

The second example will make use of several spatial-taxon rules: The aperture-frame rule, the centering rule and to small rule.

Definition 5. Cognitive relevancy attribute: Aperture Frame

Let X be the universe of discourse consisting of all pixels within the aperture of a rectangular image, such that $X_{1,1}$ is located in the upper left hand corner, pixel $X_{I,J}$ is located in the bottom left most corner and is a pixel that has non-zero fuzzy membership in at least one cognitively relevant attribute set. Suppose also that it has (high) connectivity with the outermost pixels of the image such that (i $\vee$ j)= 1, where the upper most corner of the image has indices $X_{1,1}$. This pixel is defined as having (aperture-frameness), where connectivity is a linguistic constraint whose membership decrease with connectivity with image frame decreases.

---

[8] In my implementation of the model in Barghout(2014), I implemented generalized color constraints that are functions of white anchoring (such as the Retinex model) and the relative distance between prototype colors [13] and RGB colors in the image. However, a detailed discussion is beyond the scope of this paper.

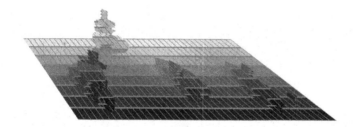

Fig. 12 Application of fuzzy color rule set (definition 4) applied to the World Color Survey. The x-y plane of the image is shown in skew to enable height, which illustrates color membership to be represented vertically.

Definition 6. Cognitive relevancy attribute: Centered

Suppose $X_{i,k}$ is a pixel that has non-zero fuzzy membership in at least one cognitively relevant attribute set. Suppose also that it has non-zero membership within circle of radius = Center, centered in the middle of the photograph. For this example, let's set the Center to be about 1/3 the width of length. Fuzzy membership is equal to 1 at center, and decreases with distance falling to zero beyond the radius.

Definition 7. Cognitive relevancy attribute: Too Small

Suppose ST_k is pixel region posited to be a hypothetical spatial-taxon. Let Size be the number of pixels included in ST_k. If Size is less then 1 percent of the total size of the photograph then dilate the magnitude of the spatial taxon membership such that μ (new) $= \sqrt{mu_k}$.

Definition 8. Fuzzy partition by cognitively relevant attributes blur and high detail

Suppose CV_k is pixel region partitioned as a fuzzy variable. Let MOMS be a set of Sobal filters tuned to spatial frequency filter bandwidths of about .01 of size of an image and three orientations: vertical, both diagonals and horizontal. Let $CV_{High}Detail$ be the sum over all the convolutions of each MOMS filter and the monochromatic image. The designer may choose a threshold. In the work presented at IPMU 2014, the threshold was set to 5 percent contrast.

Let $CV_{High}Detail$ be those pixels above frequency. Let $CV_{B}lurry = ST_k$ - $CV_{High}Detail$

Example 2. Spatial-taxon by Blurry Aperture-Frame
   Knowledge
   If CV_{Blurry} is true enough, $\wedge$ X_k is μ_{Blurry} then X $\rightarrow$ $Background_{Blurry}$
   If $CV_{ApertureFrame}$ is true enough, $\wedge$ X_k is μ_{Blurry} then X $\rightarrow$ $Background_{BlurryFrame}$
   If $CV_{ApertureFrame}$ is true enough, $\wedge$ X_k is μ_{Blurry} then X $\rightarrow$ $Background_{BlurryFrame}$
   Facts
   X_{green} is μ_{Blurry}&X_{green} is μ_{blurry}.
   X_{blue} is $\mu_{highdetail}$&X_{blue} is μ_{center}.

Fig. 13 Fuzzy partition of cognitively relevant variables blurry and green. Example of the region parsed by the blurry and green-aperture frame rule. The x-y plane of the image is shown in skew to enable height to be represented vertically. Height designates aggregated cognitively relevant variable.

X_{red} is $\mu_{highdetail}$&X_{red} is μ_{center}.
X_{orange} is μ_{blurry}&X_{orange} is μ_{center}.
Conclusion
∴ Hypothetical Spatial taxon is (not $X_{blurryFrame}$) ∧ $X_{bluehighDetail}$ ∨ $XcenterRed$ ∨ $XcenterOrange$

The rule base in this fuzzy reasoning example, contains 13 cognitively relevant fuzzy partitions: 11 colors, Blurry and Centered. It contains one spatial-taxon fuzzy partition: Too Small. The other reasoning rules were chosen to clarify the example, but the full rule base and linguistic constraints are summarized in figure 11 & 12. Note that all consequences in the inference system were fuzzy spatial-taxon information granules. In this example, all the rules had equal weights.

Requiring the consequences to all be at the spatial-taxon information granule is key to dealing with complex images. As noted earlier, the nested spatial-taxon taxonomy enables images to be segmented as a nested two-class inference problem. Getting all the information granules into the spatial-taxon level is the first step. The next step is distinguishing the correct abstraction level. This is where direct human interaction or off line consensus building as to image taxometric structure becomes useful. Also note, that though in this system all rules are used with equal weight, in the system shown in 4, the decision making system needs to decide on the optimal weights and if some hypothetical spatial-taxons should abstain from the process. Abstaining is important, because when defuzzifing in the fuzzy stage the conclusions will need to be normalized. The abstaining hypothesis should not be counted in the normalization process.

Each spatial-taxon is formed from the conjunction of visual-taxometric antecedents, where the antecedent is a fuzzy reasoning paradigm regarding how the foreground is distinguished from the background at that level granularity.

Fig. 14 Combined hypothetical spatial-taxon membership. The x-y plane of the image is shown in skew. Height, combined fuzzy membership, is vertical. The combined fuzzy membership is the weighted sum of memberships as parsed by the blurry-rule, green-aperture frame rule, too-small rule and center rule.

3 Autonomous Decision Making: Hypothetical Spatial-Taxons

As discussed earlier, spatial-taxons are the granular level best suited for decision making about image segmentation. In section 2, we detailed methods for inferring hypothetical spatial-taxons and how to elicit user feedback about the hierarchical relationships between spatial-taxons within nested taxometric structure. In this section we focus on the issue of combining hypothetical spatial-taxons. We start by discussing an example of a hypothetical spatial-taxon combination that incorrectly infers a foreground containing a hole within it rather than an object laying upon it. We then discuss the general case autonomous decision making process that uses an estimation of the utility and attentional resource requirements to make an educated guess on how best to combine and weigh hypothetical spatial-taxons.

Figure 15 illustrates the autonomous process of deciding on the best hypothetical taxon set and appropriate weighting of the natural-vision-processing-system shown in figure 10, box C. The autonomous decision making algorithm

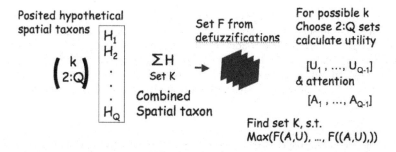

Fig. 15 Autonomous decision making system, as used Visual Taxometric Approach to Image Segmentation Using Fuzzy-Spatial Taxon Cut

(box C in figure 15) has two functions: the selection the set of hypothetical spatial-taxons for consideration and the weighting of each spatial-taxon in the fuzzy inference system.

The autonomous system decides on the hypothetical spatial-taxon set and appropriate weighting by iterating through various combinations of hypothetical spatial-taxons, to infer the defuzzifed spatial-taxon that would result from each combination, and to score the output for each combination. This enables posits to abstain.[9]. The score is a combination of spatial-taxon utility and the attentional resource requirement of the hypothetical spatial-taxon combination. The optimal set is chosen such that it maximizes utility and minimizes attentional resources.

3.1 Example: Whole Object or Hole Within an Object?

The next example shows the fuzzy spatial-taxon membership returned from the natural-vision-processing engine on a photograph of a ladybug on a yellow daisy. This image was run on the same engine whose results were presented at the vision science society annual meeting in 2013 and at IPMU 2014. [10] [4] In this engine, hypothetical spatial-taxon inference systems were chosen to simulate meaningfulness cues, known composition styles that the domain expert expects the segmentation system to encounter. The meaningfulness

Fig. 16 Linguistic constraints derived by decision making system from meaningfulness cues. On the left is the original image. [29] On the right is spatial-taxon membership (height) inferred by autonomous decision making system. The x-y plane of the image is shown in skew. Height, fuzzy membership, is vertical. Note that both the black background and ladybug are at zero or close-to-zero membership. The system grouped the ladybug with the background, incorrectly inferring a hole in the flower.

---

[9] The idea to allow psychological detectors to abstain from contributing information to the system was suggested to me by Lotfi Zadeh in 2006, personal communication.

spatial-taxon by blurry-aperture-frame inference described in example 2, is an example of a meaningfulness spatial-taxon. The engine also included a meaningfulness spatial-taxon inference that responded to an image composition consisting of an object in front of a plain uniformed color background.

As you can see, the system incorrectly inferred that the ladybug was a hole [27] in the flower! Why?

As explained in section 2.4, the system needed to decide how to combine the non-mutually exclusive cognitively relevant color antecedents to infer the spatial-taxon consequence. The ladybug has white eyes, a black head and black dots on it's wings. The red wings have bright areas (where light is reflected off the wing) and areas in shadow. Since cognitively relevant variables are not mutually exclusive, the bright pixels have membership in red, pink and possibly white. The shadowed pixels have membership in red, gray and possibly black (or brown).[10] What appears to us humans as a coherent ladybug, has several possible visual grammatically correct colors interpretations. Each hypothetical spatial-taxon inference system inferred the color most correct for it's grammar [9] [5]. A hypothetical spatial-taxon inference that assumes a uniform color $\neg(spatialtaxon)$ background as black would included these desaturated pixels as possibly $\neg(spatialtaxon)$. Thus the combined aggregate designates these pixels as $\neg(spatialtaxon)$. In other words, system infers the flower as having a hole in it rather than it being a whole foreground composed of a daisy with an object on it.

3.2 Deciding Spatial-Taxon Weight Combination

We now discuss how to choose an optimal set of non-mutually exclusive hypothetical spatial-taxons and how to choose optimal weighting for combining these hypothesis.

The utility function used to score the posited spatial-taxon was inspired by a seminal study of pictorial object naming [24] that found that objects were identified first at an "entry point "level of abstraction. Curious as to the whether the scene-architecture had an "entry level "region, I began

---

[10] This example illustrates the power of using fuzzy sets to regulate uncertainty with respect to meaning. There is little uncertainty regarding the classification of a pixel that has the color characteristics of prototypical red. But what does it mean for a pixel to be desaturated? Is the desaturated pixel meaningful because of properties intrinsic to the ladybug such as loss of pigmentation due to disease or genetic variation? Or is the color desaturation meaningful because of extrinsitc conditions caused by the lighting environment? Conventional probability methods would treat this a Bayesian problem, attempting to enumerate the likelihood of all possible events correlated with the desaturated pixels. Fuzzy logic, however, enables the system to entail the uncertainty with respect to the meaning of the desaturated pixel with symbolic logic. In short, fuzzy logic enables us to navigate uncertainty without having to specify each and every possible causal event with a prior probability.

experimenting with the visual taxometric approach [6] [5]. As discussed in section 2.2, I found that the frequency at which people labeled spatial-taxons correlated with the abstraction rank with the nested-scene-hierarchy. This suggested an underlying power-law, such as Zipf-law might be underlying human spatial-taxon choice.

The Zipfian result suggested an underlying cognitive law of least effort similar to that found in other cognitive processes [15]. Thus the utility function is inspired by the law of least effort. I define it operationally over an ordinal scale such that entry-level had the most utility, super-ordinate the next highest utility and all sub-ordinate decrease utility as a function of abstraction. This is a soft restriction, with granularity at abstraction levels. Use of attentional resources was also defined on an ordinal scale with granularity at the number of hypothetical spatial-taxons possible in the natural-vision processing engine. It's constrained to be inversely related to the number of significant spatial-taxon combination sets above threshold, where threshold was defined in terms of sub-population variance verses variance of the sub-population with the lowest within-group variance.

Definition 9. Attentional Resource Requirement

Suppose there are K hypothetical spatial-taxons under consideration. Suppose a high quality segmentations have similar centers and similar contours. Also suppose that similarity is defined as per C.L. Chang (2014). Registration artifacts, where the same correct pixel match is chosen by two or more hypothetical spatial-taxons, but are mislabeled as a false-hit or incorrect rejection, are common. Thus contour similarity is defined as the interval-valued intersection of spatial-taxons with uncertainty α $k1,kNo$ for each hypothetical contour in the set K.

Assume that attentional resource load increase with the number of potential spatial-taxons the system has to monitor.[11] The attentional load is then the cardinality of the set of similar hypothetical spatial-taxons under consideration.

More formally attentional load = fuzzy cardinality = Under consideration count of Similar(Hypothetical Spatial taxons))).

Definition 10. *For $[^k_{2:Q}]$ defuzzifications calculate utility and attention-resources-requirement where*

$$Utility(\Phi) = \int \int_\Phi hypothetical - spatial - taxon - utility(\Phi)d\Phi \qquad (1)$$

---

[11] Vision scientists have devoted considerable study to understanding how channels (psychophysical correlate of specialized neural receptive field structures) detect signals within complex scenes and the presence of noise. [5] [12][39] In this chapter, I use a simplified version of signal detection theory and assume that the attention increases with the number of receptive fields being monitored. A more nuanced implementation of signal detection theory would require multiple rules, mechanisms and possibly a large field of irrelevant channels.

$$Attentional_r esources(\Phi) = \int \int_\Phi Attentional - inference - load(\Gamma)d\Phi \quad (2)$$

Let A be a fuzzy set defined on a universe of Φ discrete meaningfulness cues $\Phi = [\Phi_1, \Phi_2, ..., \Phi_a]$ defined on the universe of discourse of images expected to be encountered by segmentation system. Let ST_1 be the spatial-taxon definition in definition 1. The spatial-taxon 'cut 'results from the defuzzification that optimizes $F(U,A)$ where U the utility function defined to me most similar to meaningful cues chosen by the designer and A is the attentional load. In this case the attentional load increases as the variance of error between spatial taxons. Calculation of spatial-taxon error will be described in the section on performance metrics. The crisp conclusion is normalized to lie between zero and one. Spatial-taxon threshold is chosen according to use-case. In this system the threshold was set to 0.5.

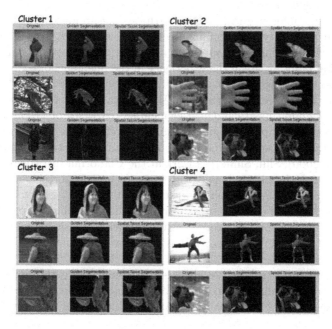

Fig. 17 Results presented at IPMU 2014, which used the autonomous decision making system

Results for the autonomous system described were presented at IPMU 2014. To familiarize the reader, I show the results again here along with the weights of the spatial-taxon meaningfulness cues.

Cluster 1		Cluster 2	
Linguistic Hedge	Meaningfulness Cue	Linguistic Hedge	Meaningfulness Cue
abstain	Blurry	some	Blurry
some	Color Surround	abstain	Color Surround
very	Connected Taxon Color	very	Connected Taxon Color
abstain	Wall-like Background	some	Wall-like Background

Cluster 3		Cluster 4	
Linguistic Hedge	Meaningfulness Cue	Linguistic Hedge	Meaningfulness Cue
very	Blurry	low	Blurry
very	Color Surround	some	Color Surround
abstain	Connected Taxon Color	some	Connected Taxon Color
some	Wall-like Background	not	Wall-like Background

Fig. 18 Meaningfulness cues along with linguistic hedges

4 Consensus Building

Now that we have discussed the automated decision-making process, we will turn to an interaction process that enables iteration.

It's one thing to say[12] that humans choose a spatial-taxon, according to a series of fuzzy inference rules, emulate those rules, and decide among them by maximizing utility and minimizing effort; it's another to say *when* humans make the choice. Since this approach attempts to emulate human thinking, the decision point at which the experience of the phenomenology of a specific scene organization is relevant.

Vision scientist Ken Nakayma (2012) of Harvard University has suggested that "we abandon fixed canonical elementary particles of vision as well as a corresponding simple-to-complex cognitive architecture for vision." Studies of postdiction, retrospective modulation of feature extraction (Eagleman and Sejnowski, 2000), Shin Shimojo's work at the California Institute of technology suggesting the phenomenological sequence human perceive which visual events occur are not isomorphic to the physical sequence of neural correlate events which caused them. In other words, by the time a human makes a decision regarding the spatial-taxon scene organization, he/she may have revised memory into causal "story" consistent with how the human mind thinks the world works.

If the human interaction we elicit is not isomorphic with sequence of the feature extraction phase, than dynamical feedback may be required to modify fuzzy inference of the cognitively relevant variables. I refer to this as backward causation.

From a design perspective, backward causation allows us to re-set the posited weights, which in turn provides an improved segmentation.

---

[12] For the sake of clarity, I'm assuming that visual-taxometrics provides a strong enough theory of cognition to justify mimicking. There is much back and forth about using computer models from A.I. to suggest psychological theories, but until these have been tested psychophysically, they are strongly speculative.

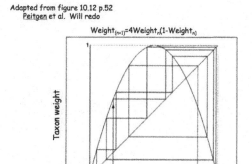

$Weight_{(n-1)} = 4Weight_n(1 - Weight_n)$

Fig. 19 Illustration of mixing behavior: Adaptation of figure 10.12, p521 from Chaos and Fractals [31]. This illustration of mixing behavior of the quadratic iteration. This is analogous to iterating weighting functions in the decision phase. (*A*) Initial weight of Aperture Rule, yields the weight for that rule in the next inference cycle. (*B*) After the 11th iteration the system is clearly filling in the unit interval, not converging. it. (*C*) The equation used to illustrate this point is the logistics equation, however, the visual-taxometric systems will be similarly nonlinear.

Re-setting the early visual processes - modeled as cognitively relevant fuzzy constraints is the next step. Human survey and/or reverse search query should yield several clusters - all of which could be potential spatial-taxons. For these images, it's possible that the image composition is not at all like the meaningfulness cues designed for by the domain expert. For these cases, the fuzzy partitions that were designed to simulate early human visual processes should be adapted. Barghout (2003, 2014a) shows examples where contrast transduction was altered to simulate color constancy illusions. This was done by altering contrast transduction, amplifying color memberships for pixels close to the potential spatial-taxon and decreasing fuzzy membership for cognitively relevant variables that support the non-selected spatial-taxon. This process, meta-iteration, simulates very low level processing that enables humans to alter their perception in favor of a visual insight.

Highly non-linear dynamical systems, such as the backward causation process described here are capable of deterministic chaos.[13] Though we want an iterative system capable of deterministic chaos, we need a decision system with stable behavior.

---

[13] Andrey Kolmogorow of the former Soviet Union and the American mathematician Stephen Smale started classifying such natural phenomenon of by the early 60s.

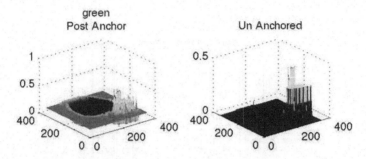

Fig. 20 Color constancy, a visual grammar rule applied during backward causation, enables phenomenological color to be invariant under different illuminations. The left (green post-anchor) fuzzy membership of green brings up the membership of the white background from zero, enabling the possibility of those pixels being included in red-inference. The chart uses dark red to indicate membership close to one and dark blue as membership close to zero.

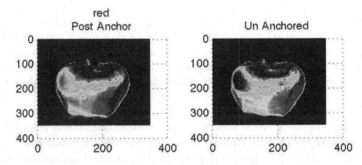

Fig. 21 Example of backward causation: In the red apple example (example 1), we designed the hypothetical spatial-taxon as Red. Color constancy (red post-anchor) applied to red inference rule, removes the dark blue (zero) membership. This enables the formerly zero red-membership pixels to now be grouped as part of the spatial-taxon. Notice that the reduction in luminance in of the bright spot on the apple, as illustrated in post Red Post-Anchor. The chart uses dark red to indicate membership close to one and dark blue as membership close to zero.

There is much quality work on chaotic non-linear systems. I point the interested reader to Chaos and Fractals, Peitgen, Jurgens and Saupe for an in depth study, and to Chaos by James Glick for an overview. For the purposes of iterative decision making, I'm borrowing two key concepts: Ljapunov exponents and very low precision arithmetic (also known as grid arithmetic).

As pointed out by Peitgen et al, an artifact of very-low-precision arithmetic is periodicity for a system that if iterated at high precision would exhibit erratic or chaotic behavior. This works to our advantage, since fuzzy granulation is a form of low-precision-arithmetic. Thus when designing your system, be aware that increasing the size of the information granules may stabilize an otherwise unstable system.

Figure 19 illustrates an ergostic system, which though it eventually converges, would be extremely challenging to iterate. If mixing behavior seems to be occurring after several iterations, try dampening the nonlinearities in the function that changes the weights. Set your halting procedure to stop after it's cycled through that number of states or sooner if the Ljapunov exponent is high. Increase granularity of the arithmetic grid to stabilize an unstable system - though use this sparingly.

4.1   User Interaction for Meta-Iteration

User interaction serves two purposes. First, the user can select the center of the spatial-taxon of interest, which enables the system to reset its parameters to favor hypothetical spatial-taxons with similar centers. Second, after sampling many users we can determine the rank frequency distribution spatial-taxons for the nested spatial-taxon hierarchy for that image. Users tend to agree on the centers of spatial-taxons, but disagree about the borders [11] [5]. Depending on the depth of the image taxonomy, clear clusters will emerge at the centers of spatial-taxons.

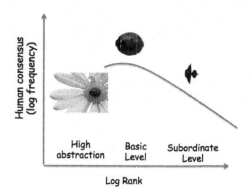

Fig. 22 Users typically choose (or have the highest consensus at) the primary level spatial-taxon, in this case the ladybug. The superordinate, in this case the ladybug on the flower is chosen less often. The subordinate level spatial-taxon, in this case the ladybug 'face ', is occasionally chosen.

The natural vision processing system discussed in section 3, assumed default maximum utility occurred at spatial-taxon width 66 percent of the total image width. After eliciting enough user feedback to infer two or three levels of the spatial-taxon hierarchy, this assumption can be replaced with one that assumes each spatial-taxon a local utility maximum. Figure 20 illustrates a typical spatial-taxon rank frequency diagram that would be expected by human subjects.

Reverse image queries provide information about the number of objects in the image and the type of image. If, for example, the search query is "Mary and Bob "then we can assume that there will be a foreground spatial-taxon with at least two children taxa (one for Mary and one for Bob), a deep hierarchy and smaller sized spatial-taxons. The query "close-up of a humming bird "indicates that the background may be blurry. This enables the system favor the burry hypothetical $\neg(spatialtaxon)$.

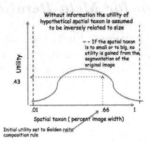

Fig. 23 Utility a function of the inverse square of the width. The default logic assumes that the utility of the hypothetical spatial-taxon inversely related to length of the width. The red dotted line shows the very very low utility of a segmentation to is too small or too large. Once the system obtains feedback, the default logic is circumscribed in favor of a more contextually relevant utility function.

Designing methods for measuring human feedback is notoriously difficult. Fields such as psychophysics, psychology, library science and human factors all share significant bodies of work that design experimental methods to tease out the important human metrics. I point the interested reader to: Introduction to the Taxometric Method [32], Introduction to Psychophysics and of course a text on Statistics.

A few pointers to keep in mind when designing such a method are:

– Design metrics that distinguish between taxa (categorical) and dimensional data and incorporate consistency testing across several methods.
– Include metrics that assume human criterion will vary. Signal detection theory from psychophysics provides useful methods for handling this issue.
– When applying performance data, be sure that the machine metrics correlate with the human measures of performance - across several naive human subjects.

Spatial-taxon are taxa, which means they vary categorically in the dimension of abstraction. This can make it tricky to measure performance of nested taxons that overlap in space. Take for example the apple shown in figure 7. The center of the apple is displaced from the leaf. Thus a human metric that measure performance in terms of agreement between the center of these spatial-taxons will yield meaningful results.

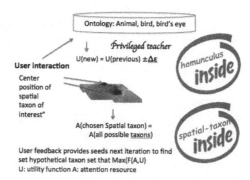

Fig. 24 Human interaction system for improving utility calculation. The user choses the center of the spatial taxon of interest. This increases the utility of the hypothetical spatial-taxons, which in turn adjust the rule weights.

Fig. 25 Alternative decision segmentation. Ladybug is grouped with aperture-frame and is thus considered a 'hole '. For this to occur the, the decision system asked the Spatial-taxon as center rule to abstain.

Meta-iteration repeats the method of backward causation, but optimizes the reparameterization to support the visual grammar of spatial-taxons inferred by human iteration. Figure 20 and figure 20 show the fuzzy memberships of green and red after reparameterization in support of $\neg(spatialtaxon)$ being white and the application of color constancy to within inferred spatial-taxons.

When including human interaction in a segmentation system to implement meta-iteration and/or backward causation take the followings into consideration while designing the system.

5 Conclusion

This chapter described iterative fuzzy-decision-making for image segmentation using spatial-taxon information granules as the primary level of decision making. Important points are as follows:

- Assuming a universe of discourse consisting of a hierarchy of nested-spatial-taxons enables us to approach image segmentation as a closed universe problem. It operationalizes the solution into an iterative fuzzy two-class inference problem. Each operation infers a distinct level of abstraction within the nested spatial-taxon scene taxonomy originally assumed.
- Domain-specific knowledge based on human vision provides the basis for these fuzzy inference systems. These inferences are not monotonic, because the support for a specific belief set does not always grow as evidence is accumulated. Common perceptual phenomena, such as color constancy, may cause a system to change it's mind regarding the belief supported by accumulated. This is particularly true for cases where the phenomenological sequence of perceptual events is not isomorphic to the sequence physical events driving perception.
- Automated decision making regarding scene organization involves at least two steps: (1) maximizing utility and minimizing attentional resources among multiple possible combinations of hypothetical spatial-taxons (2) implementing reverse-causation parameter adjustment of fuzzy-inference rules.
- Eliciting user feedback by exploiting the human tendency to agree on the center of spatial-taxons, enables dynamical feedback. Since the dynamical system is highly nonlinear, care must be taken when designing the feedback to avoid unstable behavior.

Acknowledgements. I thank Dr. Christopher Tyler (Smith-Kettlewell Eye Research Institute) for his mentoring in human vision, computational-psychophyics and for his help with this chapter. I thank Dr. Lotfi Zadah (Berkeley Initiative for Soft Computing (BISC), EECS U.C. Berkeley) for his mentoring and suggestions, Dr. Stan Klein (Vision Science, U.C. Berkeley) for discussing the similarity between Quantum duality and observer(homunculus)-sensory duality. I thank Ana DaSilva, Colin Rhodes, Janet Kramer and Kennan Rossi for the edits and comments. I also thank Janet Kramer and Yurik Reigel for their illustrations.

Bibliographical and Historical Notes

- In his 2010 paper, On Concept Algebra For Computing With Words, Yingxu Wang defines a concept as:

  "a basic cognitive unit to identify and/or model a concrete entity in the real world and an abstract subject in the perceived world. "

  I define concepts slightly differently. My own view is that the basic cognitive unit is defined as a phenomenological unit. Phenomenology, the consciousness of what is perceived by a person as directed toward an object from a particular point of view, may not correspond to physical structures in the world. A phenomenological variable exhibits uncertainty as to meaning, not it's occurrence in the real world. Fuzzy logic provides a well developed methodology for handling uncertainty with respect to meaning, making it well suited for phenomenological inference.

- In this chapter, I avoid the controversy of defining information. Many consider the 1948 Bell Labs technical paper titled A Mathematical Theory of Communication by C.E. Shannon as the start of the modern field of information theory. In his paper, Shannon defined information in terms of the carrying capacity of the system transporting messages, not in terms of the semantics (meaning) of the message or the phenomenology of the human. Thus information, as he described in that paper, is not closed loop because it requires a person (homunculus) to interpret the meaning. The second paragraph of Shannon's paper makes this distinction clear.

> "The fundamental problem of communication is that of reproducing at one point either exactly or approximately a message selected at another point. Frequently, the messages have *meaning*: that is they refer to or are correlated according to some system with certain physical or conceptual entities. These semantic aspects of communication are irrelevant to the engineering problem. "
>
> -C.E. Shannon, A Mathematical Theory of Communication. [37]

Recent work in Information Theory [28] describes a phenomenological experience as an irreducible conceptual structure. According to this theory, integrated information theory, the example shown in Figure 5 has integrated information as specified by fusion of binocular inputs that can not be reduced to it's components without losing the phenomenology.

- Stanley A. Klein, discusses the importance of addressing the role of the observer (homunculus) in his chapter, Will Robots See, [25] in distinguishing Weak A.I., Strong A.I. and Cognitivism.

References

1. Bargiela, A., Pedrycz, W.: Granular computing: an introduction. Springer (2003)
2. Bargiela, A., Pedrycz, W.: Toward a theory of granular computing for human-centered information processing. IEEE Transactions on Fuzzy Systems 16(2), 320–330 (2008)
3. Bargiela, A., Pedrycz, W.: Granular computing: an introduction. Springer (2003)
4. Barghout, L.: Visual Taxometric Approach to Image Segmentation Using Fuzzy-Spatial Taxon Cut Yields Contextually Relevant Regions. In: Laurent, A., Strauss, O., Bouchon-Meunier, B., Yager, R.R. (eds.) IPMU 2014, Part II. CCIS, vol. 443, pp. 163–173. Springer, Heidelberg (2014)
5. Barghout, L.: Vision: Global Perceptual Context Changes Local Contrast Processing Updated to include computer vision techniques. Scholars Press (2014) ISBN-10: 3639709624, ISBN-13: 978-3639709629

6. Barghout, L.: Empirical Data on the Configural Architecture of Human Scene Perception using Natural Images. J. Vis. 9(8), 964 (2009), doi:10.1167/9.8.964
7. Barghout, L.: System and Method for Edge Detection in Image Processing and Recognition. WIPO WO/2007/044828 (2006),
   https://www.google.com/patents/WO2007044828A3
8. Barghout, L.: Linguistic Image Label Incorporating Decision Relevant Perceptual, Semantic, and Relationships Data, USPTO, 20080015843 (2007),
   https://www.google.com/patents/US20080015843
9. Barghout, L., Lee, L.: Perceptual information processing system, USPTO patent application number: 20040059754 (2003),
   http://www.google.com/patents/US20040059754
10. Barghout, L., Sheynin, J.: Real-world scene perception and perceptual organization: Lessons from Computer Vision. Journal of Vision 13(9) (July 24, 2013)
11. Barghout, Winter, Riegal: Empirical Data on the Configural Architecture of Human Scene Perception and Linguistic Labels using Natural Images and Ambiguous figures. In: VSS 2011 (2011)
12. Barghout Stein, L., Tyler, C.W., Klein, S.A.: Partioning mechanisms of masking: contrast transducer versus multiplicative noise. In: Rogowitz, B.E., Pappas, T.N. (eds.) Proceedings of SPIE. Human Vision and Electronic Imaging II, vol. 3016 (1997)
13. Berlin, B., Kay, P.: Basic color terms: their universality and evolution. The University of California Press, Berkeley (1969)
14. Bloom, P.: How Children Lean the Meanings of Works, p. 82. MIT Press (2000)
15. Cancho, Sole: Zipf's law and random texts. Advances in Complex Systems 5(1), 1–6 (2002)
16. Chang, C.L.: Fuzzy-logic based programming. World Scientific, Singapore (1985)
17. Chen, S.-J., Chen, S.-M.: Fuzzy risk analysis based on the ranking of generalized trapezoidal fuzzy numbers. Applied Intelligence 26(1), 1–11 (2007)
18. Chen, S.-M., Wang, J.-Y.: Document retrieval using knowledge-based fuzzy information retrieval techniques. IEEE Transactions on Systems, Man and Cybernetics 25(5), 793–803 (1995)
19. Chen, S.-M.: Weighted fuzzy reasoning using weighted fuzzy Petri nets. IEEE Transactions on Knowledge and Data Engineering 14(2), 386–397 (2002)
20. Deng, Y., Manjunath, B., Shin, H.: Color image segmentation. In: IEEE Computer Society Conference on Computer Vision and Pattern Recognition, vol. 2 (1999)
21. Granzier, J.: Color Constancy Explained. Thesis Vrige Universiteit Amsterdam, ISBN 97 890 86591442
22. Grycuk, R., Gabryel, M., Korytkowski, M., Scherer, R., Voloshynovskiy, S.: From Single Image to List of Objects Based on Edge and Blob Detection. In: Rutkowski, L., Korytkowski, M., Scherer, R., Tadeusiewicz, R., Zadeh, L.A., Zurada, J.M. (eds.) ICAISC 2014, Part II. LNCS, vol. 8468, pp. 605–615. Springer, Heidelberg (2014)
23. James, W.: Principles of psychology, p. 403. Holt, New York (1890)
24. Jolicoeur, Gluck, Kosslyn: Pictures and names: making the connection. Cognitive Psychology 16, 243–275 (1984)
25. Klein, S.A.: Will robots see? In: Spatial Vision in Humans and Robots: The Proceedings of the 1991 York Conference on Spatial Vision in Humans and Robots. Cambridge University Press (1993)

26. Nelson, K.: Constraints on Word Meaning? Cognitive Development 3, 221–246 (1988)
27. Nelson, R., Palmer, S.E.: Of holes and wholes: The perception of surrounded regions. Perception-London 30(10), 1213–1226 (2001)
28. Pedrycs, W., Bargiela, A.: Granular Clustering: A Granular Signature of Data. IEEE Transactions on Systems, Man and Cybernetics - Part B: Cybernetics 32(2) (2002)
29. Ladybug on yellow daisy. Photosbyflick, http://www.flickr.com/photos/17773534N03/3611852338/
30. Prasad, S., Kumar, P., Sinha, K.P.: Grayscale to Color Map Transformation for Efficient Image Analysis on Low Processing Devices. In: El-Alfy, E.-S., Thampi, S.M., Takagi, H., Piramuthu, S., Hanne, T. (eds.) Advances in Intelligent Informatics. AISC, vol. 320, pp. 9–18. Springer, Heidelberg (2015)
31. Peitgen, H.-O., Saupe, D., Jurgens, H.: Chaos and Fractals: New Frontiers of Science. Springer (2004) ISBN-10: 0387202293
32. Ruscio, J., Haslam, N., Ruscio, A.: Introduction To Taxometric Method Lawrence Eelbaum Associates (2006)
33. Russell, S.: Unifying Logic and Probability: A New Dawn for AI? In: Laurent, A., Strauss, O., Bouchon-Meunier, B., Yager, R.R. (eds.) IPMU 2014, Part I. CCIS, vol. 442, pp. 10–14. Springer, Heidelberg (2014)
34. Russell, S., Peter, N.: Artificial Intelligenc, A modern approach, 3rd edn. Pearson Education, Prentice Hall (2010)
35. Shi, J., Malik, J.: Normalized Cuts and Image Segmentation. IEEE TPAMI 22(8) (2000)
36. Simon, H.: The Architecture of Complexity. Proceedings of the American Philosophical Society 106(6), 467–482 (1962)
37. Shannon, C.E.: A Mathematical Theory of Communication. The Bell System Technical Journal 27, 623–656 (1948)
38. Treisman, A.M.: Strategies and models of selective attention. Psychological Review 76(3), 282–299 (1969)
39. Tyler, C.W., Chen, C.-C.: Signal detection theory in the 2AFC paradigm: attention, channel uncertainty and probability summation. Vision Research 40(22), 3121–3144 (2000)
40. Vapnik, V.: Invited Speaker. IPMU Information Processing and Management of Uncertainty in Knowledge-Based Systems (2014)
41. Wang, L.-X.: Generating Fuzzy Rules by Learning from Examples. IEEE Transactions on Systems, Man and Cybernetics 22(6) (1992)
42. Wang: On Concept Algebra for Computing with Words (CWW). International Journal of Semantic Computing 4(3) (2010)
43. Wertheimer: Laws of Organization in Perceptual Forms (partial translation) Ellis, W.B. (ed.) A Sourcebook of Gestalt Psychology, pp. 71–88. Harcourt Brace (1938)
44. Zadeh, L.A.: Fuzzy sets. Inform. and Control 8, 338–353 (1965)
45. Zadeh, L.A.: Probability measures of fuzzy events. J. Math. Anal. Appl. 23, 421–427 (1968)
46. Zadeh, L.A.: Similarity Relations and Fuzzy Orderings. Information Sciences 3, 177–200 (1971)
47. Zadeh, L.: Outline of a new approach to the analysis of complex systems and decision processes. IEEE Trans. Syst. Man & Cybern. SMC-3 (1973)

48. Zadeh, L.A.: Fuzzy sets and information granularity. In: Gupta, M., Ragade, R., Yager, R. (eds.) Advances in Fuzzy Set Theory and Applications, pp. 3–18. North-Holland Publishing Co., Amsterdam (1979)

49. Zadeh, L.A.: A theory of approximate reasoning. In: Hayes, J., Michie, D., Mikulich, L.I. (eds.) Machine Intelligence, vol. 9, pp. 149–194. Halstead Press, New York (1979)

50. Zadeh, L.A.: Possibility Theory and Soft Analysis. In: Proc. of AAAS Symposium on Soft Data Analysis (1980)

51. Zadeh, L.A.: Toward a theory of fuzzy information granulation and its centrality in human reasoning and fuzzy logic. Fuzzy Sets and Systems 90, 111–127 (1997)

52. Zadeh, L.: Generalized theory of uncertainty (GTU) - principal concepts and ideas. Computatiional Statistics & Data Anaylysis 51, 15–16 (2006)

53. Zadeh, L.: Toward a Restriction-centered Theory of Truth and Meaning (RCT). Information Sciences 248 (2013)

54. Berkeley Segmentation Database,
    http://www.eecs.berkeley.edu/Research/Projects/CS/vision/bsds

Group Decision Making in Fuzzy Environment – An Iterative Procedure Based on Group Dynamics

Mahima Gupta[*]

Abstract. Group decision making (GDM) has become a necessity to seek a solution to real life complex problems. The complexity of the problem is due to multiple aspects of any problem such as social, political and economical that is perceived differently by multiple actors (members) due to their diverse, often conflicting evaluation system. In order to reach consensus in the group, members tend to change their opinions guided by the views of other members in the group. In this paper, we have given a methodology that obtains group's consensus view by finding the shift in the members' opinions as dictated by group's dynamics i.e. their importance and support in the group. The members' preferences for the alternatives are elicited using linguistic terms by comparing pairs of alternatives. Also, importance values of a member as perceived by others in the group are taken in linguistic terms. We have developed a Fuzzy Inference System that gives a rule base for the likely shift in the members' opinions given the group dynamics. The methodology proceeds iteratively to calculate likely shift in the members' opinions till the time consensus in the group reaches a predefined threshold value.

Keywords: Group Decision Making, Linguistic Expression, Pair –Wise Preferences, Group Dynamics, Fuzzy Inference System.

1 Introduction

GDM is an approach to problem solving by using information provided by multiple decision makers. In many real life decision making problems, a solution

Mahima Gupta
Department of Operations, Great Lakes Institute of Management, Chennai, India
e-mail: 19.mahima@gmail.com

[*] Corresponding author.

© Springer International Publishing Switzerland 2015
W. Pedrycz and S.-M. Chen (eds.), *Granular Computing and Decision-Making*,
Studies in Big Data 10, DOI: 10.1007/978-3-319-16829-6_13

is sought under a group setting, due to the complexity and importance of problem at hand. The GDM is used in various domains such as information retrieval, investment and planning etc. [1]. In order to obtain the group's view, one common approach is to aggregate individual's opinions to arrive at the group's view [2]. In this approach, it is possible that aggregated view (group view) deviates from some members' views to a large extent and thereby leads to high discontent among the members. In our work, we obtain the group's consensus view as a result of an iterative procedure wherein the members change their views depending on their importance and support of their views in the group. The members would be influenced by other members whom they consider important in the group. Thus they are amenable to shift their opinion if their views find less acceptability in the group. The challenges to obtain group's consensus view for such approach are as follows:

1)   The members give vague expressions of their preferences by comparing a pair of alternatives.
2)   In the group, the members give different importance to each other's views depending on their common interest or confidence on other's knowledge or expertise.   Their perception of importance of other members in the group is given in fuzzy terms.
3)   The members change their views depending on their importance and support of their views in the group. The resultant shift of each member is to be calculated considering both the factors.

In our work we propose an iterative procedure to solve GDM problems using Fuzzy MCDM techniques. The member's preferences for alternatives by comparing a pair of alternatives are elicited using fuzzy linguistic approach. Further, their views about the importance of other members in the group are also taken in linguistic terms.   Using the concept of similarity between two linguistic tuples, we obtain support to the views of each member. The importance of each member in the group is obtained by fuzzy aggregation of the importance accorded to that member by other members in the group. Next, we design a Fuzzy inference system that calculates the probable shift in the opinions of members in the group given the current group dynamics i.e. their importance and support in the group. Thus iteratively, making changes in the members' views, our methodology obtains the group's view. Graphically, our methodology can be described as shown in fig.1 below.

There exist many variants of GDM problem in the literature [3], [4], [5]. In general, the methodologies focus on aggregation of views of individual members in the group given in exact and complete terms. In [4], [5], [6], the members are asked to evaluate the complete set of alternatives either by attribute wise utility value or their pair wise preferences in numeric, linguistic or ordinal scale. This is deviant from real life situations where the members do not have expertise or interest to evaluate entire set of alternatives and decisions are based on incomplete preference knowledge. In [7], [8], [9], the group members are asked to submit a

---

INPUT

Member's preferences in linguistic terms (By comparing alternatives pair wise)
Importance of a member as perceived by other members in the group

GROUP DYNAMICS

Importance of a member in the group $IMP(e_i)$
Support to the member's views in the group $s_{l,m}(e_i)$

FIS

If importance is High, Support is High,
then the member's likely shift in the opinion is significantly low

OUTPUT

Group's consensus view

Fig. 1 Methodology's description

partial preference list of the alternatives. Though the methodologies given in [7], [8], [9] take into account the partial preference information in precise form, and arrive at consensus, these cannot be implemented for real life decision making situations as realistic data or information very often are imprecise, or in fuzzy terms. There are works where the members' imprecise information in GDM is incorporated using fuzzy sets [10], [11] . The other dimension of decision making, the members' changing preferences, group decision making in dynamic environment [8], [12], [13], [14] has been discussed in the literature. The GDM has been studied extensively in the literature with the focus on different aspects of decision making such as input information, group characteristics and output (pair wise or ranked list of alternatives). To the best of our knowledge, we could not find any work that arrives at group's consensus by considering group dynamics. In our work, we have given a methodology to obtain group's view in an exploratory way where the members change their views depending on their importance and support in the group.

Section 2 describes the problem statement and linguistic framework of members' expressions. In section 3, formalization of the concepts of group dynamics such as importance and support of the members in the group is explained. In section 4, construction of Fuzzy Inference System is explained. In section 5, the algorithm for obtaining group's view is outlined. In section 6, the procedure is explained with the help of a numerical example. Some concluding remarks are made in section 7.

2   Elicitation of Members' Preferences

Consider a group decision making problem where the members give their preferences by comparing the pairs of alternatives. In general, many aspects of the evaluation of alternatives cannot be assessed in quantitative form, but rather in a qualitative one with vague or imprecise knowledge. In such cases, a better approach is to use linguistic assessments   instead of numerical ones. The linguistic approach facilitates them to express their preferences of alternatives in the event of their imprecise or insufficient knowledge about the problem or inability to discriminate explicitly one alternative over the other. In linguistic approach, a term set consisting of a finite number of linguistic terms with their appropriate descriptors and semantics are defined.   The cardinality of term set depends on the level of granularity and distinctness in members' expressions. The semantics of these terms are defined as fuzzy sets whose membership functions map the domain values to the semantics of the linguistic terms [15] [16]. The linguistic term set $S = \{s_0, s_1 \ldots s_g\}$ is a finite and totally ordered discrete term set where s_r represents a possible value of a linguistic variable such that

i.   The set of linguistic terms are ordered such that $s_r > s_t;\, if\, r > t$

ii.   There is a negation operator (Neg) such that Neg $(s_r) = s_{g-r}$

iii.   The maximization operator Max satisfies that Max $(s_i, s_j) = s_i$; if $s_i \geq s_j$

iv.   The minimization operator Min satisfies that Min $(s_i, s_j) = s_i$; if $s_i \leq s_j$

Following [17] we have taken a term set S of 9 linguistic terms to express the members' pair wise preferences. The descriptors and semantics of 9 terms are shown in figure 2 below.

$S = (s_0=$IMP$,s_1=$NLG$,s_2=$VL$,s_3 =$L$,s_4=$M$,s_5=$H$,s_6=$VH$,s_7=$SH$,s_8=$EH$)$

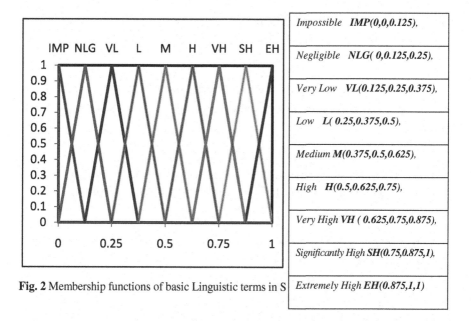

	Impossible **IMP(0,0,0.125)**,
	Negligible **NLG(0,0.125,0.25)**,
	Very Low **VL(0.125,0.25,0.375)**,
	Low **L(0.25,0.375,0.5)**,
	Medium **M(0.375,0.5,0.625)**,
	High **H(0.5,0.625,0.75)**,
	Very High **VH (0.625,0.75,0.875)**,
	Significantly High **SH(0.75,0.875,1)**,
	Extremely High **EH(0.875,1,1)**

Fig. 2 Membership functions of basic Linguistic terms in S

The fuzzy linguistic approach allows the member's expression to be represented as one of the linguistic terms and make the computations directly on their membership functions. This approach has certain limitations: a linguistic expression may not match exactly to any of these terms; secondly, in computational procedure when we perform retranslation step as an approximation process to express the result (a value in domain) in original term set, it may lead to lack of accuracy.   In order to overcome above shortcomings, the 2-tuple linguistic computational model is introduced, by treating the linguistic domain as continuous but keeping the linguistic basics (linguistic terms and semantics). In this model, linguistic information is represented by a linguistic tuple that consists of a pair of values namely (s, α), where $s \in S$ is a linguistic term $and\ \alpha \in [-0.5, 0.5)$ a numerical value representing the symbolic translation. For instance, if we have linguistic termset as {Very Low, Low, Medium, High, Very High} and a member gives her preference as "above medium" or "somewhat high", the expression can be represented as (M,α) where α signifies the difference of information between the linguistic term "Medium" and the linguistic statement "Above Medium".

2.1   Linguistic 2-Tuple and Its Transfomation to a Numeric Value in [0,g] or [0,1]

Let (s_k, α) be a linguistic 2-tuple ($s_k \in S = \{s_0, \ldots \ldots s_g\}$ and $\alpha \in [-0.5, 0.5)$). We have its numerical equivalent $\beta \in [0, g]$ and a fractional value $v \in [0,1]$ that supports the information represented by it through a transformation process

[17], [18]. Let (s_k, α) be a linguistic 2-tuple ($s_k \in S = \{s_0, \ldots \ldots s_g\}$ and $\alpha \in [-0.5, 0.5)$). Its numerical equivalent $\beta \in [0, g]$ is obtained as

$$\beta = k + \alpha$$

Further, to obtain a fractional value $v \in [0,1]$, first we find a pair of 2-tuples $\{(s_h, 1 - \gamma), (s_{h+1}, \gamma)\}$ where $h = trunc(\beta); \gamma = \beta - h$ and calculate v as shown below.

$$v = CV(s_h) * (1 - \gamma) + CV(s_{h+1}) * (\gamma)$$

Where $CV(.)$ is a function providing a characteristic value.
Example: For a tuple $(H, 0.3)$ in linguistic termset in fig. 2, we have

$$\beta = 5 + 0.3 = 5.3$$

Correspondingly we have a pair of 2-tuples as $\{(H,0.7),(VH,0.3)\}$ with $h = trunc(5.3) = 5; \gamma = 5.3 - 5 = 0.3$

$$v = 0.625 * 0.7 + 0.75 * 0.3 = 0.6625$$

2.2 Linguistic Preference Relation

Suppose 'q' members $E = \{e_1, e_2, \ldots \ldots e_q\}$ evaluate 'n' alternatives $\mathcal{A} = \{a_1, a_2 \ldots a_n\}$ in the GDM problem. For a set of alternatives $\mathcal{A} = \{a_1, a_2 \ldots a_n\}$, the preference information of pair wise comparison for a member is represented on the set $S = \{s_0, \ldots \ldots s_g\}$ of linguistic terms where the tuple $s_{ij} = (s, \alpha)$; ($s \in S = \{s_0, \ldots \ldots s_g\}$ and $\alpha \in [-0.5, 0.5)$) estimates the degree of linguistic preference of the alternative a_i over a_j.

Particularly, $s_{ij} = (s_{g/2}, 0)$ indicates indifference between a_i and a_j, $s_{ij} > (s_{g/2}, 0)$ indicates that a_i is preferred a_j and $s_{ij} < (s_{g/2}, 0)$ that a_j is preferred a_i [19]. A linguistic matrix $T = (s_{ij})_{n \times n}$ is known a 2-tuple linguistic preference relation (LPR), if

$$\beta_{ii} = \frac{g}{2};$$

$$\beta_{ij} + \beta_{ji} = g$$

A 2-tuple LPR is said to be additively consistent if $\forall i, j, k \in N$, we have

$$\beta_{ij} + \beta_{jl} + \beta_{li} = 1.5g \text{ or } \beta_{ij} = \beta_{il} + \beta_{lj} - 0.5g$$

However, in real life situation LPR may not be additively consistent. In our work, we have not assumed the relations to be additively consistent.

In our work k^{th} member's initial pair wise linguistic preferences for the alternatives $P^k(a_1, a_2)$ are converted into their numerical equivalents. The kth member's preference for the pair (a_l, a_m) in numerical terms is denoted as $e_k^0(a_l, a_m)$.

3   Group Dynamics

In many approaches to GDM, aggregated decision is taken as consensus decision of the group, assumed to be acceptable to all the members. If a member's views differ from group's consensus view to a large extent, it might arouse discontent in her and lead to unstable decisions for the group. In order to have a stable decision, consensus should be evolved among the members by leading them to change their views as guided by group dynamics. Depending on their importance and support in the group, the members will change their views to facilitate group's consensus view. The members having low support as well as low importance in the group are more amenable to changes in comparison to the members with high importance and high support in the group. Thus we need to determine the shift in the members' opinion determined by their importance and support in the group. The methodology to calculate importance of the members and their support in the group is described below.

3.1   *Importance in the Group*

The members give importance to other group members depenpending on their perceived knowledge of others' expertise or knowledge. Further, importance can be ascribed due to mutual trust, similar tastes or past interactions. In our work, we elicit the member's importance as perceived by other members in the group in linguistic terms. Suppose the members use the following linguistic termset S_{imp} to give importance of members in the group.

S_{imp}={(VL,(0,0,0.25)),

L,(0,0.25,0.5)),(M,(0.25,0.5,0.75)),(H,(0.5,0.75,1)),(VH,(0.75,1,1))}

For instance member e_i's importance as perceived by the member e_j $IMP(e_i|e_j)$ is VH (say). Similarly, we have the member's importance expressed by all other members in the group. The member's overall importance in the group is obtained by aggregating his/her importance perceived by other members in the group. Depending upon the context of the problem, we may define the member's importance in the group as perceived by *some*, *at least a few* or *most* of the members in the group. We use concept of fuzzy linguistic quantifier,to obtain the member's importance in the group as shown in equation (1) below. The readers are referred to the works [20], [21] for the concepts on linguistic quantifiers and the associated aggregation.

$$IMP(e_i) = Q\ (IMP(e_i|e_1), IMP(e_i|e_2), IMP(e_i|e_3) \ldots \ldots IMP(e_i|e_q)) \ldots \ (1)$$

In equation (1), Q could be any linguistic quantifier.

Example 3.1: Suppose for a member e_1 we have associated importance as

$$IMP(e_1|e_2) = L, IMP(e_1|e_3) = H, IMP(e_1|e_4) = M$$

These linguisic expressions are converted to their fractional counterparts as discussed in section 2.

$$IMP(e_1|e_2) = 0.25, IMP(e_1|e_3) = 0.75, IMP(e_1|e_4) = 0.5$$

If we define over all importance of the member in the group as her imporance percieved by '*most*' of the members in the group, the weights are calculated in accordance with the quantifier 'Q=most' whose membership function is defined in equation (2).

$$\mu_{most}(x) = \begin{cases} 0, & x < 0.3 \\ \frac{x-0.3}{0.5}, & 0.3 \le x < 0.8 \\ 1, & x \ge 0.8 \end{cases} \quad \dots \quad (2)$$

The individual values are aggregated by calculating the weights using formula in equation (3).

$$w_i = Q\left(\frac{i}{n}\right) - Q\left(\frac{i-1}{n}\right) \ for \ i = 1,2 \dots n. \quad \dots \quad (3)$$

Since over all importance of member e_1 is as perceived by most of other members in the group.

$$IMP(e_1) = most(IMP(e_1|e_2), IMP(e_1|e_3), IMP(e_1|e_4))$$

There fore for n=3, we have $W = (w_1 = 0.06, w_2 = 0.66, w_3 = 0.28)$

After obtaining the weights contextually, we have aggregated value using OWA aggregation of a member's perceived importance by most of the members in the group. The definition for OWA aggregation procedure is given below:

An OWA operator of dimension n is a mapping $F: R^n \to R$ with associated weighting vector $W = (w_1, w_2 \dots w_n)$ with $w_i \ge 0$ and $w_1 + w_2 + \cdots . w_n = 1$
Such that

$$F_W(x_1, x_2 \dots x_n) = \sum_{i=1}^{n} w_i y_i$$

where y_i is i^{th} largest of $(x_1, x_2 \dots x_n)$.

Using $W = (w_1 = 0.06, w_2 = 0.66, w_3 = 0.28)$, we have the importance of member e_1 using OWA aggregation principle as

$$IMP(e_1) = 0.06 \times 0.75 + 0.66 \times 0.5 + 0.28 \times 0.25 = 0.445.$$

3.2 Support in the Group

The members' views may conflict with some and be supported by others. In our work , we calculate similarity between members' views to calculate their support in the group. Suppose for a pair of alternatives say (a_l, a_m), we have two members giving the preferences as $P^i(a_l, a_m) = s_u$ and $P^j(a_l, a_m) = s_v$, similarity between two members for the pair of alternatives (a_l, a_m) [22] is obtained as:

$$sim_{l,m}(e_i, e_j) = 1 - \frac{|(v-u)|}{g} \quad \ldots \quad (4)$$

The support of a member in the group for a pair of alternatives (a_l, a_m) is defined as aggregation of the similarity of an individual's opinion with other members in the group.

$$s_{l,m}(e_i) = Q(sim_{l,m}(e_i, e_1), sim_{l,m}(e_i, e_j) \ldots \ldots sim_{l,m}(e_i, e_q)) \quad \ldots (5)$$

Example 3.2: Suppose for a pair of alternatives (a_1, a_2) we have members' expressions as

$$P^1(a_1, a_2) = s_2 = 'VL' \quad , \quad P^2(a_1, a_2) = s_3 = 'L' \quad , P^3(a_1, a_2) = s_3 = 'L' \quad ,$$
$$P^4(a_1, a_2) = s_7 = 'SH'$$

Using equation (4), we have

$$sim_{1,2}(e_1, e_3) = 1 - \frac{|(3-2)|}{8} = 0.875$$

Similarly we have $sim_{1,2}(e_1, e_2) = 0.875$ and $sim_{1,2}(e_1, e_4) = 0.375$

Using equation (5), we have support of $e_1's$ opinion for alternatives (a_1, a_2) in the group is

$$s(e_1) = average\left(sim_{1,2}(e_1, e_3), sim_{1,2}(e_1, e_4), sim_{1,2}(e_1, e_2)\right) = 0.7083$$

4   Fuzzy Inference System

We have designed a FIS to calculate the shift in the individual's opinions in order to reach consensus in the group. Here we have two input variables namely net importance and support in the group and one output variable i.e. shift in the opinion . For both the factors (antecedents) i.e. importance and support in the group and output variable, fuzzy sets and their membership functions were defined on the range based on their characteristic values [23], [24]. Cardinality of fuzzy sets were chosen to be either 5 or 9, reflecting the tradeoff between granularities of uncertainty and distinguishing ability of human mind. After defining input and output variables, their interrelationships are modelled in the form of linguistic if-then rules. The fuzzy sets and membership functions for the member's support, importance and likely shift in the view are given in figure 3, 4 and 5 respectively.

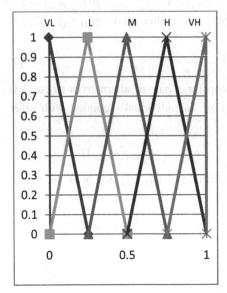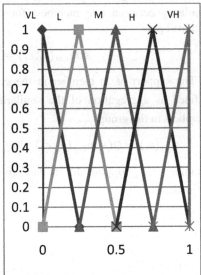

Fig. 3 Support of a member　　　　　　　　**Fig. 4** Importance of a member

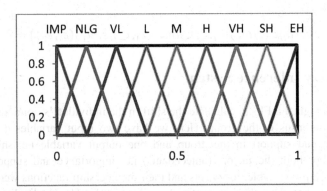

Fig. 5 Shift in the opinion

We have 25 sets of rules covering all possible compinations of input variables and resultant output i.e. shift in individual's opinion. The members with less support and importance in the group will have larger shift in their earlier opinions in comparison to the ones with high support and importance. The rules for GDM problem are given in Table 1 below.

Table 1 (Rule Base System)

	If Support is	If importance is	Then shift is		If Support is	If importance is	Then shift is
1	VL	VL	EH	14	M	H	L
2	VL	L	VH	15	M	VH	SL
3	VL	M	SH	16	H	VL	H
4	VL	H	H	17	H	L	M
5	VL	VH	M	18	H	M	L
6	L	VL	VH	19	H	H	SL
7	L	L	SH	20	H	VH	VL
8	L	M	H	21	VH	VL	M
9	L	H	M	22	VH	L	L
10	L	VH	L	23	VH	M	SL
11	M	VL	SH	24	VH	H	VL
12	M	L	H	25	VH	VH	NLG
13	M	M	M				

4.1   Evaluation of FIS

When a crisp input set of antecedents is entered into a rule, membership value of each input is calculated. Implication operator AND connects through minimum function and gives membership value of consequent (output) according to the rule. This membership value gives weight of the rule.

Example 4.1: Suppose a member e_i with importance $IMP(e_i) = 0.445$ gives her preference for the pair a_l, a_m $P^i(a_l, a_m)$ = 'L' that generates the support $s_{l,m}(e_i) = 0.7083$ for her in the group.

In order to calculate shift in her opinion, we will use rule base given in Table 1 as follows:

Rule 1

$$\min(\mu_{VL}(s_{l,m}(e_i) = 0.7083), \mu_{VL}(IMP(e_i) = 0.445\ )) = \min(0,0) = 0$$
$$= \mu_{VH}(shift_{l,m}(e_i)) = w_1 = 0$$

..

...

Rule 12:

$$\min(\mu_{M}(s_{l,m}(e_i) = 0.7083), \mu_{L}(IMP(e_i) = 0.445\ )) = \min(0.1688,0.22)$$
$$= 0.1688 = \mu_{H}(shift_{l,m}(e_i)) = w_{12} = 0.1688$$

Rule 13:

$$\min(\mu_{M}(s_{l,m}(e_i) = 0.7083), \mu_{M}(IMP(e_i) = 0.445\ )) = \min(0.1688,0.78)$$
$$= 0.1688 = \mu_{M}(shift_{l,m}(e_i)) = w_{13} = 0.1688$$

...

...

Rule 17:

$$\min(\mu_{H}(s_{l,m}(e_i) = 0.7083), \mu_{L}(IMP(e_i) = 0.445\ )) = \min(0.8332,0.22)$$
$$= 0.22 = \mu_{M}(shift_{l,m}(e_i)) = w_{17} = 0.22$$

Rule 18

$$\min(\mu_{H}(s_{l,m}(e_i) = 0.7083), \mu_{M}(IMP(e_i) = 0.445\ )) = \min(0.8322,0.78)$$
$$= 0.78 = \mu_{L}(shift_{l,m}(e_i)) = w_{18} = 0.78$$

...

Rule 20:

$$\min(\mu_{M}(s_{l,m}(e_i) = 0.7083), \mu_{H}(IMP(e_i) = 0.445\ )) = \min(0,0) = 0$$
$$= \mu_{L}(shift_{l,m}(e_i)) = w_{20} = 0$$

...

Rule 25:

$$\min(\mu_{L}(s_{l,m}(e_i) = 0.7083), \mu_{H}(IMP(e_i) = 0.445\ )) = \min(0,0) = 0$$
$$= \mu_{M}(shift_{l,m}(e_i)) = w_{25} = 0$$

Therefore, for above inputs $IMP(e_i) = 0.445$ and support $s_{l,m}(e_i) = 0.7083$, rules 12,13,17 and 18 are activated.

4.2 Defuzification Method

This FIS is used to calculate likely shift in the member's opinion $Shift_i\ (a_l, a_m)$ using centroid method [23]. Corresponding to each rule we have the consequent linguistic term in liguistic termset (Fig. 5) as an output and its weight corresponding to input values to the rule.

$$Shift_i\ (a_l, a_m) = \frac{w_1 XCV(Rule1) + w_2 XCV(Rule2)\dots w_{25} XCV(Rule25)}{w_1 + w_2 + \cdots w_{25}} \quad \dots \quad (6)$$

Where $CV(Rule12)$ denotes characteristic value of consequent of rule 12 'If Support is M and importance is L then shift is H'. For instance in our rule base CV(Rule 12), corresponding to output'H' is 0.625.

Example 4.2: For all the rules with non-zero membership, we have (H, 0.1688) (M, 0.1688) (M, 0.22) (L, 0.78) For the above scenario, $IMP(e_i) = 0.445$ and preference for the pair a_l, a_m $P^i(a_l, a_m) = $ 'L' that generates the support $s_{l,m}(e_i) = 0.7083$ for him/her in the group, the shift in the opinion is

$$Shift_i\ (a_l, a_m) = \frac{0.1688 X 0.625 + 0.1688 X 0.5 + 0.22 X 0.5 + 0.78 X 0.375}{0.625 + 0.5 + 0.5 + 0.375} = 0.4412.$$

5   Algorithm

The algorithm that iteraively generates the consensus view of the preference of the alternatives in pair wise form considering group dynamics is explained below. The algorithm proceeds from one iteration to next till the group's view i.e. average of individual views are acceptable to majority in the group. In our work we calculate dispersion of views of individual members from group's average and if it is found to be   below a predetermined threshold value, we take that average view of the members as group's consensus view. Otherwise, the algorithm calculates the possible shift in the members' opinions and correspondingly changed view of the members.

Step-1: Elicit the members' opinions giving pair wise preferences for the alternatives $P^i(a_l, a_m)$, i= 1,2…q and $\forall(a_l, a_m) \in \mathcal{A} X \mathcal{A}$ using the linguistic termscale given in fig. 2. Convert the linguistic expression into their equivalent numerical ones [0,8] and denote them as $e_i^0(a_l, a_m)$. Take the member's views for the importance of other members in the group $IMP(e_i|e_j)$. Set iteration count r =0. Set dispersion threshold $\delta = \delta_0$.

Step-2: Calculate average of the opinions of the members for each pair of alternatives denoted as $Group\_view^r(a_l, a_m)$. The group's view for the preference of a_l over a_m is taken as average of the members' views $(e_1^r(a_l, a_m), e_2^r(a_l, a_m), \dots, e_q^r(a_l, a_m))$.

Step-3: Calculate dispersion in group's opinion $Disp(r)$ as

$$= 2 * \sum_{(a_l, a_m) \in \mathcal{A} X \mathcal{A}} \sum_{e_i \in E} abs(e_i^r(a_l, a_m) - Group\_view^r(a_l, a_m))/(q * n * (n-1)) \quad (7)$$

In the above equation, we have taken pair wise relations (a_l, a_m) ; $l < m$ because $P^i(a_l, a_m)$ and $P^i(a_m, a_l)$ are reciprocal relations.

Step-4: If $Disp(r) \geq \delta_0$, Set r = r+1.Go to Step-5

Else stop. Take $Group\_view^r(a_l, a_m)$ as group's consensus view.

Step-5: Calculate support of the members in the group using equation (5).

Step-6: Also, calculate importance of the members in the group using equation (1).

Step-7: Calculate shift of each member $Shift^r_i (a_l, a_m)$ from their respective positions towards average of the opinion using the fuzzy Inference System explained in section 4.

The new opinion of the e_i for the pair of alternatives a_l, a_m is equal to

$$e_i^r(a_l, a_m) = e_i^{r-1}(a_l, a_m) + Shift^r_i (a_l, a_m) * (Group\_view^{r-1}(a_l, a_m) - e_i^{r-1}(a_l, a_m)) \ldots \tag{8}$$

Repeat the procedure for all the members and pairs of alternatives.

Step-8: Go to step 2.

6   Example

Consider a problem of 4 members in a group giving pair wise preferences for evaluation of six alternatives as shown in tables (2.1-2.4) below. The members give their preference for the alternatives using the linguistic term scale given in fig. 2.

Table 2.1 Member e_1's preferences

$P^1(a_l, a_m)$	1	2	3	4	5	6
1		VL	NLG	H	M	L
2			L	NLG	H	SH
3				VH	L	M
4					EH	VL
5						VL
6						

Table 2.2 Member e_2's preferences

$P^2(a_l, a_m)$	1	2	3	4	5	6
1		L	M	VL	M	VH
2			H	L	M	VL
3				NLG	NLG	L
4					L	H
5						SH
6						

In above GDM problem, the members give importance to others' views using linguistic term scale S_{imp} in section 3.1 in table 3 below.

Above information on members' pair wise preferences and member's preference for other members are used as an input to algorithm given in section 5. For the problem, threshold for group's dispersion is taken 0.5.

Table 2.3 Member e_3's preferences

$P^3(a_l, a_m)$	1	2	3	4	5	6
1		L	M	VL	L	EH
2			NLG	L	VH	EH
3				VL	H	NLG
4					M	VL
5						L
6						

Table 2.4 Member e_4's preferences

$P^4(a_l, a_m)$	1	2	3	4	5	6
1		SH	VH	SH	NLG	VL
2			SH	EH	M	L
3				IMP	NLG	L
4					M	L
5						VL
6						

Table 3 Members' Importance

$IMP(e_i \| e_j)$	1	2	3	4
1	-	L	VH	VL
2	L	-	M	M
3	H	VL	-	VH
4	M	H	M	-

Iteration (0)

The group's view as an average of individual's opinions is shown in table 4 below.

Table 4 Group view

$Group\_view^0(a_l, a_m)$						
	1	2	3	4	5	6
1		3.75	3.75	4	3	4.75
2			4	3.75	4.75	5
3				2.25	2.5	2.75
4					4.75	3
5						3.5
6						

Using equation (7), we have dispersions in the group's opinion as $Disp(0) = 1.5$. Since $Disp(0) > 0.5$; we move to next iteration.

Iteration (1)

We calculate support to the members using equation (5). Support to each member in the group is given in tables 5.1-5.4 below.

Table 5.1 Support to member e_1

$s_{l,m}(e_1)$	1	2	3	4	5	6
1		0.708333	0.541667	0.666667	0.833333	0.625
2			0.666667	0.541667	0.875	0.583333
3				0.375	0.75	0.791667
4					0.458333	0.833333
5						0.75
6						

Table 5.2 Support to member e_2

$s_{l,m}(e_2)$	1	2	3	4	5	6
1		0.791667	0.791667	0.666667	0.833333	0.625
2			0.666667	0.708333	0.875	0.5
3				0.708333	0.75	0.875
4					0.708333	0.666667
5						0.416667
6						

Table 5.3 Support to member e_3

$s_{l,m}(e_3)$	1	2	3	4	5	6
1		0.791667	0.791667	0.666667	0.833333	0.458333
2			0.5	0.708333	0.791667	0.5
3				0.708333	0.583333	0.708333
4					0.791667	0.833333
5						0.75
6						

Table 5.4 Support to member e_4

$s_{l,m}(e_4)$	1	2	3	4	5	6
1		0.458333	0.625	0.5	0.666667	0.541667
2			0.5	0.291667	0.875	0.583333
3				0.625	0.75	0.875
4					0.791667	0.833333
5						0.75
6						

Using equation (1), we have importance of the members as

$IMP(e_1) = 0.445$; $IMP(e_2) = 0.21$;
$IMP(e_3) = 0.53$; $IMP(e_4) = 0.39$;

With this information, we use FIS to determine shift in the members' opinions and the resultant opinions are given below in tables 6.1-6.4. The value in parenthesis gives resultant shift.

Table 6.1 Member e_1's revised opinion and (shift)

$e_1^1(a_l, a_m)$	1	2	3	4	5	6
1		3.3744 (0.4992)	3.8752 (0.4992)	3.1502 (0.5751)	3.5196 (0.4804)	5.263 (0.5896)
2			4.4249 (0.5751)	3.4199 (0.5599)	4.3486 (0.4648)	3.9503 (0.6501)
3				1.6999 (0.5599)	1.7875 (0.525)	2.884 (0.4648)
4					3.9798 (0.5599)	3.8498 (0.5751)
5						4.5489 (0.7003)
6						

Table 6.2 Member e_2's revised opinion and (shift)

$e_2^1(a_l, a_m)$	1	2	3	4	5	6
1		2.7721 (0.4412)	2.3926 (0.5064)	4.5409 (0.4591)	3.637 (0.363)	3.8274 (0.4728)
2			3.4591 (0.4591)	2.3926 (0.5064)	4.913 (0.3479)	6.0242 (0.4879)
3				3.7579 (0.5979)	2.796 (0.4077)	3.523 (0.3816)
4					6.1602 (0.5661)	2.363 (0.363)
5						2.6115 (0.4077)
6						

Table 6.3 Member e_3's revised opinion and (shift)

$e_3^1(a_l, a_m)$	1	2	3	4	5	6
1		3.242 (0.3227)	3.919 (0.3227)	2.8002 (0.4001)	3 (0.3065)	6.354 (0.5065)
2			2.4421 (0.4807)	3.2858 (0.3811)	5.5966 (0.3227)	6.5579 (0.4807)
3				2.0953 (0.3811)	3.922 (0.4312)	1.669 (0.3811)
4					4.242 (0.3227)	2.3065 (0.3065)
5						3.1776 (0.3552)
6						

Table 6.4 Member e_4's revised opinion and (shift)

$e_4^1(a_l, a_m)$	1	2	3	4	5	6
1		3.242 (0.5854)	3.919 (0.4954)	2.8002 (0.5563)	3 (0.4811)	6.354 (0.5288)
2			2.4421 (0.5563)	3.2858 (0.654)	5.5966 (0.3705)	6.5579 (0.5093)
3				2.0953 (0.4954)	3.922 (0.4316)	1.669 (0.3705)
4					4.242 (0.404)	2.3065 (0.3845)
5						3.1776 (0.4316)
6						

With the revised opinion, we calculate group's opinion $Group\_view^1(\ )$ as shown in table 7 below.

Table 7 Group view (Iteration 1)

$Group\_view^1(\ )$	1	2	3	4	5	6
1		3.6215	3.7681	3.9556	3.0297	4.725
2			3.9143	3.5797	4.784	5.13775
3				2.1669	2.5383	2.7453
4					4.6713	2.879
5						3.246
6						

At this stage dispersion in group view is $Disp(1) = 0.769$
Since it is above threshold value, we go to next iteration.

Iteration (2)

We proceed with same steps and obtain $Group\_view^2(\ )$ as shown in table 8 below.

Corresponding to above group view we have dispersion as $Disp(2) = 0.4639$. The algorithm terminates here and we take $Group\_view^2(\ )$ as final view of the group.

Table 8 Consensus Group View

$Group\_view^2(\ )$	1	2	3	4	5	6
1		3.5935	3.7707	3.947	3.0284	4.7202
2			3.874	3.547	4.807	5.1908
3				2.1611	2.5714	2.7313
4					4.6629	2.836
5						3.196
6						

7 Conclusion

In our work, we have given an algorithm to obtain group's view by an iterative procedure based on group's dynamics. The group's dynamics are captured by the members' support and importance in the group. The group's consensus decision obtained in this way is expected to be more stable as the members are induced to change their views in accordance to their importance and support in the group. We have obtained the shift in the member's opinions by considering two factors i.e. their importance and support in the group. The work can be extended by incorporating other relevant factors in group decision making such as trust and various other interrelationships among the members.

References

[1] Rodriguez, R.M., Martinez, L., Herrera, F.: A group decision making model dealing with comparative linguistic expressions based on hesitant fuzzy linguistic term sets. Information Sciences 241, 28–42 (2013)
[2] Su, Z.-X., Chen, M.-Y., Xia, G.-P., Wang, L.: An ineractive method for dynamic intuitionistic fuzzy multiple attribute group decision making. Experts System with Applications 38, 15286–15295 (2011)
[3] Kacprzyk, J., Fedrizzi, M.: Multiperson Decision Making Models Using Fuzzy Sets and Possibility Theory. Kluwer, Dordrecht (1990)
[4] Morais, D.C., de Almeida, A.T.: Group decision making on water resources based on analysis of individual rankings. Omega 40(1), 42–52 (2011)
[5] Yu, L., Lai, K.K.: A distance-based group decision-making methodology for multi-person multi-criteria emergency decision support. Decision Support Systems 51(2), 307–315 (2011)
[6] Yager, R.R.: Fusion of multi-agent preference orderings. Fuzzy Sets and Systems 117(1), 1–12 (2001)
[7] Chen, Y.-L., Cheng, L.-C.: Mining maximum consensus sequences from group ranking data. European Journal of Operational Research 198(1), 241–251 (2009)
[8] Chen, Y.-L., Cheng, L.-C.: An approach to group ranking decisions in a dynamic environment. Decision Support System 48(4), 622–634 (2010)

[9] Cook, W.D., Golany, B., Penn, M., Raviv, T.: Creating a consensus ranking of proposals from reviewers' partial ordinal rankings. Computers & Operations Research 34(4), 954–965 (2007)

[10] Guha, D., Chakraborty, D.: Fuzzy multi attribute group decision making method to achieve consensus under the consideration of degrees of confidence of experts' opinions. Computers & Industrial Engineering 60(4), 493–504 (2011)

[11] Chen, S.-M., Lee, L.-W.: Fuzzy multiple criteria hierarchical group decision-making based on interval Type-2 fuzzy sets. IEEE Transactions on Systems, Man, and Cybernetics—Part A: Systems and Humans 40(5) (September 2010)

[12] Weiss, E.N.: Using the Analytic Hierarchy Process in a dynamic environment. Mathematical Modelling 9, 211–216 (1987)

[13] Campanella, G., Ribeiro, R.A.: A framework for dynamic multiple-criteria decision making. Decision Support Systems 52, 52–60 (2011)

[14] Pérez, I.J., Cabrerizo, F.J., Viedma, E.H.: A mobile decision support system for dynamic group decision-making problems. IEEE Transactions on Systems, Man, and Cybernetics—Part A: Systems and Humans 40(6), 1244–1256 (2010)

[15] Herrera, F., Martinez, L.: A 2-tuple fuzzy linguistic representation model for computing with words. IEEE Transactions on Fuzzy Sets 8(6), 746–752 (2000)

[16] Martınez, L., Ruan, D., Herrera, F.: Computing with words in decision support systems: an overview on models and applications. International Journal of Computational Intelligence Systems 3(4), 382–395 (2010)

[17] Herrera, F., Martinez, L.: An approach for combining linguistic and numerical information based on the 2- tuple fuzzy linguistic representation model in decision-making. International Journal of Uncertainty, Fuzziness and Knowledge-Based Systems 8(5), 539–562 (2000)

[18] Martinez, L., Herrera, F.: An overview on the 2-tuple linguistic model for computing with words in decision making: Extensions, applications and challenges. Information Sciences 207(1), 1–18 (2012)

[19] Gong, Z.-W., Forrest, J., Yang, Y.-J.: The optimal group consensus models for 2-tuple linguistic preference relations. Knowledge Based Systems 37, 427–437 (2013)

[20] Yager, R.R.: Quantifier guided aggregation using OWA operators. International Journal of Intelligent Systems 11, 49–73 (1996)

[21] Yager, R.R.: Families of OWA operators. Fuzzy Sets and Systems 59, 125–148 (1993)

[22] Xu, Z.: Deviation measures of linguistic preference relations in group decision making. Omega 33(3), 249–254 (2005)

[23] Madami, E.H., Assilian, S.: An experiment in linguistic synthesis with a fuzzy logic controller. International Journal of Man-Machine Studies 7(1), 1–13 (1975)

[24] Dixit, V., Srivastava, R.K., Chaudhari, A.: Integrating materials management with project management of complex projects. Journal of Advances in Management Research 10(2), 230–278 (2013)

Fuzzy Optimization in Decision Making of Air Quality Management

Wang-Kun Chen and Yu-Ting Chen

Abstract. This study presents an optimization method in fuzzy decision making of air quality management. The optimization method presented in this chapter gives the mathematical representation to find the equilibrium point. How to obtain and express these optimal data depends on the fuzzy optimization techniques. The methodology and algorithm of fuzzy decision making process by interactive multi-objective approach and iterative optimization method are described, with the application in the process of air quality management. This paper also provides the interactive multi-objective model and iterative calculation method for the application of air quality management. First, the comparison of model output and field monitoring results was discussed, and then the experimental outcome of interactive fuzzy optimum model was presented. Secondly, the comparison of optimum decision from different decision makers was considered, and the experimental outcome of iterative fuzzy optimum model was presented. The combined approach of interactive and iterative method for fuzzy optimization model makes the decision of air quality management more accurate and pragmatic.

Keywords: Fuzzy decision making, Optimization method, Air quality management.

1   Introduction

1.1   Decision Making for Policy Maker

Decision making analysis is a very important routine task for management. Decision analyst decides what kind of strategy to take from the data they obtained

Wang-Kun Chen
Department of Environment and Property Management,
Jinwen University of Science and Technology, New Taipei City, Taiwan
e-mail: wangkun@just.edu.tw

Yu-Ting Chen
Department of Environmental Engineering,
National Cheng-Kung University, Tainan, Taiwan
e-mail: ytdantim@gmail.com

© Springer International Publishing Switzerland 2015
W. Pedrycz and S.-M. Chen (eds.), *Granular Computing and Decision-Making,*
Studies in Big Data 10, DOI: 10.1007/978-3-319-16829-6_14

every day. Their decision will affect the future operating of an enterprise and many people's welfare; therefore, such decision must be made carefully. However, decision making analysis is a very difficult job, because the data comes from many different sources, and its credibility is not the same. Finally, what is believable? This is the biggest difficulty faced by policy-makers. In addition, what kind of message is provided among large amounts of data? These factors must be considered when conducting policy analysis. Fortunately, big data analysis techniques provide us the solution in this respect. Through big data analysis, we can get a more accurate analysis result.

How to get a reliable decision is the issue which policymakers continue to consider. The results of decisions are usually "Yes "or "No". And the data support decision may come from simulation results of the theoretical model or real observed results. If there is a difference between the two, what can be trusted for the decision-makers? The analysis results from theory provide observation mechanism of detailed changes. However, if the assumptions have errors, it can easily lead to erroneous results. That's why the results of theoretical analysis need to be verified through empirical analysis.

In addition to theoretical analysis and real observations, we also often get the support of decision-making through the expert. Because the experts have many valuable implicit knowledge, these knowledge is unable to acquire by theoretical models or real observations. While how is this valuable knowledge of experts join our decision-making process? The expert knowledge is usually hidden and unknown, but appears to be very reliable. This situation is a problem faced by policy makers, in the same way, how will this knowledge be put into our decision-making process. If in another case, the opinions from different experts are not the same, what people really can believe? Is there a solution to based on the advantages of the above three methods to get the best solution for decision-makers? It is an issue to be discussed in this article.

1.2 *Review of Previous Studies*

Air quality management is a very typical problem of decision analysis, it is related to the above three dimensions. As such, how the decision analyst obtains the message from above three sources to make the best judgment become the problems they faced every day. This chapter described the methodology of the decision analysis which combines the three issues. Theoretical development and its characteristics are presented. The experiment results are also shown in the article.

Before discussing this article, first, make a review of the past research that scholars have done. The representation of observational data and theoretical results of the model exist in all walks of life. So long, many scholars presented their views. (1oldstein and Landoritz, 1977) (Gustafson et al,, 1977) Sasaki first proposed in Calculus of Variation to optimize the best of meteorological data, to improve the consistency of observational meteorological data and results from meteorological model. And because air pollution is increasingly importance for everyone, Heimbach and Sasaki apply it to the assessment of air pollution. (Heimbach and Sasaki, 1977) They deal with the question of discrepancy between air quality monitoring stations and results of diffusion model. However, due to the large number of air pollution sources, his research has not described how to deal with the problem of value inconsistency between pollution sources and monitoring station.

Liang further developed his theory of optimization so that the theoretical model of air pollution can be capable of interaction with actual observed value. The model tries to have a good balance between each other. The constrained conditions of this theory were derived so that policy makers can have good space to determine what extent to be used. Thus, their decision will be the most realistic situation. His control theory was divided into two parts, strong and weak. And a parameter was derived to represent the degree of interaction between the two. There are also a lot of practical applications done by his theory. (Liang, 1979)(Liang, 1980)(Liang and Young, 1980)

Hsieh and Liang together use the finite element method to interact the information between the results from numerical models and observational data. (Liang and Hsieh, 1980) But they did not take into consideration of the air pollution sources. Liang and Lee take the example of carbon monoxide pollution in Taipei and apply the calculus of variation method for analysis. (Liang and Lee, 1980) Liang and Lee take the example of sulfur dioxide pollution in Kaohsiung to make good assessment between the values of observations and theoretical output. (Liang and Lee, 1980) And Chen and Liang use multivariate statistical analysis to optimize the value of model and observation (Liang and Chen, 1981)

However, Sasaki with the above researchers only takes into account of comparison and interaction between observations and model results. They did not consider the participation of wisdom from experts. And no doubt the wisdom of experts is a very important part of the decision analysis process. If not applying this part, the result is bound to be something omissions in the decision-making process. How to solve the problem of expert's wisdom participation is the focus of this article. In this chapter, the author tries introducing fuzzy decision theory to have a good inter-connected among the above three. Using fuzzy mathematics and iteration procedure, the opinions of expert's can be integrated. The expertise and advantages could give full play to obtain the best strategy. (Novák, 2005) (Torof, 1970)

Nevertheless, the above research only focuses on establishing the mathematical representation of the decision making of air quality management by Calculus of Variation. And these studies are mainly on the comparison of model output results and field monitoring results. In addition, there is very little discussion about the application of granular computing in air quality management. Although there are some researches which applies the fuzzy method to forecast the air quality. It is still remain empty in the way to obtain the optimum value of the parameters in the environmental simulation models.

This study applied the fuzzy method to define the optimum value in the parameters of the model, such as the wind speed, stability in the Gaussian diffusion model. The main topics in this chapter include the air quality forecasting and decision making method is in the second section. , Framework of fuzzy optimization in decision making of air quality management is described in the third section. Finally, the application of this model is presented in section four.

2 Air Quality Forecasting and Decision Making Method

2.1 Forecasting by Monitoring

To predict future results, using current information to speculate is the most direct way. Especially when the time interval is not far away from now, the best way to do the estimation is the estimation by current situation. As shown in Figure 1, it is the results of an air quality monitoring stations for three consecutive observing days. We can use the results of this three-day observation to speculate one-day, two-day, and three-day or future value. Because the data comes from the actual observation, if all the processes are in line with the necessary procedures, then the data is reliable. So there will be not much controversy of this data. (Liang and Tsai, 1980)(Liang and Chang, 1983)(Chen, 2009)(Yuan et al, 2000a,b)

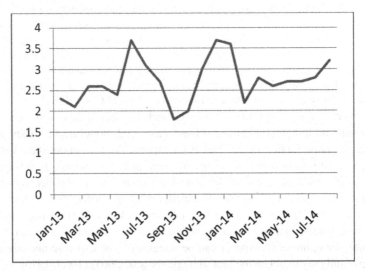

Fig. 1 An example to obtain the data by observation from a monitoring station (Source: Taipei EPD, 2014)

But when the time to speculate is far away from the observation time, say, in the case when the policymakers want to know whether typhoon will happen or not in the time next year? Or the environmental protection authorities now want to know how the air quality in Taipei is in order to decide the possibility to hold large outdoor sports without affecting the participant's health. These data this time has been inadequate clearly. Then we need the help of other ways to get more information for policymakers to analyze.

Another drawback of actual monitoring is that it is unable to simulate different scenarios. For example: the policymakers want to know what possible disasters of petrochemical plants is, when the gas explosion occurs, so as to make an emergency response plan to the local residents after gas explosion. However we cannot create a gas explosion situation for the policymakers to monitor. And most of the monitoring may not be the scenario that policymakers need to know because the situations policymakers want to know are always the situation of extreme condition. The real example such as the Kaohsiung oil pipeline gas explosion incident, Japan's Fukushima nuclear power plant accident events, both are not able to be analyzed by the actual monitoring data from a particular situation we designed and set in advance.(Lipscy et al., 2013)(Chen and Hong, 2014)

Another economy consideration is the burden from the cost. Monitoring stations usually takes a very high cost, and the information obtained in the monitoring can be used in the current time and space only. Change to other scenarios, then the monitoring information cannot be applied. Therefore, the monitoring result data are usually regarded as a verification and corroboration of decision. For example, if the environmental decision-making officials wondered the effectiveness of air pollution control measures, then they can conduct field monitoring in certain locations at certain times. The results obtained can be used as the verification of control strategy effectiveness. (Goldstein, 1977)

2.2   Forecasting by Model

Model provides us with more information than monitoring data, including time and space. For example in Figure 2, that is the distribution of air pollution in Taipei, it provides us with the distribution in each point of contamination on this map. This is what the actual monitoring cannot be done. It can also be obtained of the values in domain of different times. For example, time series analysis model, it can simulate the concentration distribution of the time for period of decision analysis. (Chang and Chang, 2002)(Chen and Wang, 2007)(Chen,2009)(Chen, 2010a,b)(Chen et al., 2008).

Model can be divided into two categories, physical models and mathematical models. The physical condition is to reduce the actual type into the laboratory, and then use the laboratory data backstep to the real field size for evaluation. The example is the water canal experiments in water conservancy engineering and the wind tunnel experiments in aeronautical engineering. (Faunae et al., 1986) (Uehara et al., 2000)(Naidu et al., 2013) Since the model is controlled in a laboratory, so it is possible to create a variety of different scenarios for simulating.

Mathematical model considers the various changes in physical and chemical factors. (Yao and Liu, 2013) If expanded to other areas, it is also possible to include social factors and economic factors. The advantage of mathematical model is that it is able to set a variety of different parameter to make the different contexts. Helping the decision-makers to get a clearer idea and make their decision more explicit. For example, in a mathematical model of socio-economic, the birth rate can be taken into account for considerations, so that the real GDP results can be forecasted. In the chemical reaction, the chemical reaction rate at different temperatures changes can be considered to know what species to be generated finally. In an air pollution dispersion model, It is also possible to include different weather conditions, including wind speed, wind direction, atmospheric stability degree, to know the distribution of air pollution under different scenarios for policy makers.

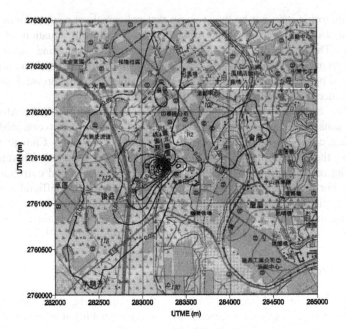

(a) An example for air quality distribution in a city

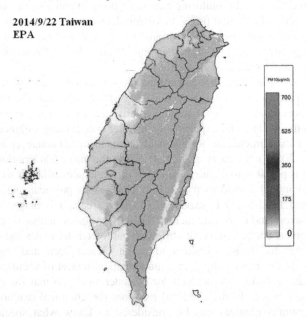

(b) Air quality distribution of Taiwan (Source: Taiwan EPA, 2014)

Fig. 2 Spatial distribution pattern predicted results by air quality model

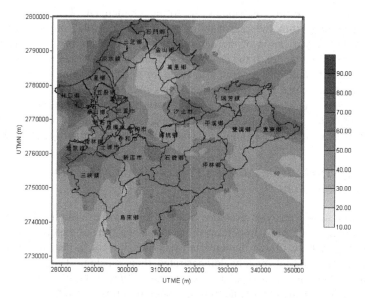

(c) Air quality map for decision making, the condition before control

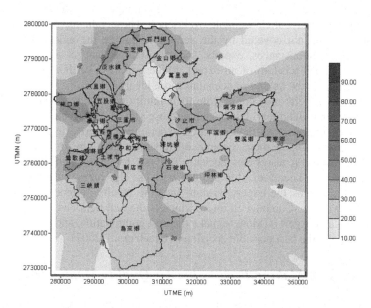

(d) Air quality map for decision making, the condition after control

Fig. 2 *(continued)*

However, the results of models are often challenged. The most common challenge comes from the reasonability of model variable settings. Since the model is the simplification of the nature phenomenon, therefore, there are many assumptions in the model. If the assumption is far from the actual situation, then it is not suitable to take into account as a basis for decision-making analysis. But in model the trend of the real situation and the simulated conditions in a particular context is showing clearly. These are the basis for decision making which can offer very effective suggestion decision-makers. So make the best use of the advantages of model is a big help for decision-makers of air quality management.

2.3   Expert Decision Making Process for Air Quality Management

Both of the above analysis provides a lot of help for us. But in the decision-making process, the most important thing is the wisdom of experts from policy makers. So we have to add the experts involving process in decision-making so that experts can be adequately demonstrated their wisdom. The expert usually have different opinions, and these opinions are different, how to integrate these opinion together is the main issue to be considered in this research.

Wisdom of experts from different levels: In the above example, the measured data and model data, although both have advantages and disadvantages, but experts may also have their valuable experience. These valuable resources cannot be observed, and not covered by the parameter among the model. Even more, under certain circumstances, experts' intuitions alone can get a good decision. That is the hidden wisdom of an expert. Make good use of this expert's hidden wisdom can make the quality of decisions more ideal.

Wisdom of experts may come from their intuition, yet it can be derived from their professional judgment. For example, those values setting for the parameters in the model, experts with their research experience can provide good advice to decision-makers. Using their recommended values also tend to get preferable simulation results. It also helps decision analysis for those who make the final decision making. Therefore, this chapter will use the fuzzy theory to derive a method which takes the expert wisdom into the results of model analysis and actual observations. Combine all three, which is the real field observations, model simulations, and expert wisdom to make the best combination, and provide policy-makers to make the best possible decisions. (Green et al., 2007)

2.4   Decision Making Process for Air Quality Management

Knowing the true value of air quality is essential for decision making of air quality management. The air pollutant uniformly distributed in the atmosphere so it is difficult to decide the representative point. Scientist has to find the way to tell the decision makers. Good quality of data help us to know the truth of environmental phenomenon, therefore the methodology to find the optimum predicting results becomes more important.

The information of air quality includes the domain of time and space. To know the detail variation, it is necessary to include time segment and space grid. Measuring by equipment is the most direct method, although the cost is very expensive. Developing the cheapest method of measurement can provide more information although it seems not possible in the near future.

The other way to obtain the air quality information is through the simulation model derived from physical, chemical, and mathematical principle. The model made the assumption according to the real situation, nevertheless there is always a bias to the reality. The simulation model plays an important role in forecasting the air quality because it provide adequate data in both time and space domain.

The concept to combine the two data system, measurement and simulation, is a complex procedure. Advantage of them should be involved so as to obtain the optimum solution. If there is no mistake in the measuring procedure, then the observed results are reliable, even if there is only one or few points in the space or time domain. The simulated results undoubtedly include the systematic errors which come from the theoretical assumption. Thus, the errors can be eliminated through properly adjustment by the observed data.

Using statistical methods to force the numerical model to approach the observed value may cause the deletion of the loss of significant physical meaning in the simulation results, although mathematically meet the requirements of optimization. This is the issue of internal consistency many scholars have repeatedly stressed in the study of objective analysis. This issue has been solved until they get a reasonable solution in the year 1958. Sasaki proposed a theoretical basis to engage objective analysis with the variational principle. It is able to maintain internal consistency of the analysis field in a variety of constraints. This method later was known as numerical variational analysis (NVA), or variational objective analysis, also known as variational optimization analysis. This method is able to combine the dynamic, energy, statistical, or empirical condition into an optimal analysis process In order to analyze the variables in weather or ocean. It provide a reliable basis for decision-making for the purposes of air quality management.

But this method has little progress in the subsequent decade. It is not until 1969, Sasaki (1969, a,b) noted characteristic feature of this method is that the constraint functions and filters. Sasaki (1970) has continuously published three articles, it laid the theoretical foundation variational analysis, and the feasibility of its use in meteorology decision analysis. Later, there are many scholars engaged in the application research of variational analysis and objective analysis. Groll (1975) uses Lewis' (1972) model to analyze the weather in Europe and found using binding conditions can really filter out short wave. Sasaki (1976) use integral condition of energy conservation to control the truncation error generated in the integral calculation of numerical weather prediction. It can avoid the errors formation of short wave and high-frequency.

The reliability of weather forecast is the basis of air quality management decisions. After the variational method can be successfully applied in the case of weather forecast, it can also be used in air quality management. The application of variational optimization method allows the interaction patterns between the model and monitored value have more physical meaning. The weighting factor of parameters in the optimization process can be determined by fuzzy expert decision-making process to make it more representative in the variational analysis.

3   Granular Computing in Air Quality Management

3.1   *Interactive and Iterative Fuzzy Optimization Model*

This chapter described the methodology of the fuzzy optimization by interactive and iterative method. Theoretical development and its characteristics are presented. The experiment results are also shown in the article. Figure 3 is the research framework of this chapter.

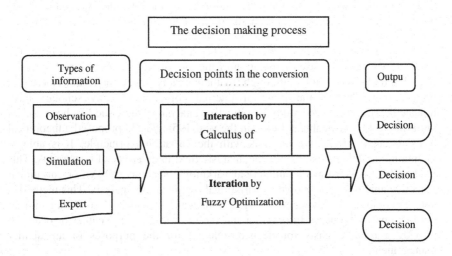

Fig. 3 Framework of fuzzy optimization in decision making of air quality management

Granular computing is the basis of fuzzy decision-making. (Dubois and Prade, 1988) (Goguen, 1967) This chapter also uses the spirits of granular computing to carry out fuzzy decision analysis for air quality management. Assume the experts and human judgment in the fuzzy decision, although very reliable, but it is often based on a lack of clearly and vague. The fuzzy mathematics tries to find a trusted basis from the semantic of expert, although it is often lack of clarity. Zadeh first derived the theoretical equations of fuzzy mathematics.(Zadeh, 1965) (Zadeh, et al. 1996).With his thesis proposed, researchers successor one after another to apply the fuzzy theory to different areas.(Chen, 2010a,b) (Chen and Cheng, 2010)(Yang and Yuan, 2003)(Yuan and Shaw,1995)

3.2   Fuzzy Optimization Method (FOM)

A fuzzy optimization method was developed in this chapter which includes the interactive and iterative stage to obtain the optimum solution of air quality management decision. In the interactive stage, the observed and simulated results are optimized through the fuzzy determination and calculus of variation method. The relative advantage of them can be considered and tuned by the decision makers. In the iterative stage, the optimum value of each parameter in the physical models is determined by the   iterative procedure of the invited experts.

This article use calculus of variation to combine and interact between the monitoring data and model results, and take advantage of fuzzy theory to link different expert opinions for the parameters in a model, using fuzzy mathematics to allocate the relative importance of each parameter, and finally obtaining the optimum simulated result.

Because there is a big gap between the different expert opinions, so we reach convergence through iteration process, and provide a clear basis for decision-makers. After each opinion inquiry, repeat the same process for expert's participation, and set minimum threshold of expert opinion which could be tolerated. When the views of the experts reached a minimum requirement of threshold, which is the convergence of expert decisions.

3.3   Mathematical Representation of Interactive Multi-objective Approach

Let{ $\tilde{c}_m|=1,2,.....,m$ }represent the observed value of M places, and { $\tilde{q}_n|=1,2,.....,n$ }be the emission amount of N sources. The distributed concentration is a function of Q_n. Assume the concentration in location (x,y) be C(x,y), then

$$C(x,y) = \sum_{n=1}^{N} f_n(x,y)Q_n \tag{1}$$

where f(x,y) is the function determined by the environmental factors such as topography, meteorology, and location, etc.

From the view point of fuzzy theory, the representative of $\tilde{c}_m$ is better than c(x,y) because it is an actual observed data . However, the value of c(x,y) provide more information than $\tilde{c}_m$ because it offers the pattern of the trend in both time and space domain. The other reason to take notice on the value from the simulated model is because the cost of observation is always very expensive. The simulated results could help us to know the phenomenon in case only the limited budget provided.

The uncertainty of simulated value comes from the emission amount of sources and the parameters of diffusion functions, f_n. The ambiguous condition should be treated by the fuzzy theory to obtain the optimum judgments. Thus, the $\acute{\Gamma}_{emi}$, $\acute{\Gamma}_{met}$, and $\acute{\Gamma}_{opt}$ were defined as the parameters which come from the influence of source

Table 1 Main factor affecting the final results of air quality assessment

No.	Symbol	name	description
1	Γ_{emi}	Source emission influence factor	Point, line, area source etc.
2	Γ_{met}	Metrological influence factor	Wind speed, wind direction, mixing height etc
3	Γ_{top}	Topographical influence factor	Elevation, surface roughness etc.

emission, meteorology, and topography. The main factors affecting the final results of air quality assessment value of them are shown in table 1.

Sasaki and Liang proposed the optimization theory to link the observed value and simulated value by the variational technique. First, considering the errors, R_{1m}, between them is

$$(R_{1m}) = C(x_n, y_m) - \tilde{C}_m$$
$$= C_m - C_m, m = 1,2,......M \tag{2}$$

where $f_n(x,y)$ is the dispersion function. Since there are M observed value and N source, in order to use the most value of observed information, it is essential to consider the errors from source emission, which means not only C_m close to $\tilde{C}_m$, but also the Q_n has to be close to $\tilde{Q}_m$. Thus

$$(R_2)_n = Q_n - Q_n, n = 1,2,.......N \tag{3}$$

where $\tilde{Q}_1, \tilde{Q}_2,.......\tilde{Q}_N$ is the observed value of source emission. The optimum solution exist in the condition of the following

$$E = \sum_{m=1}^{M} (R_1)_m^2 + \beta^2 (R_2)_n^2$$
$$= \sum_{m=1}^{M} (C_m - \tilde{C}_m)_m^2 + \beta^2 (Q_n - \tilde{Q}_n)_n^2 \tag{4}$$

where β is the weighting factor determined by the relative importance of these two factors.

If there are more parameters which influence the factor, β, then

$$\beta^2 = \beta_1^2 + \beta_2^3 + \beta_3^2 \tag{5}$$

where β_1 represent the influence of source emission, β_2 represent the influence of meteorology, β_3 represent the influence of topography.

Expanding the above equation, if there are more parameters involved in the simulation model, the general forms of equation becomes

$$\beta^2 = \beta_1^2 + \beta_2^3 + \beta_3^2 + ...+ \beta_\kappa^2 \tag{6}$$

where κ is the total number of parameters which exists in the simulation model.

therefore, the optimum value for decision making could obtained when the E value is minimum, thus

$$\delta(E) = 0 \tag{7}$$

3.4 The Interactive Calculation between the Analytical and Observed Data

In case of only one parameter, say source emission, considered in the optimization procedure, then the optimum strength of source should be determined. To obtain the optimum source emission strength, The matrix form of equation is rewritten as the following(Liang, 1979)

$$(F^F + \beta^2 I)\vec{Q} = (F^T \vec{\vec{C}} + \beta^2 \vec{\vec{Q}}) \tag{8}$$

Where

$$\vec{Q} = [Q_1, Q_2, \cdots, Q_N]^T,$$
$$\vec{\vec{C}} = [C_1, C_2, \cdots, C_M]^T,$$
$$\vec{\vec{Q}} = [Q_1, Q_2, \cdots, Q_N]^T,$$

T is the transformation of matrix, and I is the unit matrix.

The equation (8) become

$$A\vec{Q} = \vec{B}, \tag{9}$$

and

$$A = F^T F + \beta^2 I$$
$$B = F^T \vec{\vec{C}} + \beta^2 \vec{\vec{Q}}$$

The optimum emission strength can be determined by the interactive procedure as the following equation

$$\vec{Q} = A^{-1}\vec{B}, \tag{10}$$

The optimum concentration for decision making of air quality management is

$$C(x, y) = \sum_{n=1}^{N} f_n(x, y)Q_n. \tag{11}$$

Liang proposed a Strong Optimization and Weak Optimization method in decision making theory to solve the problem of differences between measured and simulated value, and how to get the best approach by interaction tradeoff. The strong interactive optimization (SINO) requires all the data consistent with the observed results. Thus the sum of errors has the minimum value to make sure all of them are closed to the observed data.

3.5 Mathematical Representation of Iterative Calculation Approach

In this paragraph, Gaussian diffusion equation will be used as an example in the air quality assessment. It is used to describe the method using three variables in the function and obtain the best decision. The best optimization of expert's opinion was obtained by fuzzy optimization method. Commonly used Gaussian diffusion equation is as follows (Stern, 1968)

$$C(x, y, z, H)$$

$$= \frac{Q}{2\pi U \sigma_x \sigma_Y} \exp\left(-\frac{y^2}{2\sigma_y^2}\right)\left\{\exp\left[-\frac{(z-H)^2}{2\sigma_y^2}\right] + \exp\left[-\frac{(z+H)^2}{2\sigma_z^2}\right]\right\} \tag{11}$$

In the above formula, Q is the amount of emissions, U is the wind speed, in addition, sigma x, y is the diffusion coefficient in the horizontal and vertical direction. Because they are independent of the function, therefore, the above equation can be removed individually, and individually treated by optimization.

Let β_1, β_2, β_3, β_4 represents the influence from emissions, wind speed, atmospheric stability, and diffusion coefficient. Then equation (6) can be written as

β^2= (source variation) + (wind speed variation) + (stability variation) + (diffusivity variation)

$$= \beta_{source}^2 + \beta_{meteorology}^2 + \beta_{stability}^2 + \beta_{diffusivity}^2$$

$$= \beta_1^2 + \beta_2^2 + \beta_3^2 + \beta_4^2 \tag{12}$$

When individually consider of their impact, the above equation can be rewritten as

$$C(x, y) = \sum_{n=1}^{N} Q_n \cdot f_{Q_n}(x, y). \tag{13}$$

$$C(x, y) = \sum_{n=1}^{N} U_n \cdot f_{U_n}(x, y). \tag{14}$$

$$C(x, y) = \sum_{n=1}^{N} S_n \cdot f_{S_n}(x, y). \tag{15}$$

$$C(x, y) = \sum_{n=1}^{N} K_n \cdot f_{K_n}(x, y). \tag{16}$$

Q_n, U_n, S_n, K_n, Q_n, U_n, S_n, and K_n in formula (13), (14), (15), (16) respectively represent the errors from the pollution emissions, wind speed, atmospheric stability, and diffusion coefficient. Qn, Un, Sn, Kn, these four key variables are part of Gaussian dispersion function, therefore, it is possible to obtain the optimized emission amount, wind speed, atmospheric stability, and diffusion coefficient.

$$C(x, y) = \sum_{n=1}^{N} Q_n \cdot f_{Q_n}(x, y). \tag{17}$$

$$C(x, y) = \sum_{n=1}^{N} U_n \cdot f_{U_n}(x, y). \tag{18}$$

$$C(x, y) = \sum_{n=1}^{N} S_n \cdot f_{S_n}(x, y). \tag{19}$$

$$C(x, y) = \sum_{n=1}^{N} K_n \cdot f_{K_n}(x, y). \tag{20}$$

3.6 Iterative Calculation from Different Decision Maker

The determination of an environmental condition can be judged by a group of expert through the fuzzy optimization procedure. The procedure is an iterative process to obtain the optimum results among a group of experts. In the beginning, the opinion from these experts is not the same, so the second time choose the most closed results and re-run the procedure. After a certain runs, the opinion comes to consistent. (Rowe and Wright, 2001) (Green et al, 2007) (Tapio, 2003)

If the iteration among the experts is very strong, then it is called "Strong Interactive Optimization (SITO), which means the results comes from the common idea of all the experts is high. The SITO method requires the results from the experts be very close. If the iteration among the experts is few, then it is called "Weakly Interactive Optimization (WITO), which means the results comes from the common idea of all the experts is low. The extent of these parameters should be determined by the expert through a fuzzy determination process as list in table 2.

Table 2 The fuzzy representation of the main parameters affecting the simulated results

type	Extremely low	very low	low	slight low	medium	slightly high	strong	very high	extremely high
symbol	A	B	C	D	E	F	G	H	I
number	1	2	3	4	5	6	7	8	9
Percent	0.1	0.2	0.3	0.4	0.5	0.6	0.7	0.8	0.9

The above values are determined by the experts of decision makers for further analysis. The decision makers can decide the value by the objective determination based on their understanding. (Rescher, 1998)

Delphi method was used in this study to make the expert advice to achieve convergence condition. (Michael et al.,1996)Delphi method, also known as an expert investigation, is a way of consultation using the communication method with experts. It requires the experts to solve the problem alone by the questionnaires, and summarizes the views of all the experts and recovered, then sorting out a comprehensive advice. (Harold et al., 2002)) Subsequently the comprehensive advice and prediction problems then were back to the experts, consultation with the experts once again. So many times of iteration over and over again, And gradually obtain more consistent approach to the decision making predictions. (Rowe and Wright,1999) (Basu, et al.,1977) (Dalkey, and Helmer ,1963)

The optimum value of the parameters applied in the model depends on the opinion expert from different view point. Thus an iteration procedure to obtain the optimum value is required. Procedure for determining the value is as the following. (Harold et.al,1975) (Adler and Ziglio,1996)

(1) Form a group of experts. Identify experts in accordance with the required knowledge of the subject. The number of experts is according to the subject size, Usually it is no more than 20 people.

(2) Raise the issue to be predicted and the requirements to all the experts, attach all the background material on the subject, and also requested experts what is the required material. Then, make a written reply from the experts.

(3) Each expert make their forecast opinions according to the material they received, and explain how they use these materials to make the suggested value.

(4) Make a summary of the views of the first time judgment of experts, tabulated chart are compared, then circulated to the experts again. Let the experts compare themselves with others of different opinions, Modify their opinions and judgments. Is also possible to collate the views of the experts, requested higher status or other experts to comment, then put these views again distributed to the experts so that they can modify their own views for reference.

(5) Collect the modified views of all the experts, summarize, and once again circulated to all the experts in order to make the second revision. By-round collection of advice and feedback expert information is the main part of Delphi method. Collect comments and feedback generally through three or four round. In time feedback to the experts, only give various opinions, but does not indicate the specific names of all opinions of experts. This process is repeated until each expert does not change views.

(6) Comprehensive handle on expert advice.

The opinion among the experts is very different. However, the final decision still necessary, therefore, the results is the aggregate solution of all these experts, as shown in the following figure.

4   Application of FOM in Air Quality Management

4.1   Comparison of Model Output Results and Field Monitoring Results

Figure 4 is the result obtained by monitored data. Because it is impossible to obtain the monitoring value in each grid point, so it is using the information of different monitoring stations and curve drawing program to get the final concentration.

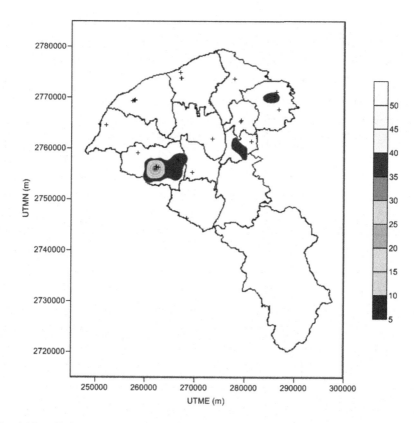

Fig. 4 Air pollution distribution using the interpolate data of monitoring station

The monitoring results provide us the concentration in the exact point. On the other hand, Gaussian modes can obtain the different concentrations distribution at respective grid points. This is a mathematical model with analytical solution. We can also use different numerical models to get similar distribution pattern.

4.2 Influences of Interactive Fuzzy Optimum Method

The results from the simulated results have much useful information. In some condition, it is not necessary to force the data to consistent with the observed data. This situation could be regarded as the "weak interactive optimization (WINO), which means the interaction between the observed data and simulated data is very weak. The air quality model is a Gaussian type air quality model, and there are five stations in this study, if all the results are fitted by the stations, then the results may lost its physical meaning, just because of the purpose to fit the measured data.

Table 3 Determine the relative importance weighting by iteration method among the experts

No. of expert	First round			Second round			Third round		
Parameters in the model	β_1	β_2	β_3	β_1	β_2	β_3	β_1	β_2	β_3
A	0.7	0.2	0.1	0.75	0.2	0.05	0.67	0.25	0.09
B	0.6	0.3	0.1	0.65	0.25	0.1	0.70	0.20	0.10
C	0.8	0.1	0.1	0.75	0.1	0.15	0.69	0.10	0.21
D	0.6	0.2	0.2	0.72	0.1	0.18	0.70	0.11	0.19
E	0.4	0.4	0.2	0.55	0.35	0.1	0.58	0.32	0.10
F	0.55	0.4	0.05	0.65	0.3	0.05	0.65	0.30	0.05
G	0.7	0.25	0.05	0.72	0.23	0.05	0.70	0.2	0.1
Average	0.6214	0.2642	0.1142	0.6842	0.2185	0.0971	0.67	0.2114	0.12

Because the SINO forced the simulated value to be equal to the value of monitoring station, so it has to adjust the value of diffusion equation in emissions or other parameters. It may result in the circumstances of negative emissions, or the wind speed is less than 0. This method provides a mechanism for policy makers to decide the weighting by themselves. The results of strong iterative optimization are shown in figure (a) in this figure. Apart from the previous researches which only apply the monitored data and simulated results to decide the optimum value, our method includes the expert opinion the uncertainty of parameter value. So it reserves the physical meaning of the model and offer a more reasonable explanation of our decision.

4.3   *Influences of Iterative Fuzzy Optimum Method*

The optimum value for decision from different decision makers is usually not the same. In order to get the optimum values in line with the physical significance. it is suggested to determine the best value beta with experts iteration.

Which is the optimum value of the different parameters can be obtained by the fuzzy decision procedure. In this model, for example, after expert discussions, the set of weights were   0.2,0.3,0.4,and 0.1。

4.4   Experimental Outcome of Fuzzy Optimum Model

Take the information of Figure 4 as the basis and applying the fuzzy optimization method to obtain the results of figure 5. This result is in line with the actual monitoring data obtained after correction interaction.

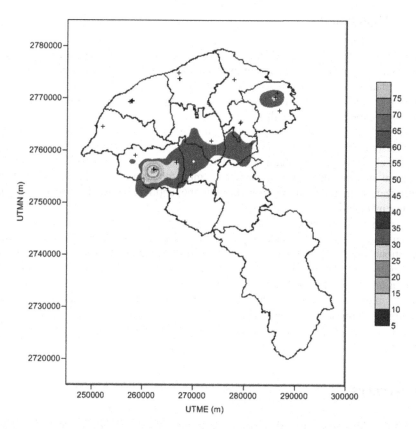

Fig. 5 Results of air pollution distribution using interactive correction between actual value and monitored values

4.5   Limitations and Future Research Needs

In this study, the fuzzy variational optimization theory was deduced and applied in the simulation of an air quality management. From preliminary results it seems this method can really be used in the decision making process of air quality management. Not only the value can be verified by the actual observations, but also keep the physical meaning in the air quality dispersion model. The results of air quality diffusion model after the handle of variational optimization method still can clearly describe the behavior of air pollutant dispersion. Therefore, the conclusions of policy become more reliable.

Owing to the limit time and resource, this research cannot provide more detail results of various parameters. The future study should focus on the different parameters' characteristic such as the wind speed, temperature, and diffusion coefficient etc.

5   Conclusions

This chapter has described an optimization model for decision making in air quality management which can combine the observed value and simulated output. The variational calculus was used to maintain the physical meaning of the model. The optimized value of the parameters in the physical model was determined by the fuzzy expert decision method.

The optimization method gives the mathematical representation to find the equilibrium point. Yet, how to obtain and express these optimal data depends on the fuzzy optimization techniques. The methodology and algorithm of fuzzy decision making process by interactive multi-objective approach and iterative optimization method are described, with the application in the process of air quality management.

A numerical experiment was presented in this chapter, a physical diffusion model was used to calculate the air pollutant concentration, and then some variational equations were derived. Then the calculated results were corrected by the monitored data with the variation calculus. The experts determine the weighting factor of each parameter. Therefore, the final decision was determined by the interactive process between the simulated and monitored results. The combined approach of interactive and iterative method for fuzzy optimization model makes the decision of air quality management more accurate and pragmatic.

References

1. Adler, M., Ziglio, E. (eds.) Gazing Into the Oracle: The Delphi Method and Its Application to Social Policy and Public Health. Kingsley Publishers, London (1996) ISBN 978-1-85302-104-6
2. Basu, S., Schroeder, R.G.: Incorporating Judgments in Sales Forecasts: Application of the Delphi Method at American Hoist & Derrick. Interfaces 7(3), 18–27 (1977), doi:10.1287/inte.7.3.18
3. Chang, D.C., Chang, Y.C.: Application of fuzzy cluster algorithms in pollutant's spatial distribution of Kaohsiung coast. Master thesis, National Sun-Yat Sen University, Kaohsiung, Taiwan (2002)

4. Chen, W.K.: Study of the Pattern Recognition of Particulate Concentration by Fuzzy Time Series and Neural Network Analysis. Paper presented at the 2009 Cross Strait Conference on Aerosol Science and Technology (2009)

5. Chen, W.K.: An Approach to Pattern Recognition by Fuzzy Category and Neural Network simulation. Paper presented at the 2010 International Conference on Machine Learning and Cybernetics, ICMLC (2010a)

6. Chen, W.-K.: Environmental applications of granular computing and intelligent systems. In: Pedrycz, W., Chen, S.-M. (eds.) Granular Computing and Intelligent Systems. ISRL, vol. 13, pp. 275–301. Springer, Heidelberg (2011b)

7. Chen, W.K., Wang, P.: Numerical Modeling of Gas-phase Kinetics in Formation of Secondary Aerosol. China Particuology 5(4), 267–273 (2007), doi:10.1016/j.cpart.2007.05.002

8. Chen, W.K., Cheng, C.S.: Application of Grey Theory in Predicting the Disaster Loss of Typhoon. In: 2012 Global Chinese Conference on Environment and Energy, Hsinchu, Taiwan (2012)

9. Chen, W.K., Hong, Y.H.: Causal Analysis of Kaohsiung Gas Explosion Incident by Event History Method. In: Sustainable 2014 Development and Environmental Conference, Chungli, Taiwan (2014)

10. Chen, W.K., Lin, C.C., Yu, C.H.: Establishment of the Pollution Simulation and Particulate Control Technology to the Exposed Surface Area–An Assessment Techniques Combining the Satellite Image Interpretation and Air Quality Models. Journal of Environmental Protection 31 (2008)

11. Dalkey, N., Helmer, O.: An Experimental Application of the Delphi Method to the use of experts. Management Science 9(3), 458–467 (1963)

12. Dubois, D., Prade, H.: Fuzzy Sets and Systems. Academic Press, New York (1988)

13. Fraunie, P., Beguier, C., Parachivoiu, I., Brochier, G.: Water channel experiments of dynamic stall on Darrieus wind turbine blades. Journal of Propulsion and Power 2(5), 445–449 (1986), doi:10.2514/3.22927(1986)

14. Goguen, J.A.: L-fuzzy sets. Journal of Mathematical Analysis and Applications 18, 145–174 (1967)

15. Goldstein, I.F., Landoritz, I.: Analysis of Air Pollution Patterns in New York City. I. Can on station represent the Large Metropolitan Area? II. Can one Aerometric Station Represent the Area Surrounding it? Atmos. Envir. 11, 47–57 (1977)

16. Green, K.C., Armstrong, J.S., Graefe, A.: Methods to elicit forecasts from groups: Delphi and prediction markets compared. Foresight: The International Journal of Applied Forecasting 8, 17–20 (2007)

17. Gustafson, S.A., Kortanek, K.O., Sweigart, J.R.: Numerical Optimization Techniques in Quality Modeling: Objective Interpolation Formulus for the Spatial Distribution of Pollution Concentration. J. Appl. Meteor. 16, 1243–1255 (1977)

18. Linstone, H.A., Torof, M.: The Delphi Method: Techniques and Applications. Addison-Wesley, Reading (1975) ISBN 978-0-201-04294-8

19. Linstone, H.A., Torof, M. (eds.) The Delphi Method:Techniques and Applications, Murray Turoff and Harold Linstone, TOC III.B.3. The National Drug-Abuse Policy Delphi: Progress Report and Findings to Date, IRENE ANNE JILLSON (2002), http://is.njit.edu/pubs/delphibook/ch3b3.html

20. Heimbach, J.A., Sasaki, Y.: A Variational Technique for Mesoscale Objective Analysis of Air Pollution. J. Appl. Meteor. 16, 1243–1255 (1977)
21. Liang, W.J.: The Optimization theory of Air Pollution Assessment Annual Report of the Institute of Physics. Academia Sinica 9 (1979)
22. Liang, W.J., Lee, C.T.: Optimization Assessment of Sulfur Dioxide Pollution and Monitoring Network in Kaohsiung Area. Journal of the Chinese Institute of Engineers 3(2), 105–115 (1980)
23. Liang, W.J., Lee, K.T.: Optimization Assessment of Carbon Monoxide Pollution and Monitoring Network in Taipei City. Journal of the Chinese Society of Mechanical Engineers 1(1) (1980)
24. Liang, W.J., Tsai, F.C.: Research on the Time Series Model for Prediction of the Sulfur Dioxide Pollution in Taipei City. Annual Report of the Institute of Physics, Academia Sinica 10 (1979)
25. Liang, W.J., Young, P.H.: Theory and Application of the Aggregate and Average Optimization in Air Pollution Assessment. Annual Report of the Institute of Physics, Academia Sinica 10 (1980)
26. Liang, W.J., Hsieh, C.S.: Objective Analysis of the Finite Element Method. Annual Report of the Institute of Physics, Academia Sinica 10 (1980)
27. Liang, W.J.: The Variational Optimization of Wind Field for the Estimation of Vertical Velocity. Proc. Natl. Sci. Counc. ROC. 4(3), 249–259 (1980)
28. Liang, W.J., Chen, W.K.: Application of Model Output Statistics in Air Quality Assessment. Annual Report of the Institute of Physics, Academia Sinica 11 (1981)
29. Liang, W.J., Chang, R.T.: Study on the Time Series Model for the Parameters in Atmospheric Diffusion Model. Annual Report of the Institute of Physics, Academia Sinica 13, 1979 (1983)
30. Lipscy, P.Y., Kushida, K.E., Incerti, T.: The Fukushima Disaster and Japan's Nuclear Plant Vulnerability in Comparative Perspective. Environmental Science & Technology dx.doi.org/10.1021/es4004813 I Environ. Sci. Technol. 47, 6082–6088 (2013)
31. Turoff, M.: The Design of a Policy Delphi. Technological Forecasting and Social Change 2, 2 (1970)
32. Naidu, M.K., Tian, Z.F., Medwell, P.R., Birzer, C.H.: An Investigation of Cool Roofing on Urban Street Canyon Air Quality. In: 20th International Congress on Modelling and Simulation, Adelaide, Australia, December 1-6 (2013), http://www.mssanz.org.au/modsim
33. Novák, V.: Are fuzzy sets a reasonable tool for modeling vague phenomena? Fuzzy Sets and System 156, 341–348 (2005)
34. Rescher: Predicting the Future. State University of New York Press, Albany (1998)
35. Rowe, Wright: The Delphi technique as a forecasting tool: issues and analysis. International Journal of Forecasting 15(4) (October 1999)
36. Rowe, Wright: Expert Opinions in Forecasting. Role of the Delphi Technique. In: Armstrong (ed.) Principles of Forecasting: A Handbook of Researchers and Practitioners. Kluwer Academic Publishers, Boston (2001)
37. Sasaki, Y.: Proposed inclusion of time variation terms, observed theoretical, in numerical variational objective analysis. J. Meteor. Soc., 115–124 (1969)

38. Sasaki, Y.: Some basic formalisms in numerical variational analysis. Mon. Wea. Rev. 98, 875–883 (1970a)
39. Sasaki, Y.: Numerical variational analysis formulated under the determined by long wave equations and a low pass fielter. Mon. Wea. Rev. 884–898 (1970b)
40. Sasaki, Y.: Numerical variational analysis with weak constraints to surface analysis of severe storm gust. Mon. Wea. Rev. 98, 89 (1970c)
41. Sasaki, Y.: Variational design of finite difference scheme for problem of an integral constraints. J. Comp. Phy. 278, 21–270 (1976)
42. Stern, A.C.: Air Pollution, pp. 425–463. Academic Press, New York (1968)
43. Tapio, P.: Disaggregative Policy Delphi: Using cluster analysis as a tool for systematic scenario formation. Technological Forecasting and Social Change 70(1), 83–101 (2003), http://dx.doi.org/10.1016/S0040-1625(01)00177-9
44. Taipei Environmental Protection Department, Historical Air Quality data of Taipei City (2014),
   http://www.tldep.taipei.gov.tw/C_INDEX/ENVIR/air_5.asp
45. Taiwan Environmental Protection Administration, Historical Air Quality data of Taiwan (2014),
   http://taqm.epa.gov.tw/taqm/zh-tw/YearlyDataDownload.aspx
46. Uehara, K., Murakami, S., Oikawa, S., Wakamatsu, S.: Wind tunnel experiments on how thermal stratification affects flow in and above urban treet canyons. Atmospheric Environment, volume 34, Issue 10 2000, 1553–1562 (2000), doi:10.1016/S1352-2310(99)00410-0
47. Yang, H.Y., Yuan, C.S.: The correlation of the visibility variation with weather patterns and meteorological factors in the south of Taiwan. In: 96th AW&MA Annual Meeting, San Dieago, California (June 2003)
48. Yao, J., Liu, W.: Nonlinear Time Series Prediction of Atmospheric Visibility in Shanghai. In: Pedrycz, W., Chen, S.-M. (eds.) Time Series Analysis, Model. & Applications. ISRL, vol. 47, pp. 385–399. Springer, Heidelberg (2013)
49. Yuan, C.S., Lee, C.G., Chang, J.C., Liu, S.H., Yuan, C., Yang, H.Y.: Correlation of Atmospheric Visibility with Chemical Composition and Size Distribution of Aerosol Particles in Urban Area. In: 93rd Air and Waste Management Association Annual Meeting, Salt Lake City, Utah (June 2000a)
50. Yuan, C.S., Lee, C.G., Liu, S.H., Chang, F.T.: Innovative Measurement of Visual Air Quality and Its Correlation with Meteorological Factors and Air Pollutants in Subtropics. presented at Seventh International Conference on Atmospheric Sciences and Applications to Air Quality and Exhibition, Taipei (November 2000b)
51. Yuan, Y., Shaw, M.J.: I nduction of fuzzy decision trees. Fuzzy Sets and Systems, 125–139 (1995)
52. Zadeh, L.A.: Fuzzy sets. Information and Control 8(3), 338–353 (1965)
53. Zadeh, L.A., et al.: Fuzzy Sets, Fuzzy Logic, Fuzzy Systems. World Scientific Press (1996) ISBN 981-02-2421-4

Author Index

Subject Index

Printed in the United States
By Bookmasters